Advances in Experimental Medicine and Biology

Volume 1054

Editorial Board
IRUN R. COHEN, *The Weizmann Institute of Science, Rehovot, Israel*
ABEL LAJTHA, *N.S. Kline Institute for Psychiatric Research, Orangeburg, NY, USA*
JOHN D. LAMBRIS, *University of Pennsylvania, Philadelphia, PA, USA*
RODOLFO PAOLETTI, *University of Milan, Milan, Italy*
NIMA REZAEI, *Tehran University of Medical Sciences, Children's Medical Center Hospital, Tehran, Iran*

More information about this series at http://www.springer.com/series/5584

Jeffrey E. Plowman
Duane P. Harland
Santanu Deb-Choudhury
Editors

The Hair Fibre: Proteins, Structure and Development

Editors
Jeffrey E. Plowman
AgResearch Ltd.
Lincoln, New Zealand

Duane P. Harland
AgResearch Ltd.
Lincoln, New Zealand

Santanu Deb-Choudhury
AgResearch Ltd.
Lincoln, New Zealand

ISSN 0065-2598   ISSN 2214-8019 (electronic)
Advances in Experimental Medicine and Biology
ISBN 978-981-10-8194-1   ISBN 978-981-10-8195-8 (eBook)
https://doi.org/10.1007/978-981-10-8195-8

Library of Congress Control Number: 2018937518

© Springer Nature Singapore Pte Ltd. 2018
This work is subject to copyright. All rights are reserved by the Publisher, whether the whole or part of the material is concerned, specifically the rights of translation, reprinting, reuse of illustrations, recitation, broadcasting, reproduction on microfilms or in any other physical way, and transmission or information storage and retrieval, electronic adaptation, computer software, or by similar or dissimilar methodology now known or hereafter developed.
The use of general descriptive names, registered names, trademarks, service marks, etc. in this publication does not imply, even in the absence of a specific statement, that such names are exempt from the relevant protective laws and regulations and therefore free for general use.
The publisher, the authors and the editors are safe to assume that the advice and information in this book are believed to be true and accurate at the date of publication. Neither the publisher nor the authors or the editors give a warranty, express or implied, with respect to the material contained herein or for any errors or omissions that may have been made. The publisher remains neutral with regard to jurisdictional claims in published maps and institutional affiliations.

Printed on acid-free paper

This Springer imprint is published by the registered company Springer Nature Singapore Pte Ltd.
The registered company address is: 152 Beach Road, #21-01/04 Gateway East, Singapore 189721, Singapore

Preface

This book is the child of two revolutions and the death of an empire. The first revolution is glorious and inspiring. This is the "omics" revolution – that surge of new molecular genomic, proteomic and lipidomic technologies which are now, with frightening ease, liberating new information about keratin fibre chemistry and biology, fibres development and the entangled control mechanisms by which hair phenotype is determined. The second revolution is the "internet" revolution. The inclusion of data, discoveries and processing tools on the internet has been inexorable and transformative, allowing analyses and uptake of findings that we could only dream of a few decades ago. What these first two revolutions are delivering is an ability to solve complex problems across disciplines, and it is here that the fall of the empire that was wool research is felt.

We, the editors of this book, somehow survived this fall, and more than a decade ago took the opportunity to diversify from wool, to encompass hair from a wide range of mammals, including humans. Keratin fibre, and especially hair follicle, research is a wide, multi-focused and multi-discipline arena, with plethoric differences in approaches and nomenclatures. One particular problem we frequently encountered over the past decade has been new generations of researchers who are largely unaware of research carried out prior to the year 2000, because this work is largely off-line, and often using wool research terminology unfamiliar to, for example, bio-medical dermatologists.

It has also been a long time since a comprehensive book was published on animal hairs and follicles. For our research on fibres, follicles and keratin proteins, we found ourselves frequently reaching for a small number of older books and reviews. These include David Parry and Peter Steinert's book on intermediate filament structure; the Jollès, Zahn and Höcker edited book on the formation and structure of human hair; William Montagna's textbooks from the 1960s and 1970s; and Don Orwin's 1979 review, "the cytology and cytochemistry of the wool follicle". Meanwhile, the exciting new research is scattered among various journals. Many of the researchers involved in this area have long since retired and, while some are still active in their retirement, we felt it was important to capture their research in one volume.

It was for these reasons that we have felt that the time was long overdue for an updated book on the subject of keratins and fibre development that drew together all these threads of research. We hope that this book will be as useful as a learning tool and reference to readers as it has been to us.

Lincoln, New Zealand
November 2017

Jeffrey E. Plowman
Duane P. Harland
Santanu Deb-Choudhury

Contents

Part I Introduction to Hair and Follicles

1 **Fibre Ultrastructure** 3
Jeffrey E. Plowman and Duane P. Harland

2 **The Follicle Cycle in Brief** 15
Jeffrey E. Plowman and Duane P. Harland

Part II Hair Proteins

3 **Diversity of Trichocyte Keratins and Keratin Associated Proteins** 21
Jeffrey E. Plowman

4 **Evolution of Trichocyte Keratins** 33
Leopold Eckhart and Florian Ehrlich

5 **Evolution of Trichocyte Keratin Associated Proteins** 47
Dong-Dong Wu and David M. Irwin

6 **Structural Hierarchy of Trichocyte Keratin Intermediate Filaments** 57
R. D. Bruce Fraser and David A. D. Parry

7 **Trichocyte Keratin-Associated Proteins (KAPs)** 71
R. D. Bruce Fraser and David A. D. Parry

Part III Hair Development

8 **Introduction to Hair Development** 89
Duane P. Harland

9 **Environment of the Anagen Follicle** 97
Duane P. Harland

10 **Development of Hair Fibres** 109
Duane P. Harland and Jeffrey E. Plowman

11 **Macrofibril Formation** 155
Duane P. Harland and A. John McKinnon

Part IV Hair Chemistry and Thermodynamics

**12 Crosslinking Between Trichocyte Keratins
and Keratin Associated Proteins** 173
Santanu Deb-Choudhury

13 The Thermodynamics of Trichocyte Keratins 185
Crisan Popescu

14 Oxidative Modification of Trichocyte Keratins 205
Jolon M. Dyer

Index... 219

Part I

Introduction to Hair and Follicles

Fibre Ultrastructure

Jeffrey E. Plowman and Duane P. Harland

Contents

1.1	The Cuticle	4
1.1.1	The Fibre Surface Cuticle Lipids	5
1.1.2	The Exocuticle	6
1.1.3	The Endocuticle	7
1.2	**The Cortex**	7
1.3	**The Medulla**	11
1.4	**Other Components of the Fibre**	12
1.4.1	The Cell Membrane Complex	12
1.4.2	The Cytoplasmic/Nuclear Remnant	12
References		12

Abstract

Mammalian hair fibres can be structurally divided into three main components: a cuticle, cortex and sometimes a medulla. The cuticle consists of a thin layer of overlapping cells on the surface of the fibre, constituting around 10% of the total fibre weight. The cortex makes up the remaining 86–90% and is made up of axially aligned spindle-shaped cells of which three major types have been recognised in wool: ortho, meso and para. Cortical cells are packed full of macrofibril bundles, which are a composite of aligned intermediate filaments embedded in an amorphous matrix. The spacing and three-dimensional arrangement of the intermediate filaments vary with cell type. The medulla consists of a continuous or discontinuous column of horizontal spaces in the centre of the cortex that becomes more prevalent as the fibre diameter increases.

Keywords

Fibre surface cuticle lipids · Exocuticle · Endocuticle · Cortex · Medulla · Cell membrane complex · Cytoplasmic/nuclear remnant

J. E. Plowman (✉) · D. P. Harland
AgResearch Ltd., Lincoln, New Zealand
e-mail: Jeff.Plowman@agresearch.co.nz;
Duane.Harland@agresearch.co.nz

Hair, along with milk production, is a definitive phenotypic trait for mammals. Hair appears to have evolved from scales of synapsid reptiles around 200 million years ago [1, 2], and there is also a common ancestral link between reptilian claws and mammalian hairs [3]. Because hair is such a well-established trait from the dawn of mammalian evolution, most aspects of hair biology are highly conserved across all mammals including general patterns of fibre morphology, hair assembly in a follicle, keratin sequence homology and follicle cycling.

Consequently, hair fibres, and those fine-diameter hairs referred to as wool, are structurally divided into three main components (Fig. 1.1). The surface is defined by a thin sheath of cuticle cells that wraps around and protects the fibre from environmental damage. Inside these can be found the cortical cells, the proteins of which determine the mechanical properties of the fibre. The cuticle constitutes around 10% of the fibre on a weight basis, while in fine fibres, such as those from merino sheep, the cortex makes up the remaining 86–90% [4]. A third structural component, the medulla, is a feature of many specialised fibres such as those from rodents, and occurs also in human hairs and wool as fibre diameter increases. Not all fibres have a medulla, but it is found close to the centre of the cortex.

1.1 The Cuticle

The cuticle is composed of a series of flattened cells that in merino are approximately 20×30 μm by 0.7 μm [5]. These cells overlap both longitudinally and peripherally to form a sheath with an exposed edge, known as the scale edge, that points to the distal end or tip of the fibre (Fig. 1.2a, b). This arrangement is such that each cell is also in contact with the cortex. It is thought that the overlapping of cells in the vicinity of the scale edge is important in assisting the control of water ingress and egress and hence in maintaining normal fibre function and durability [6–8].

Ultrastructural studies of stained ultrathin sections by transmission electron microscopy (TEM) have shown that the cuticle has a laminated structure of which the lipid layer on the surface of the cuticle, the exocuticle a-layer, rest of the exocuticle and endocuticle are the primary components

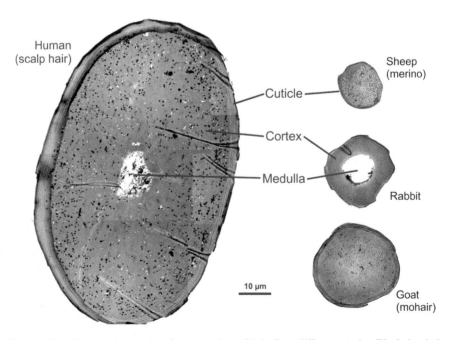

Fig. 1.1 Transmission electron micrographs of cross sections of hairs from different species. Black dots in human hair are melanin granules

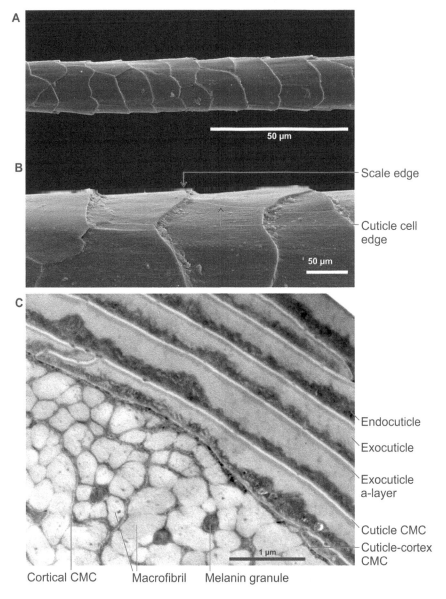

Fig. 1.2 The hair fibre cuticle. (**a**) Scanning electron micrograph of merino wool fibre showing overlapping scale pattern (left of figure is toward skin surface). (**b**) Higher magnification view showing external features, note debris and wool grease caught along scale edges. (**c**) Transmission electron micrograph of cross section of human scalp hair showing multiple overlapping cuticle layers and main features

(Fig. 1.2c). Additional studies have revealed a surface band, the latter generally referred to as the fibre cuticle surface membrane (FSCM) or epicuticle (Fig. 1.3a). This component was originally identified after treatment with chlorine and water when a portion of membrane was observed to be raised above the fibre surface [9].

1.1.1 The Fibre Surface Cuticle Lipids

The surface of the cuticle of wool has been found to be composed of a class of lipids resistant to extraction with lipid solvents but that could be released by a mild treatment with alcohols under

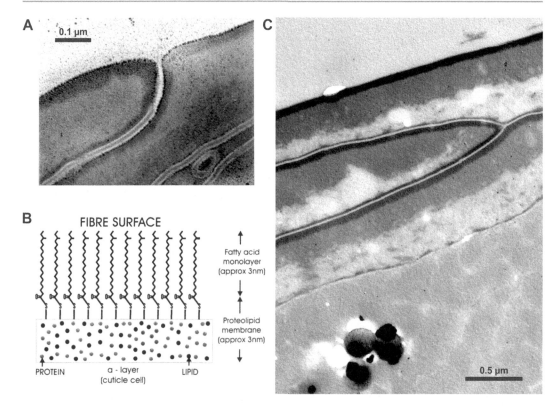

Fig. 1.3 Hair fibre surface. (**a**) Transmission electron micrograph of a cross-section of a wool fibre cuticle in which chemical staining has caused the extent of the fibre cuticle surface membrane (FSCM) or epicuticle to be highlighted as dark grains. (**b**) The model for the arrangements of proteins and lipids at the surface of the fibre cuticle. In this model there is a monolayer of fatty acids of which 18-methyleicosanoic acid is the major constituent. These acids are covalently linked through thioester bonds to the underlying proteins. (**c**) Micrograph of a cross-section through human cuticle stained to highlight the exocuticle a-layer. ((**b**) reprinted from Micron, 1997, Jones and Rivett [6], with permission from Elsevier)

alkali conditions [10, 11]. The composition of this lipid component was found to consist of a mixture 16- 18- and 21-carbon fatty acids of which the latter was found to make up about half of the total bound fatty acids. Using a combination of NMR and mass spectrometry this 21-carbon fatty acid has subsequently been identified as 18-methyleicosanonic acid (MEA) [11]. Support for the notion that these lipids are associated with the fibre surface has come from studies that show that the proportion of both the bound fatty acids and MEA decreases as the fibre diameter increases [12]. Thus the current view of the surface of wool and hair fibres is that the long chain fatty acids are covalently linked as thioesters to a heavily crosslinked protein membrane to form a hydrophobic barrier at the surface of the cuticle cell (Fig. 1.3b) [6, 12].

1.1.2 The Exocuticle

Located below the proteolipid complexes, based on distinct differences in staining intensity, the exocuticle is generally considered to consist of two separate components (Fig. 1.3c): the a-layer and the exocuticle itself. Of these, the uppermost layer, the a-layer, is approximately 100 nm thick [7] and has a high cysteine content (1 in every 2.7 residues is cysteine) [13]. Recent studies have identified a number of cuticle ultra-high sulfur proteins in the a-layer [14]. Though the a-layer is resistant to extraction via chemical means and some enzymes, part of it can be digested by mixtures of papain and dithiothreitol. This has led to the suggestion that this intractability is due to the action of transglutaminases in the surface membranes extending their action to the underlying a-layer to catalyse isopeptide bonds such as ε-amino-(γ-glutamyl)lysine.

The exocuticle itself is of variable thickness and appears to be generally amorphous [7] and it has a significantly lower sulfur content than the a-layer [13, 15, 16]. The total protein content of the exocuticle is unknown but studies with transgenic wool have indicated that KAP5 ultra-high sulfur proteins are found in it [17], while immunolabelling techniques have demonstrated the presence of both KAP5.1 and KAP10.1 in the exocuticle [18]. Likewise, immunolabelling approaches have shown that the Type I keratins K32 and K35 and the Type II keratins K82 and K85 are also expressed in it [19–21]. The exocuticle is readily dissolved when treated with papain/dithiothreitol, or is extractable when chaotropes such as urea are used in conjunction with the reductant tris(2-carboxyethyl)phosphine which suggest that this region of the cuticle is devoid of isopeptide bonds [14].

1.1.3 The Endocuticle

Located below the exocuticle is the endocuticle, a second amorphous layer of variable thickness. It also has a markedly lower sulfur content than the exocuticle [15, 16]. It has been found to contain high levels of acidic and basic amino acids but is devoid of isopeptide bonds [7] and as such is readily digestible by enzymes. Based on its general appearance the endocuticle appears to have been derived from developing cell cytoplasm and cytoplasmic components (nucleus and cell organelles).

1.2 The Cortex

The mature cortex in wool (Fig. 1.4) and hair fibres is noted for its spindle-shaped cells between 80 μm and 115 μm in length [22] (Fig. 1.5). Characteristically the ends of cortical cells are

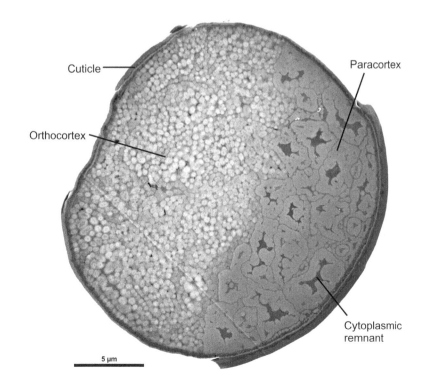

Fig. 1.4 Transmission electron micrograph of a cross section of a merino wool fibre illustrating the two main classes of cortex cell, which in fine, high curl wool differ in chemistry and internal morphology. The differences are often less extreme in other species and in higher diameter wool fibres

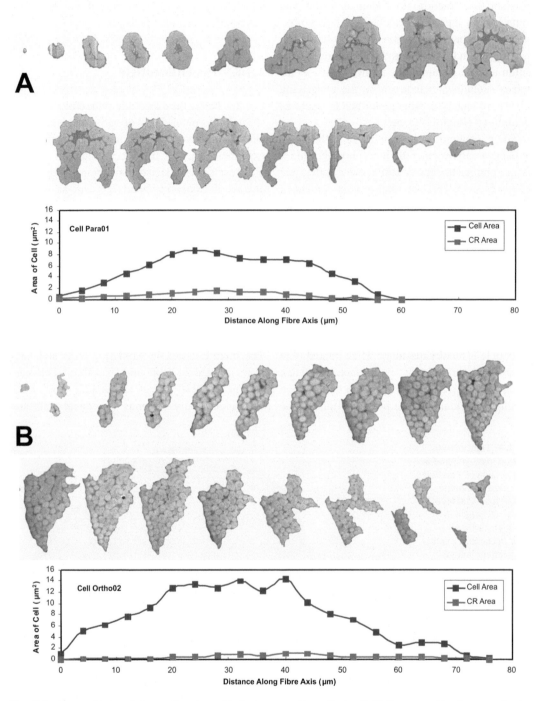

Fig. 1.5 Along-cell morphology of cortex cells from merino wool fibres as measured from transmission electron micrographs. Examples of a (**a**) paracortical and (**b**) an orthocortical cell. Cell cross-sectional area and cytoplasmic remnant areas are plotted

polyhedral as their fingertip projections interdigitate with neighbouring cells [23]. These cells are notable for being aligned in the same direction as the fibre longitudinal axis [23].

In the merino wool cortex, three types of cells have traditionally been recognized – ortho, meso and para – where they are typically clumped into groups separated in a bilateral arrangement across the fibre, the orthocortical cells on the outside of the crimp curve wave [24, 25]. Orthocortical cells appear to be the most predominant type in merino fibres, followed by paracortical cells, while mesocortical cells are a relatively minor component. These three cell types were originally identified by their different uptake of methylene blue stain [26–28]. Chemical differences between orthocortical and paracortical cells in wool has been confirmed using electron dispersive spectroscopy mapping of sulfur and iodine distribution in the fibres, which revealed a higher sulfur content in the paracortex, while iodine, which reacts with tyrosine, predominates in the orthocortex [29].

On an ultrastructural scale, macrofibrils are a composite material of an array of intermediate filaments (IFs) embedded within matrix. The arrangement of the filaments within the macrofibril varies considerably, but in fine diameter wool is well correlated with cell types (Fig. 1.6) [23, 30]. The close relationship between cell type and macrofibril structure in fine diameter wools, has become a *de facto* definition applied to hairs from many species. In fine wools, orthocortical cells tend to contain macrofibrils which are cylindrical columns composed of IFs arranged along a central linear axis, around which the filaments are helically pitched with a linear increase in helical angle for filaments in the radial dimension. It is as if there is one central straight IF surrounded by six filaments which are gently wound around the core, the next concentric ring of filaments wound gently around the inner shell, and so on to the macrofibril periphery. A term borrowed from liquid crystal studies describes the architecture perfectly as a double-twist arrangement (Sect. 2.4).

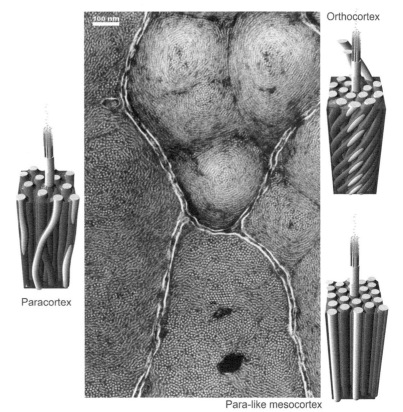

Fig. 1.6 Macrofibril types in wool cortex cells. Transmission electron micrograph of a wool fibre cross section at the boundary between paracortical, orthocortical and mesocortical cells, next to which are illustrations of the three-dimensional arrangement of IFs that are characteristic for the macrofibrils of each type (confirmed by electron tomography)

Paracortical cells have more widely spaced filaments, which are all roughly parallelly aligned along the fibre axis. Mesocortical cells always contain sizable regions in which IFs are highly aligned and packed laterally into a tight hexagonal array with a spacing intermediate between ortho- and paracortical cells. However, mesocortical macrofibrils can appear similar to those of the orthocortex or to those of the paracortex, leading to some to use the terms "ortho-like mesocortex" and "para-like mesocortex" [31]. Furthermore, in human scalp hair [32, 33], non-sheep wool fibres such as goats, alpaca and rabbits [34], deer hairs [35], higher-diameter sheep fibres (25–50 μm) and lustre mutant wools (Harland and Plowman, unpublished results), a single cortical cell can contain macrofibrils of different architectures. Many of these appear to be intermediate between orthocortical and paracortical type arrangements but are not necessarily mesocortical (lacking the tight hexagonal local packing).

In double-twist macrofibrils, irrespective of what cell type they are found within, the intensity of the helical twist can vary, and low-intensity double-twist macrofibrils are common to both orthocortical and paracortical cells of wool (Fig. 1.7) (see Chap. 11, Sect. 11.1). Macrofibrils in which IFs have mesocortical hexagonal pack-

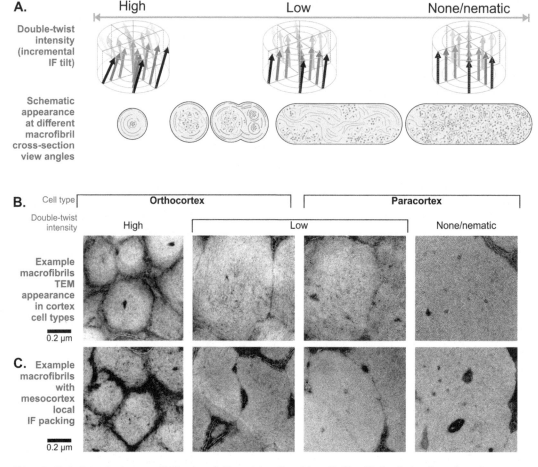

Fig. 1.7 Twist intensity in macrofibrils of wool fibres. (**a**) Theoretical model for understanding double-twist intensity of macrofibrils and key to the appearance of TEM projections of macrofibrils of differing intensity. (**b**) Examples of macrofibrils across the intensity continuum found in cells identified as being from the orthocortex or paracortex. Specific angle data cannot easily be extracted from TEM projections, but a clear impression can be made between high, low and no tilt morphologies. (**c**) The same continuum observed in mesocortical cells

ing appear to vary in the same way, suggesting that mesocortex is not an intermediate between orthocortex and paracortex as originally thought.

1.3 The Medulla

The medulla is located in the centre of the fibre and consists of a vertical column of horizontal spaces in the centre of the cortex that can be either fragmented or continuous [36]. The medulla is usually absent in fine fibres, such as the 20 μm fibres typically found in merino wool, but becomes more prevalent as the fibre diameter increases. It has a high lipid content compared to the rest of the fibre and is poor in disulfide bonding. Trichohyalin is one of the major protein components of the medulla and there is evidence that it is hardened by isopeptide bond-mediated formation of trichohyalin

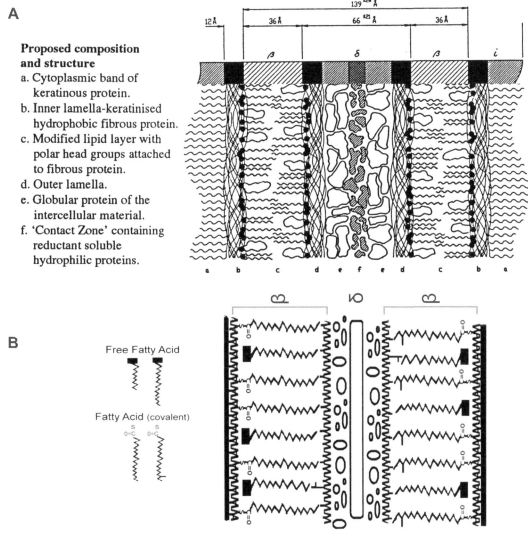

Fig. 1.8 The cell membrane complex (CMC). A schematic of the proposed structure of (**a**) the cortex-cortex CMC of wool, and (**b**) the proposed structure of the cuticle-cuticle CMC of human hair. ((**a**) modified after an original, featured in Bryson, Herbert [41]; (**b**) reprinted from J. Soc. Cosmetic Chem., 2009, Robbins [42], with permission from the Society of Cosmetic Chemistry)

molecules [37]. This appears to occur through a process whereby up to 90% of the arginine residues in trichohyalin are converted into citrulline during differentiation and that transglutaminases catalyse the formation of g-peptide bonds [38]. Morphologically it has a porous structure maintained by 'spongy keratin' and some air-filled vacuoles.

1.4 Other Components of the Fibre

1.4.1 The Cell Membrane Complex

The cell membrane complex (CMC) forms intercellular connections between the cells within the cortex and between them and the cuticle (Fig. 1.8). The CMC ranges in thickness from 25 nm to 28 nm with a centrally located δ-layer, 15–18 nm in thickness, which appears to be composed of a mixture of proteins and polysaccharides [39, 40]. Either side of this are the β-layers, 5 nm in thickness, which appear to be modified plasma membranes.

1.4.2 The Cytoplasmic/Nuclear Remnant

Each cortical cell has a nuclear or cytoplasmic remnant derived primarily from the degraded nucleus of the living cell. These are small, elongated inclusions approximately 40 μm long and 0.5–1 μm in diameter, situated in the centre of the cell.

References

1. Maderson, P. F. A. (2003). Mammalian skin evolution: A reevaluation. *Experimental Dermatology, 12*(3), 233–236.
2. Alibardi, L. (2006). Structural and immunocytochemical characterization of keratinization in vertebrate epidermis and epidermal derivatives. *International Review of Cytology, 253*, 177–259.
3. Eckhart, L., et al. (2008). Identification of reptilian genes encoding hair keratin-like proteins suggests a new scenario for the evolutionary origin of hair. *Proceedings of the National Academy of Sciences of the United States of America, 105*, 18419–18423.
4. Bradbury, J. H. (1973). The structure and chemistry of keratin fibers. *Advances in Protein Chemistry, 27*, 111–211.
5. Bradbury, J. H., & Leeder, J. D. (1970). Keratin fibres. IV. Structure of cuticle. *Australian Journal of Biological Science, 23*, 843–854.
6. Jones, L. N., & Rivett, D. E. (1997). The role of 18-methyleicosanoic acid in the structure and formation of mammalian hair fibres. *Micron, 28*(6), 469–485.
7. Swift, J. A. (1999). Human hair cuticle: Biologically conspired to the owner's advantage. *Journal of Cosmetic Science, 50*(1), 23–47.
8. Jones, L. N. (2001). Hair structure anatomy and comparative anatomy. *Clinics in Dermatology, 19*(2), 95–103.
9. von Allwörden, K. (1916). Die eigenschaften der schafwolle und eine neue untersuchungsmethodezum nachweis geschädiger wolle auf chemischem wege. *Angewandte Chemie, 29*, 77–78.
10. Logan, R. I., et al. (1989). Analysis of the intercellular and membrane lipids of wool and other animal fibers. *Textile Research Journal, 59*, 109–113.
11. Negri, A. P., Cornell, H. J., & Rivett, D. E. (1991). The nature of covalently bound fatty acids in wool fibres. *Australian Journal of Agricultural Research, 42*(8), 1285–1292.
12. Negri, A. P., Cornell, H. J., & Rivett, D. E. (1993). A model for the surface of keratin fibers. *Textile Research Journal, 63*(2), 109–115.
13. Swift, J. A. (1997). Morphology and histochemistry of human hair. In P. Jollès, H. Zahn, & H. Höcker (Eds.), *Formation and structure of human hair* (pp. 149–175). Basel: Birkhäuser Verlag.
14. Bringans, S. D., et al. (2007). Characterization of the exocuticle a-layer proteins of wool. *Experimental Dermatology, 16*(11), 951–960.
15. Jones, L. N., Kaplin, I. J., & Legge, G. J. F. (1993). Distributions of protein moieties in α-keratin sections. *Journal of Computer Assisted Microscopy, 5*(1), 85–88.
16. Hallegot, P., & Corcuff, P. (1993). High resolution spatial maps of sulphur from human hair sections; an EELS study. *Journal of Microscopy, 172*, 131–136.
17. MacKinnon, P. J., Powell, B. C., & Rogers, G. E. (1990). Structure and expression of genes for a class of cysteine-rich proteins of the cuticle layers of differentiating wool and hair follicles. *Journal of Cell Biology, 111*(6), 2587–2600.
18. Jones, L. N., et al. (2010). Location of keratin-associated proteins in developing fiber cuticle cells using immunoelectron microscopy. *International Journal of Trichology, 2*(2), 89–95.
19. Langbein, L., et al. (1999). The catalog of human hair keratins. I. Expression of the nine type I members in the hair follicle. *Journal of Biological Chemistry, 274*(28), 19874–19884.

20. Langbein, L., et al. (2001). The catalog of human hair keratins. II. Expression of the six type II members in the hair follicle and the combined catalog of human type I and II keratins. *Journal of Biological Chemistry, 276*(37), 35123–35132.
21. Yu, Z., et al. (2009). Expression patterns of keratin intermediate filament and keratin associated protein genes in wool follicles. *Differentiation, 77*(3), 307–316.
22. Kulkarni, V. G., Robson, R. M., & Robson, A. (1971). Studies on the orthocortex and paracortex of Merino wool. *Applied Polymer Symposium, 18*, 127–146.
23. Rogers, G. E. (1959). Electron microscopy of wool. *Journal of Ultrastructure Research, 2*(3), 309–330.
24. Kaplin, I. J., & Whiteley, K. J. (1978). An electron microscope study of fibril: Matrix arrangements in high and low crimp wool fibres. *Australian Journal of Biological Science, 31*, 231–240.
25. Harland, D. P., Vernon, J. A., Woods, J. L., Nagase, S., Itou, T., Koike, K., Scobie, D. A., Grosvenor, A. J., Dyer, J. M., & Clerens, S. (2018). Intrinsic curvature in wool fibres is determined by the relative length of orthocortical and paracortical cells. *The Journal of Experimental Biology, 221*(6), jeb172312.
26. Horio, M., & Kondo, T. (1953). Crimping of wool fibers. *Textile Research Journal, 23*(6), 373–387.
27. Mercer, E. H. (1953). The heterogeneity of the keratin fibers. *Textile Research Journal, 23*(6), 388–397.
28. Swift, J. A. (1977). The histology of keratin fibers. In R. S. Asquith (Ed.), *Chemistry of natural protein fibers* (pp. 81–146). London: Wiley.
29. Jones, L. N., et al. (1990). Elemental distribution in keratin fibre/follicle sections. In *Proceedings of the 8th International Wool Textile Research conference*. Christchurch: Wool Research Organisation of New Zealand.
30. Caldwell, J. P., et al. (2005). The three-dimensional arrangement of intermediate filaments in Romney wool cortical cells. *Journal of Structural Biology, 151*(3), 298–305.
31. Orwin, D. F. G., Woods, J. L., & Ranford, S. L. (1984). Cortical cell types and their distribution in wool fibres. *Australian Journal of Biological Science, 37*, 237–255.
32. Bryson, W. G., et al. (2009). Cortical cell types and intermediate filament arrangements correlate with fiber curvature in Japanese human hair. *Journal of Structural Biology, 166*(1), 46–58.
33. Harland, D. P., et al. (2014). Three-dimensional architecture of macrofibrils in the human scalp hair cortex. *Journal of Structural Biology, 185*(3), 397–404.
34. Thomas, A., et al. (2012). Interspecies comparison of morphology, ultrastructure and proteome of mammalian keratin fibres of similar diameter. *Journal of Agricultural and Food Chemistry, 60*(10), 2434–2446.
35. Woods, J. L., et al. (2011). Morphology and ultrastructure of antler velvet hair and body hair from red deer (Cervus elaphus). *Journal of Morphology, 272*(1), 34–49.
36. De Cassia Comis-Wagner, R., et al. (2007). Electron microscopic observations on human hair medulla. *Journal of Microscopy, 226*, 54–63.
37. Harding, H. W., & Rogers, G. E. (1971). (γ-glutamyl) lysine cross-linkage in citrulline-containing protein fractions from hair. *Biochemistry, 10*, 624–630.
38. Rogers, G. E. (1989). Special biochemical features of the hair follicle. In G. E. Rogers, P. J. Reis, K. A. Ward, & R. C. Marshall (Eds.), *The biology of wool and hair* (pp. 69–85). London/New York: Chapman and Hall.
39. Swift, J. A., & Bews, B. (1974). The chemistry of human hair cuticle-II: The isolation and amino acid analysis of the cell membranes and A-layer. *Journal of the Society of Cosmetic Chemistry, 25*, 355–366.
40. Orwin, D. F. (1971). Cell differentiation in the lower outer sheath of the Romney wool follicle: A companion cell layer. *Australian Journal of Biological Science, 24*(5), 989–999.
41. Bryson, W. G., et al. (1995). Characterisation of proteins obtained from papain/dithiothreitol digestion of merino and romney wools. In *Proceedings of the 9th International Wool Textile research conference*, Biella, Italy.
42. Robbins, C. R. (2009). The cell membrane complex: Three related but different cellular cohesion components of mammalian hair fibers. *Journal of the Society of Cosmetic Chemistry, 60*(4), 437–465.

The Follicle Cycle in Brief

Jeffrey E. Plowman and Duane P. Harland

Contents

2.1 Introduction 15
2.2 Telogen (The Sleeping Follicle) 16
2.3 Anagen (The Growing Follicle) 17
2.4 Catagen 17
References 17

Abstract

This chapter presents a very succinct overview of the cyclic biology of the hair follicle as it transitions from the quiescent telogen stage to the anagen stage in which hairs are actively produced before regressing through the catagen stage to telogen.

Keywords

Hair follicle · Skin · Hair cycle · Telogen · Anagen · Catagen

2.1 Introduction

The chapters within this section of the book are primarily focused on the "growing hair". However, the follicles within which hairs develop also undergo a cyclical renewal in which entirely new hairs are produced. In mammal species which have evolved in regions that undergo seasonal variations in temperature we see this renewal and replacement of hairs as moulting of summer and winter coats. At the level of the individual follicle, there is a cyclic transformation which progresses through three stages of dramatic change in morphology and metabolic activity (Fig. 2.1). The follicle cycle starts after the growth of the first hair during the embryological and post-partum morphogenesis of the skin. The basal part of the new follicle undergoes apoptosis-driven regression (catagen) ending in a period of relative quiescence (telogen). Then stem cells activate, and the base of the follicle develops anew, in a process reminiscent of its initial morphogenesis. This growth and renewal results in a new hair (anagen), which eventually stops growing, and the cycle repeats. Much of our knowledge of the details of the follicle cycle, and its regulation, have come from rodent studies, because rodent body hair follicles transition through the cycle rapidly and follow a

J. E. Plowman (✉) · D. P. Harland
AgResearch Ltd., Lincoln, New Zealand
e-mail: Jeff.Plowman@agresearch.co.nz;
Duane.Harland@agresearch.co.nz

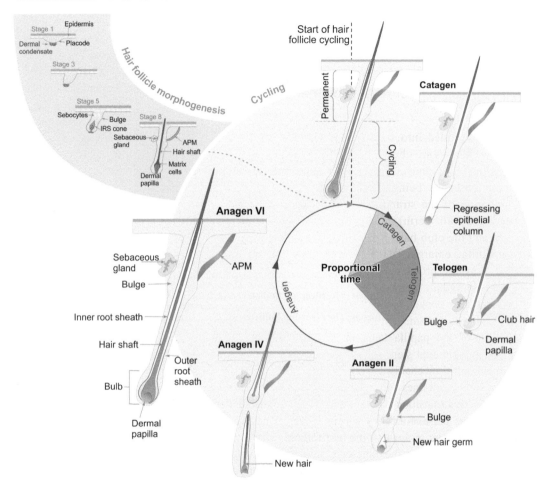

Fig. 2.1 Summary of the development and cycle of the hair follicle (based on mouse). (Minor modification of, Current Biology, 2009, Schneider, Schmidt-Ullrich [3], with permission from Cell Press). *APM arrector pili* muscle

predictable temporal pattern across their skin. Human scalp hair and wool, in contrast, have less predictable cycling, and both have extended anagen phases that can last for years. Readers with a specific interested in the field of dermatology which focuses on unpicking the genomic control of follicle cycling are directed to the references in this chapter as a good starting point for further elucidation [1–3].

2.2 Telogen (The Sleeping Follicle)

At this stage of the cycle (following catagen) the follicle is at its smallest, with a tight cluster of dermal fibroblasts (what remains of the dermal papilla) resting immediately below the bulge that contains a population of epithelial stem cells, and which marks the lower extremity of the permanent part of the follicle. The papilla is extracellular, matrix poor and its fibroblasts have very little cytoplasm, nevertheless they are largely the same as those of the anagen follicle. The epithelial cells in the lower region of the follicle show little or no DNA or RNA synthesis, nor do they synthesise anagen proteins like trichohyalin and trichocyte keratins, but they do synthesise K14 which anchors the telogen hair fibre. Embedded in the epithelial sac is the club hair or (telogen) shaft, a dead fully keratinised hair, which has a brush-like base attached to the two-layered epidermal outer root sheath sac. Interaction between the stem cells and the dermal papilla cells are responsible for triggering the end of telogen.

2.3 Anagen (The Growing Follicle)

Anagen is divided into six subphases (I–VI). Very few changes are observed in the follicle in proceeding into anagen I, one of the few changes observed being the thickening and prolongation of the strand of epidermal cells (called the hair germ) between the dermal papilla and the club hair. In anagen II the dermal papilla enlarges and becomes partly enclosed by the proliferating keratinocytes of the developing hair matrix.

In very early anagen (subphase I–III) the dermal fibroblasts becomes stratified to begin the formation of the papilla which becomes surrounded by epithelial stem cells which together form the control mechanism that will determine the order of the cell lines within the fibre and root sheath [4]. This transition is marked by a change in expression levels of a raft of genes, the most well studied of which is the down regulation of bone morphogenic protein, typically referred to simply as BMP.

In anagen III the appearance of a cone of inner root sheath (IRS) is the first sign of the new fibre to come. At the same time, the base of the follicle continues to grow downward into the skin. The next stage is characterised by the appearance of a hair shaft, entirely surrounded by all three layers of the IRS (Henle's, Huxley's and cuticle), that reaches up to the middle of the follicle.

At this point, the new hair and vanguard of new IRS may come into contact with the club hair from the previous cycle. This old hair can either be pushed out (in a process termed exogen), or the new hair can grow alongside the old, which may later be lost simply because it is not as well anchored into the skin as the growing hair.

In the final stage of anagen III the bulb and dermal papilla complete their extension downwards, coming to rest, in the human scalp, just within the subcutis.

During anagen IV the growing hair shaft and IRS reach a position just below the insertion point of the sebaceous gland, the tip of the hair shaft finally entering the hair canal in anagen V and finally emerges from the epidermis in anagen VI.

2.4 Catagen

The first outward sign of catagen is the withdrawal of the papilla fibroblast projections from the basement membrane that separates the dermal papilla cells from the stem cells in the bulb that give rise to the hair and IRS. The dermal papilla and bulb start to shrink. At this stage, the club hair forms, developing a brush-like hair base along with the three layers of the developing secondary hair germ capsule (ORS). As catagen progresses the lower follicle shrinks and withdraws as an epithelial strand from around the panniculus carnosus leaving behind a ball-shaped dermal papilla in the subcutis. Within the epithelial strand the keratinocytes become apoptotic. In the follicle, epithelial cells production of trichohyalin, transglutaminase I and desmoglein cease. By the end of catagen the germ capsule and club hair base have moved back into the dermis, while the dermal papilla is back at the boundary between the dermis and cutis.

References

1. Müller-Röver, S., et al. (2001). A comprehensive guide for the accurate classification of murine hair follicles in distinct hair cycle stages. *Journal of Investigative Dermatology, 117*(1), 3–15.
2. Alonso, L., & Fuchs, E. (2006). The hair cycle. *Journal of Cell Science, 119*(3), 391–393.
3. Schneider, M. R., Schmidt-Ullrich, R., & Paus, R. (2009). The hair follicle as a dynamic miniorgan. *Current Biology, 19*(3), R132–R142.
4. Yang, H., et al. (2017). Epithelial-mesenchymal microniches govern stem cell lineage choices. *Cell, 169*, 483–496.

Part II

Hair Proteins

Diversity of Trichocyte Keratins and Keratin Associated Proteins

3

Jeffrey E. Plowman

Contents

3.1	Introduction	22
3.1.1	Keratins	22
3.1.2	Keratin Associated Proteins (KAPs)	22
3.2	**Keratin Families**	24
3.2.1	Type I Keratins	24
3.2.2	Type II Keratins	26
3.3	**KAP Families**	26
3.3.1	The High Sulfur Proteins	26
3.3.1.1	The KAP1 Family	27
3.3.1.2	The KAP2 Family	28
3.3.1.3	The KAP3 Family	28
3.3.2	The Ultra-High Sulfur Proteins	28
3.3.2.1	The KAP4 Family	28
3.3.2.2	The KAP5 Family	29
3.3.3	The High Glycine-Tyrosine Proteins	29
3.3.3.1	The KAP6 Family	29
3.3.3.2	The KAP7 Family	29
3.3.3.3	The KAP8 Family	29
3.4	**Summary**	30
	References	30

Abstract

Wool and hair fibres are primarily composed of proteins of which the keratins and keratin associated proteins (KAPs) are the major component. Considerable diversity is known to exist within these two groups of proteins. In the case of the keratins two major families are known, of which there are 11 members in the acidic Type I family and 7 members in the neutral-basic Type II family. The KAPs are even more diverse than the keratins, with 35 families being known to exist when the KAPs found in monotremes, marsupials and other mammalian species are taken into consideration. Human hair and wool are known to have 88 and 73 KAPs respectively, though this number rises for wool when polymorphism within KAP families is included.

J. E. Plowman (✉)
AgResearch Ltd., Lincoln, New Zealand
e-mail: Jeff.Plowman@agresearch.co.nz

Keywords

High sulfur proteins · Ultra-high sulfur proteins · High glycine-tyrosine proteins

3.1 Introduction

Proteins constitute the major component of wool and hair fibres, representing 98% of the total fibre by weight [1]. Of these, keratins and their associated proteins make up 85% of the total protein content of the fibre [2]. Hair keratin proteins were first fractionated into two classes in 1934 based on their fractional salting-out properties after s-carboxymethylation [3]. Those with lower sulfur content than whole wool were designated SCMK-A and those with higher sulfur content SCMK-B. The former group became known for a time as low sulfur proteins, while the latter were more commonly referred to as high sulfur proteins.

3.1.1 Keratins

The low sulfur fraction was later subject to further fractionation using non-equilibrium two-dimensional polyacrylamide gel electrophoresis (2D-PAGE), which saw its separation into three components labeled 5, 7 and 8 [4]. Of these component 7 was resolved into three subcomponents (7a, 7b 7c) and component 8 into five subcomponents (8a, 8b, 8c-1, 8c-2 and 8c-3), although 8c-3 was later shown to be an artifact of the separation system [5]. Subsequent studies saw a further classification of these proteins. Those of lower molecular weight and higher acidity, component 8, were later designated the Type I intermediate filament proteins (IFPs), while the higher molecular weight proteins, components 5 and 7, were classified as the neutral-basic Type II IFPs [6]. This process of renaming keratins continued and in 1997 a new nomenclature was introduced in an attempt to unify the system and include keratins from other species, thus in this system the Type I keratin component 8c-1 was renamed K1.1 [7].

In parallel with this an attempt was made to introduce a comprehensive system for naming epithelial keratins in 1982 based on 2D-electrophoresis and sodium-dodecylsulfate polyacrylamide gel electrophoresis (SDS-PAGE) whereby the neutral-basic Type II keratins were labeled K1-K8 and the acidic Type I keratins K9-19 [8]. Alongside this was another system for the α-keratins of human hair where Ha stood for the acidic Type I keratins and Hb for the basic Type II keratins (H standing for hair), followed by a number [9–11]. With the addition of a prefix 'h' for human [12] or 'o' for ovine [13] the nomenclature could be adapted to name keratins in other species. This system was finally revised in 2006 with the introduction of a new consensus nomenclature for keratin genes and proteins that accommodated both functional and pseudogenes and provided room for new genes from other mammalian species [14]. Current known epithelial and trichocyte or hair keratins for wool and human hair are listed in Tables 3.1 and 3.2.

3.1.2 Keratin Associated Proteins (KAPs)

The introduction of amino acid analysis enabled a further sub-division of the SMCK-B fraction into high sulfur proteins (HSP) and ultra-high sulfur proteins (UHSP), based on whether their cysteine content was below or above 30 mol% (Table 3.3) [15]. Amino acid analysis also led to the identification of a third class of proteins in wool, one that was low in sulfur but rich in glycine and tyrosine, the so-called high glycine-tyrosine proteins (HGTP) [16, 17]. Together these three classes of proteins came to be known for a time as the matrix proteins because of their localization in the matrix material between the intermediate filaments of the fibre.

Subsequent attempts to fractionate the high sulfur group of proteins and identify subcomponents led to further improvement in our understanding of this class and also to a proliferation of new protein names. The use of fractional precipitation with ammonium sulfate solutions resulted in the definition of two fractions SCMK-B1 and SCMK-B2 [18], with subsequent sub-fractionation of the B2 group by DEAE-cellulose into a further three components by

Table 3.1 Type I epithelial and trichocyte keratins (Schweizer et al. [14])

Gene designation	Protein designation	Location
KRT9	K9	Epithelial cells
KRT10	K10	Epithelial cells
KRT11	K11	Epithelial cells
KRT12	K12	Epithelial cells
KRT13	K13	Epithelial cells
KRT14	K14	Epithelial cells
KRT15	K15	Epithelial cells
KRT16	K16	Epithelial cells
KRT17	K17	Epithelial cells
KRT18	K18	Epithelial cells
KRT19	K19	Epithelial cells
KRT20	K20	Epithelial cells
KRT23	K23	Epithelial cells
KRT24	K24	Epithelial cells
KRT25	K25	Epithelial cells
KRT26	K26	Epithelial cells
KRT27	K27	Epithelial cells
KRT28	K28	Epithelial cells
KRT31	K31	Fibre cortex
KRT32	K32	Fibre cuticle
KRT33a	K33a	Fibre cortex
KRT33b	K33b	Fibre cortex
KRT34	K34	Fibre cortex
KRT35	K35	Fibre cuticle and cortex
KRT36	K36	Fibre cortex
KRT37	K37[a]	Fibre cortex vellus hairs only
KRT38	K38	Fibre cortex
KRT39	K39	Fibre cuticle and cortex
KRT40	K40	Fibre cuticle and cortex

[a]Not found in wool

Table 3.2 Type II epithelial and trichocyte keratins (Schweizer et al. [14])

Gene designation	Protein designation	Location
KRT1	K1	Epithelial cells
KRT2	K2	Epithelial cells
KRT3	K3	Epithelial cells
KRT4	K4	Epithelial cells
KRT5	K5	Epithelial cells
KRT6	K6	Epithelial cells
KRT6A	K6A	Epithelial cells
KRT6B	K6B	Epithelial cells
KRT6C	K6C	Epithelial cells
KRT7	K7	Epithelial cells
KRT8	K8	Epithelial cells
KRT71	K71	Epithelial cells
KRT72	K72	Epithelial cells
KRT73	K73	Epithelial cells
KRT74	K74	Epithelial cells
KRT75	K75	Epithelial cells
KRT76	K76	Epithelial cells
KRT77	K77	Epithelial cells
KRT78	K78	Epithelial cells
KRT79	K79	Epithelial cells
KRT80	K80	Epithelial cells
KRT81	K81	Fibre cortex
KRT82	K82	Fibre cuticle
KRT83	K83	Fibre cortex
KRT84	K84	Dorsal tongue filiform papilla
KRT85	K85	Fibre cuticle and cortex
KRT86	K86	Fibre cortex
KRT87	K87[a]	Fibre cortex

[a]Only found in wool

chromatography; their names being shortened to B2A, B2B and B2C [19]. In parallel with these studies, column electrophoresis was used to fractionate the high sulfur component into four fractions labeled 1-IV of which only bands I and III proved to contain protein [20, 21]. Subsequent separation of these by gel filtration on Sephadex G100 resulted in the SCMK-B1 band being split in two, the BIB component being found to be equivalent to the B2 family of proteins. The SCMK-BIII peak was also split by gel filtration into two bands BIIIA and BIIIB. Thus, for a time a hybrid nomenclature system was in place where these HSPs were known as B2, BIIIA and BIIIB. Further separation of BIIIB by DEAE-cellulose led to three fractions BIIIB2, BIIIB3 and BIIIB4 [21–23].

Table 3.3 Differences in cysteine, glycine and tyrosine content of the different classes of keratins and keratin associated proteins

Group	Cysteine	Glycine	Tyrosine
Keratin	3–6.5%	2–8.5%	1–4%
High sulfur	11–28%	2–5%	1–2%
Ultra-high sulfur	29.5–38%	3–28%	0–3%
High glycine-tyrosine	0–20%	23–38%	13–29%
Other	7–9%	4–14%	0.5–9%

The HGTPs were also further sub-divided into Type I and II sub-classes by ion-exchange chromatography [24]. The Type Is were found to have a moderate percentage of glycine and tyrosine, and had only two components, C2 and F [25]. In contrast, the Type II family, which had a higher percentage of these two amino acids, was thought to contain up to ten individual components [7, 26], though only one had been fully sequenced by 1993 [27]. The final group of proteins to be sequenced were the UHSPs, two cuticle UHSPs being sequenced in 1990 and 1994 [28, 29] and two cortical UHSPs in 1994 and 1995 [30, 31].

The increasing diversity of the matrix proteins, coupled with their non-uniform naming, led to the proposal for a nomenclature for them using the abbreviation KAPm.nxpL for the protein and *KRTAPm.nxpL* for the gene [7, 32], where "m" denotes a family or unique protein, "n" denotes a variant, "p" denotes a pseudogene and "L" stand for "like". This nomenclature divides the KAPs of all species into families and further into family members based on similarities in their amino acid sequences. Historically, then, what was originally called SCMK-B became SCMK-B2, then HS-B2A and then KAP1.1 for the protein and *KRTAP1.1* for the gene. The relationship between the old and new nomenclature is illustrated in Table 3.4.

More recently the number of KAP families and proteins within those families in human hair has increased through gene sequencing [33, 34]. This has resulted in a total of 88 KAP genes in human hair, plus a total of 17 pseudogenes (non-coding genes) (Table 3.5). As a result there has been a concomitant increase in the number of KAP families and individual proteins in wool [32]. For wool this has amounted to a total of 71 proteins in 22 families but this does not take into account polymorphisms, a subject more extensively studied in sheep, where both single nucleotide polymorphism (SNP) and length polymorphism have been observed [35–40]. Polymorphism has been less well studied in human hair but a number of examples are known including some in the KAP1 family [41] and KAP4 family [33].

The rapid increase in comparative genomic data in recent times has further assisted the search for KAPs in other species [42]. From this it has been possible to identify several new subfamilies, specifically 28–35 (Table 3.6), while, based on their phylogeny subfamilies, 14 and 15 were grouped into subfamily 15 and subfamily 22 was combined with subfamily 19. Based on their amino acid profiles KAP29 and KAP31 belong to the HSP group and KAP28, KAP30, KAP32 and KAP33 are UHSPs. Interestingly a number of the HSPs have a relatively low cysteine content but a high serine content.

3.2 Keratin Families

3.2.1 Type I Keratins

The acidic Type I keratins range in size from 403 to 471 residues in length. A total of 11 Type I keratins have been identified in human hair [12, 43], while 10 Type I keratins have been found in wool [44]. Of these K31, K33a, K33b, K34 K36 and K38 are found exclusively in the cortex of human hair and wool, K32 exclusively in the cuticle, while K35, K39 and K40 are found in both the cuticle and cortex. K37, however, is found only in the cortex of human vellus hair. K31, K33a, K33b, K34, K36, K38 and K39 are also found in the medulla of wool [44] and human hair [45] (Chap. 10, Fig. 10.10a, b). Of these proteins electrophoretic studies have demonstrated that the major components of the cortex of wool are K31, K33a, K33b and K34 [46]. These four proteins exhibit the highest degree of homology, sequence alignments showing that up to 92% of

Table 3.4 Old and new ovine keratin associate protein nomenclature

Gene designation	New protein designation	Old protein designation
KRTAP1.1	KAP1.1	B2A
KRTAP1.2	KAP1.2	B2B
KRTAP1.3	KAP1.3	B2C
KRTAP1.4	KAP1.4	B2D
KRTAP2.3	KAP2.3	BIIIA3
KRTAP2.4	KAP2.4	BIIIA3A
KRTAP3.2	KAP3.2	BIIIB2
KRTAP3.3	KAP3.3	BIIIB3
KRTAP3.4	KAP3.4	BIIIB4
KRTAP6.1	KAP6.1	HGT Type II
KRTAP6.2	KAP6.2	HGT Type II component 5
KRTAP7.1	KAP7.1	HGT Type I component C2
KRTAP8.1	KAP8.1	HGT Type I component F

Table 3.5 Known human ovine and caprine keratin associated proteins (**Gong et al.** [32] #14390)

Class	Subfamily	Sheep	Goat	Human	Location
High sulfur	KAP1	4 (29)	1	4 (2)	Cortex
	KAP2	3	1	5 + 1	Cortex
	KAP3	3 + 1	2	3 + 1	Cortex
	KAP10	1		11 + 1	Cortex
	KAP11	1	1	1	Cortex
	KAP12	1	1	4 + 1	Cuticle
	KAP13	2	1	4 + 2	Cortex/Cuticle
	KAP15	1	1	1	Cortex/Cuticle
	KAP16	1	4	1	–
	KAP23	–		1	Cortex/Cuticle
	KAP24	1		1	Cuticle
	KAP25	–		1	–
	KAP26	1		1	Cuticle
	KAP27	1		1	–
Ultra-high sulfur	KAP4	27		11 + 1 (10)	Cortex
	KAP5	4 + 1 (6)		12 + 2	Cuticle
	KAP9	7	1	7 + 1	Cortex
	KAP17	–		1	Cuticle
High glycine-tyrosine	KAP6	4 (4)	6	3	Cortex
	KAP7	1 (2)	1	1	Cortex
	KAP8	2 (3)	2	1 + 2	Cortex
	KAP16	2			
	KAP19	4		7 + 4	Cortex/Cuticle
	KAP20	–		2	Cortex
	KAP21	2		2 + 1	Cortex/Cuticle
	KAP22	–		1	–
Total		73 + 2 (44)	22	88 + 17 (12)	

Numbers after "+" represent pseudogenes
Numbers in brackets are genetic variants of family members
Human KAP16.1 is a HSP, whereas the goat and sheep KAP16.1s are HGTPs

Table 3.6 KAPs in other species [42]

Gene	Species
KRTAP28	Opossum, Platypus, Dog, Rat, Mouse, Monkeys, Apes
KRTAP29	Opossum, Platypus, Dog, Rat, Mouse, Monkeys, Apes
KRTAP30	Rat, Mouse
KRTAP31	Rat, Mouse
KRTAP32	Platypus
KRTAP33	Platypus
KRTAP34	Rat, Mouse
KRTAP35	Mouse

their residues are identical and 96.5% similar. These four Type I keratins are also notable for having similar amino acid profiles and one that differs from those of K35 and K38 (Table 3.7). Of particular note is the fact that levels of glycine are higher in K35 and K38 than the major Type I keratins, more specifically, while the levels of asparagine are lower in K35 and K38. More significant differences become apparent when focusing in on the head domain, specifically there are more alanine and glycine residues in head domain of K35 and K38 and less cysteine residues than in the other four Type I keratins. In the case of glycine there are 4 or 5 residues in the head domain of K31, K33a, K33b and K34, whereas there are 17 in K35 and 12 in K38, which means that glycine residues constitute between 5% and 9% of residues of the head domain, as opposed to 18% for K35 and 11.5% for K38. Likewise, there are only two alanine residues in the head groups of K31 and K33a, K33b and K34, but 10 in K35 and 7 in K38.

Table 3.7 Amino acid profiles of the major cortical Type I keratins

Amino acids	K31	K33a	K33b	K34	K35	K38	K39	K40
Alanine	5.82	4.96	5.69	5.16	7.57	6.19	7.04	7.67
Arginine	7.76	7.94	7.67	8.35	6.68	5.75	4.97	5.12
Asparagine	7.03	7.19	7.67	6.63	4.23	4.20	5.38	5.35
Aspartic acid	4.12	3.97	3.96	3.93	4.45	4.65	3.93	5.58
Cysteine	6.06	5.70	4.46	4.18	5.35	5.97	5.18	7.67
Glutamic acid	11.40	11.41	11.14	11.55	10.02	10.40	10.14	11.40
Glutamine	6.55	6.94	7.43	7.13	6.01	5.97	7.25	6.28
Glycine	2.91	2.97	3.22	1.72	5.57	5.09	3.31	3.26
Histidine	0.48	0.99	0.50	0.98	0.89	2.21	2.28	0.70
Isoleucine	3.88	3.72	3.47	3.69	3.12	4.42	6.00	3.26
Leucine	11.89	12.15	12.13	12.29	12.03	12.61	12.01	12.79
Lysine	2.91	2.97	2.72	3.19	4.01	3.98	4.76	2.56
Methionine	0.24	0.24	0.25	0.49	1.56	0.88	2.07	0.47
Phenylalanine	1.94	2.23	1.98	1.97	2.00	1.77	1.04	1.63
Proline	4.12	3.47	3.47	2.95	3.79	3.98	3.31	3.26
Serine	8.25	8.18	8.66	12.04	10.02	9.29	7.45	9.53
Threonine	5.58	5.45	5.69	5.16	4.45	5.75	7.25	6.51
Tryptophan	0.48	0.49	0.50	0.49	0.67	0.22	0.62	0.47
Tyrosine	2.42	2.23	2.97	2.95	2.45	1.11	2.48	2.09
Valine	6.06	6.69	6.44	5.16	5.12	5.31	3.52	4.42

3.2.2 Type II Keratins

The neutral basic Type II keratins range in size from 479 to 507 residues in length, all with similar amino acid profiles (Table 3.8). A total of five Type II keratins have been identified in human hair [47] and seven in wool [44]. Of these K81, K83, K85 and K86 are found in human hair and wool, K87 also being found in the latter. K82 is found in the cuticle, with K85 in both the cuticle and cortex of both hair and wool. K84 is only found in the cytoskeletal extracts of the human tongue [47] (Chap. 10, Fig. 10.10a, b). In wool K81, K83, K85 and K86 make up the major constituents [46]. There is a high degree of homology in this family also; sequence alignments showing the number of identical residues being as high as 93% and 96% similar.

3.3 KAP Families

3.3.1 The High Sulfur Proteins

The original definition of the HSP family of proteins was defined as those proteins having levels of cysteine less than 30 moles%, a total of 12 HSP subfamilies fitting into this category (Table 3.5). The KAP1, KAP2, KAP3 and

Table 3.8 Amino acid profiles of the major cortical Type II keratins

Amino acids	K81	K83	K85	K86	K87
Alanine	8.72	8.13	9.18	9.36	7.72
Arginine	6.71	7.52	7.78	7.37	7.31
Asparagine	4.92	4.67	4.19	4.18	4.80
Aspartic acid	3.80	3.45	3.59	3.59	2.51
Cysteine	5.37	6.50	4.19	5.58	3.13
Glutamic acid	10.96	9.95	9.98	9.96	9.81
Glutamine	4.92	4.87	4.19	3.98	4.80
Glycine	6.94	8.33	6.98	6.97	7.52
Histidine	0.45	0.40	0.99	0.60	1.46
Isoleucine	4.25	4.26	4.39	4.38	4.38
Leucine	10.07	8.94	8.58	8.96	8.14
Lysine	5.59	4.87	5.18	5.38	6.47
Methionine	0.89	0.81	0.99	0.80	0.84
Phenylalanine	1.79	2.84	2.59	2.19	2.09
Proline	2.01	2.43	2.19	2.39	1.67
Serine	8.50	8.13	10.17	9.56	11.90
Threonine	3.80	4.06	4.19	4.18	4.59
Tryptophan	0.22	0.20	0.39	0.20	0.42
Tyrosine	2.68	2.43	3.19	2.99	3.76
Valine	7.38	7.11	6.98	7.37	6.68

KAP10 subfamilies are notable for having cysteine contents between 20 and 30 moles% but in other subfamilies it is lower than 20 moles% ranges and is as low as 7 moles% for KAP13.1 and KAP15.1 and 8 moles% for KAP24.1 and KAP26.1 (Table 3.9). Curiously, in these proteins the serine content is high, 25.7 moles% for KAP13.1 and 22.4 moles% for KAP24.1, while in the KAP10, KAP11 and KAP15 families the serine content is between 17 and 19 moles%.

3.3.1.1 The KAP1 Family

The four members of this highly conserved HSP subfamily differ as a result of length polymorphism involving 30 base pairs, which translates into the sequence SIQTSCCQPT [48–51]. As a result the proteins range in size from 151 to 181 residues, with anything from two to five of these decapeptide repeats (Fig. 3.1). KAP1.2 also differs from the other members of the family in the apparent loss of five residues from the C-terminus. The

Table 3.9 Amino acid profiles of selected proteins from some of the major wool HSP families

Amino acids	KAP1.1	KAP2.2	KAP3.3	KAP11.1	KAP13.1	KAP15.1
Alanine	2.94	2.29	3.06	0.63	0.61	0.74
Arginine	3.52	10.68	3.06	6.33	5.52	4.44
Asparagine	–	–	3.06	0.63	3.07	6.67
Aspartic acid	0.58	1.52	2.04	1.90	1.23	0.74
Cysteine	22.35	24.42	17.34	12.66	7.36	7.41
Glutamic acid	2.94	1.52	2.04	1.90	1.23	0.74
Glutamine	8.82	5.34	4.08	6.33	5.52	5.19
Glycine	8.92	4.58	2.04	5.70	9.82	11.11
Histidine	–	–	2.04	0.63	1.84	0.74
Isoleucine	5.29	1.52	3.06	3.80	1.23	2.22
Leucine	1.76	2.29	8.16	3.80	7.36	4.44
Lysine	–	–	1.02	0.63	–	–
Methionine	–	–	–	–	–	–
Phenylalanine	1.17	2.29	2.04	1.27	8.59	11.11
Proline	10.58	14.50	15.30	8.23	6.13	5.93
Serine	14.70	9.16	12.24	19.99	25.77	19.26
Threonine	10.58	10.68	11.22	12.03	6.75	8.15
Tryptophan	0.58	0.76	1.02	0.63	–	–
Tyrosine	2.35	1.52	2.04	3.16	4.91	5.93
Valine	2.94	6.87	5.10	10.76	3.07	5.19

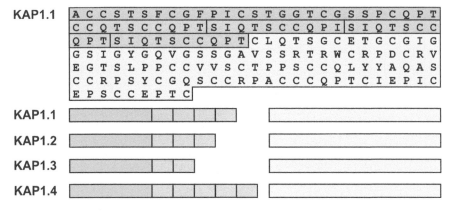

Fig. 3.1 The primary structure of a typical KAP1 protein, specifically KAP1.1, showing the sequence homology in its four decapeptide repeats (blue). The head region is coloured green and the tail region yellow. Other members of the family differ in the number of decapeptide repeats, KAP1.2, KAP1.3 and KAP1.4 having 3, 2 and 5 repeats, respectively

wool subfamily is also noted for a series of highly conserved, non-repetitive repeats that in KAP1.1 run from residues 1–11. 58–83, 101–122 and 124–157. Sequence alignments show that up to 90% of the amino acids are identical and 92% similar.

There is also a high degree of length polymorphism in the family, in particular in KAP1.1, with KAP1.1α having five such decapeptide repeats, KAP1.1β four and KAP1.1γ three [52]. Of the other members of the KAP1 family nine single nucleotide polymorphisms (SNPs) have been found for KAP1.3, of which two were silent [52] and nine have also been observed in KAP1.4 [36] and KAP1.2 [37].

In human hair the KAP1 family is expressed in the keratogenous zone of the developing follicle (Chap. 10, Fig. 10.10c) [53]. In wool there is a degree of variability in the expression of this subfamily. KAP1.3 and KAP1.4 are expressed in all animals in all breeds, while the expression of KAP1-1 variable across sheep with a breed and between breeds and KAP1.2 appears to be primarily expressed in merino wool [54].

3.3.1.2 The KAP2 Family

Though no ovine genes have been identified for this HSP family [32] it is thought to have at least 11 components of length 130–132 residues [7], though only three protein sequences have been identified to date: KAP2.2 [7], KAP2.3 and KAP2.4 [55].

Studies of human KAP2-1 have shown that this protein is predominantly expressed in the keratinization zone of the hair shaft (Chap. 10, Fig. 10.10c) [53, 56]. Like the KAP1s this subfamily is highly homologous, 93% of the sequence being identical and similar between KAP2.3 and KAP2.4.

3.3.1.3 The KAP3 Family

A total of four proteins have been described for this cortical based HSP family, with sequences of length 94–97 residues, but only three complete sequences are known, KAP3.2, KAP3.3 and KAP3.4 [22, 23, 57]. A fourth protein KAP3.1 has been identified but is considered to be a minor component, representing less than 10% of the total KAP3 protein fraction

[57]. Of the three principal elements of the family KAP3.3 and KAP3.4 are more closely related to each other, differing by only four amino acid residues, and are quite distinct from KAP3.2, differing from it by 27 residues. Even then there are differences between the gene and protein sequences, the gene sequence being one residue shorter. The offending sequence involves three cysteines at positions 46–48 as determined by amino acid sequencing, while the gene sequence predicts only two. One other interesting feature of the sequence of this family is that it appears to have two distinct regions, KAP3.2 having 6.5% sulfur content between residues 1–47 and 1.9% between residues 48–97 [22]. In human hair the KAP3 subfamily are expressed in the keratogenous zone of the follicle cortex (Chap. 10, Fig. 10.10c) [53]. The family is also highly homologous, with KAP3.3 and KAP3.4 exhibiting 96% identity and 98% similarity.

3.3.2 The Ultra-High Sulfur Proteins

Though the ultra-high sulfur subfamily was defined as those proteins having cysteine contents higher than 30 moles%, not all proteins fit exactly into this category. For instance 18 members of the 27-member KAP4 family have cysteine contents between 28 and 30 moles% (Table 3.10). The KAP4 and KAP5 subfamilies are also high in serine, while the KAP5 subfamily has a high level of glycine too.

3.3.2.1 The KAP4 Family

This UHSP family is the largest among the sheep KAPs comprising a total of 27 members ranging in length from 54 to 310 residues (Zhidong Yu, unpublished results). Approximately 80% of the protein is composed of five amino acids, cysteine, serine, proline, arginine and threonine, with cysteine making up between 28.7 and 31.2 mol%. Studies of the expression of this family in wool have demonstrated that it is found exclusively in the cortex and is expressed only on the paracortical side (Chap. 10, Fig. 10.10c) [30, 58].

Table 3.10 Amino acid profiles of selected proteins from the wool UHSP families

Amino acids	KAP4.2	KAP5.1	KAP9.2
Alanine	–	1.64	2.41
Arginine	9.95	1.09	4.47
Asparagine	–	–	0.69
Aspartic acid	0.47	–	–
Cysteine	29.38	30.76	31.62
Glutamic acid	0.47	–	2.06
Glutamine	4.27	1.09	7.56
Glycine	3.79	28.02	10.00
Histidine	0.47	–	1.37
Isoleucine	3.32	0.54	1.03
Leucine	0.95	–	1.03
Lysine	–	5.49	1.72
Methionine	–	0.54	–
Phenylalanine	–	–	1.03
Proline	11.37	4.39	11.34
Serine	21.33	20.32	11.68
Threonine	7.11	–	15.46
Tryptophan	–	–	–
Tyrosine	2.37	–	1.37
Valine	4.74	6.02	2.75

3.3.2.2 The KAP5 Family

This sheep UHSP family is found exclusively in the cuticle and has five known members, including one pseudogene, KAP5.3 (Chapter 10, Figure 10.10c) [7], though only three sequences are reported on the international databases. In sheep they range in length from 181 to 197 amino acids and are highly basic. In terms of cysteines these range in content from 29.5 moles% in KAP5.5 to 31.6 moles% in KAP5.4. They are also very high in both glycine and serine, having 26–28 moles% of the former and on average 21 moles% of the latter. Their primary structure follows the pattern of glycine-rich and cysteine-rich repeats.

3.3.3 The High Glycine-Tyrosine Proteins

At least six subfamilies of HGTPs are known in wool (Table 3.5), the glycine content ranging from 22 to 37 moles%, while the tyrosine content varies between 12 and 28 moles% (Table 3.11). Included among the HGTP subfamilies are two of the lowest molecular weight proteins in wool that have currently been named, KAP16.1 and KAP16.2, based on their sequence homology with the goat KAP16 family (KAP16.1 in human hair is a HSP). Additional members of the HGTP family include the KAP19 family of six members and KAP20.1, KAP21.1 and KAP22.1 [59].

3.3.3.1 The KAP6 Family

This HGTP family, found exclusively in the cortex (Chap. 10, Fig. 10.10c), was thought to contain as many as ten members in wool, though to date few sequences exist. These are KAP6.1, KAP6.2 and a variant of KAP6.2 that has resulted from the deletion of 12 residues, Leu^{61} to Pro^{72}, from its sequences. Five polymorphic variants of KAP6.1 have also been reported [38]. Of these KAP6.1D appears identical to KAP6.1, while KAP6.1A and KAP6.1C appear to be identical, differing in five amino acids from KAP6.1. KAP1.6B has a deletion running from S27-to G46, while KAP6.1E differs in 9 amino acids 9 deletions from KAP6.1.

True to their name the family is high in glycines (31–32 residues) and tyrosines (17–18 residues). They are relatively high in their cysteine content (8–10 residues). Comprising 62–84 residues they are among the smallest known KAPs. They are also noted for some short glycine-X repeats, such segments in silk proteins being associated with a β-strand. The glycine residue, lacking a side chain also imparts a high degree of conformational flexibility onto the chain, which would allow the tyrosine side-chains to line up in the manner of a glycine-loop protein [60].

3.3.3.2 The KAP7 Family

The KAP 7 HGTP subfamily currently has only one member. It is a basic protein, 84 residues long. It has only two short repeats of six amino acids but one notable feature is a region 18 residues long in the N-terminal half lacking in both glycine and tyrosine.

3.3.3.3 The KAP8 Family

The KAP 8 HGTP subfamily has only two members KAP8.1 and KAP8.2, the former being basic and the latter with one glutamic acid residue

Table 3.11 Amino acid profiles of selected proteins from some of the major wool HGTP families

Amino acids	KAP6.1	KAP7.1	KAP8.1	KAP16.1	KAP19.3
Alanine	–	1.19	3.28	–	
Arginine	4.87	4.76	3.28	5.36	6.94
Asparagine	1.21	4.76	1.64	–	2.78
Aspartic acid	–	–	–	–	–
Cysteine	10.97	5.95	6.56	–	6.94
Glutamic acid	–	–	–	–	–
Glutamine	–	–	–	–	–
Glycine	37.80	22.62	22.95	35.71	33.33
Histidine	–	1.19	–	–	1.39
Isoleucine	–	–	–	–	–
Leucine	6.09	5.95	3.28	8.33	4.17
Lysine	–	–	–	–	–
Methionine	–	–	–	–	–
Phenylalanine	2.43	10.71	9.84	7.14	13.89
Proline	–	7.14	6.56	1.79	1.39
Serine	14.63	14.29	13.11	12.50	13.89
Threonine	–	4.595	3.28	–	1.39
Tryptophan	–	1.19	3.28	–	–
Tyrosine	21.95	11.90	18.03	28.57	13.89
Valine	–	2.38	4.92	–	–

is acidic. This is one of the smallest proteins in wool, KAP8.1 having 61 residues and KAP8.2 62 residues. In these proteins the glycine and tyrosine residues are concentrated in the middle of protein and there are unique 10-residue segments at the N- and C-terminal ends that lack glycine and contain only one tyrosine residue. The only repeat motif is that of GYG that occurs three times.

3.4 Summary

Thus, it is apparent that there is considerable diversity in both keratins and KAPs, though just why is not clear. Some keratins are clearly critical in the initial assembly of intermediate filaments in wool and hair fibres (Chap. 10, Sect. 10.3.2). This has been demonstrated in the almost complete absence of human hair in hair and nail ectodermal dysplasia where a mutation leads to a premature termination in the K85 gene and the expression of a truncated protein [61]. In the case of the, even more diverse, KAPs the situation is less clear but at least one suggestion for their diversity has been provided in Chap. 5, Sect. 4.

References

1. Shorland, F. B., & Gray, J. M. (1970). The preparation of nutritious protein from wool. *British Journal of Nutrition, 24*, 717.
2. Gillespie, J. M., & Goldsmith, L. A. (1983). The structural proteins of hair: Isolation, characterization, and regulation of biosynthesis. In L. A. Goldsmith (Ed.), *Biochemistry and physiology of the skin* (pp. 475–510). Oxford: Oxford University Press.
3. Goddard, D. R., & Michaelis, L. (1934). A study on keratin. *Journal of Biological Chemistry, 106*, 605–614.
4. Crewther, W. G., & Lennox, F. G. (1975). Wool research in the division of protein chemistry, CSIRO. *Proceedings of the Royal Society of New South Wales, 108*(3 &4), 95–110.
5. Crewther, W. G., et al. (1980). The microfibrillar proteins of α-keratin. In D. A. D. Parry & L. K. Creamer (Eds.), *Fibrous proteins: Scientific, industrial and medical aspects* (pp. 151–159). London: Academic Press.
6. Powell, B. C. (1996). The keratin proteins and genes of wool and hair. *Wool Technology and Sheep Breeding, 44*(2), 100–118.
7. Powell, B. C., & Rogers, G. E. (1997). The role of keratin proteins and their genes in the growth, structure and properties of hair. In P. Jolles, H. Zahn, & H. Hoecker (Eds.), *Formation and structure of human hair* (pp. 59–148). Basel: Birkhäuser Verlag.

8. Moll, R., Franke, W. W., & Schiller, D. L. (1982). The catalog of human cytokeratins: Patterns of expression in normal epithelia, tumors and cultured cells. *Cell, 31*, 11–24.
9. Heid, H. W., Werner, E., & Franke, W. W. (1986). The complement of native α-keratin polypeptides of hair-forming cells: A subset of eight polypeptides that differ from epithelial cytokeratins. *Differentiation, 32*, 101–119.
10. Rogers, M. A., et al. (1998). Characterization of a 190-kilobase pair domain of human type I hair keratin genes. *Journal of Biological Chemistry, 273*(41), 26683–26691.
11. Rogers, M. A., et al. (2000). Characterization of a 300 kbp region of human DNA containing the type II hair keratin gene domain. *Journal of Investigative Dermatology, 114*(3), 464–472.
12. Langbein, L., et al. (1999). The catalog of human hair keratins. I. Expression of the nine type I members in the hair follicle. *Journal of Biological Chemistry, 274*(28), 19874–19884.
13. Plowman, J. E., et al. (2006). Wool keratins – The challenge ahead. *Proceedings of the New Zealand Society of Animal Production, 66*, 133–139.
14. Schweizer, J., et al. (2006). New consensus nomenclature for mammalian keratins. *Journal of Cell Biology, 174*(2), 169–174.
15. Gillespie, J. M., & Broad, A. (1972). Ultra-high sulphur proteins in the hairs of the Artiodactyla. *Australian Journal of Biological Science, 25*, 139–145.
16. Stein, W. H., & Moore, S. (1948). Chromatography of amino acids on starch columns; separation of phenylalanine, leucine, isoleucine, methionine, tyrosine and valine. *Journal of Biological Chemistry, 176*, 337–365.
17. Harrap, B. S., & Gillespie, J. M. (1963). A further study on the extraction of reduced proteins from wool. *Australian Journal of Biological Science, 16*, 542–557.
18. Crewther, W. G. (1975). Primary structure and chemical properties of wool. In *Proceedings of the 5th International Wool Textile Research conference*. Aachen, Germany.
19. Lindley, H., & Elleman, T. C. (1972). The preparation and properties of a group of proteins from the high-sulphur fraction of wool. *Biochemical Journal, 128*, 859–867.
20. Swart, L. S., Joubert, F. J., & Strydom, A. J. C. (1969). The apparent microheterogeneous nature of the high-sulfur proteins of à-keratins. *Textile Research Journal, 39*, 273–279.
21. Haylett, T., Swart, L. S., & Parris, D. (1971). Studies on the high-sulphur proteins of reduced merino wool. Amino acid sequence of protein SCMKB-IIIB 3. *Biochemical Journal, 123*(2), 191–200.
22. Haylett, T., & Swart, L. S. (1969). Studies on the high-sulfur proteins of reduced Merino wool. Part III. The amino-acid sequence of protein SCMKB-IIIB2. *Textile Research Journal, 39*, 917.
23. Swart, L. S., & Haylett, T. (1971). Studies on the high-sulphur proteins of reduced merino wool. Amino acid sequence of protein SCMKB-IIIB 4. *Biochemical Journal, 123*(2), 201–210.
24. Gillespie, J. M., & Darskus, R. L. (1971). Relation between the tyrosine content of various wools and their content of a class of proteins rich in tyrosine and glycine. *Australian Journal of Biological Science, 24*, 1189–1197.
25. Gillespie, J. M. (1991). The structural proteins of hair: isolation, characterisation and regulation of biosynthesis. In L. A. Goldsmith (Ed.), *Physiology, biochemistry and molecular biology of the skin* (Vol. 1, 2nd ed., pp. 625–659). New York: Oxford University Press.
26. Powell, B. C., & Beltrame, J. S. (1994). Characterization of a hair (wool) keratin intermediate filament gene domain. *Journal of Investigative Dermatology, 102*(2), 171–177.
27. Fratini, A., Powell, B. C., & Rogers, G. E. (1993). Sequence, expression, and evolutionary conservation of a gene encoding a glycine/tyrosine-rich keratin-associated protein of hair. *Journal of Biological Chemistry, 268*(6), 4511–4518.
28. MacKinnon, P. J., Powell, B. C., & Rogers, G. E. (1990). Structure and expression of genes for a class of cysteine-rich proteins of the cuticle layers of differentiating wool and hair follicles. *Journal of Cell Biology, 111*(6), 2587–2600.
29. Jenkins, B. J., & Powell, B. C. (1994). Differential expression of genes encoding a cysteine-rich keratin family in the hair cuticle. *Journal of Investigative Dermatology, 103*(3), 310–317.
30. Fratini, A., et al. (1994). Dietary cysteine regulates the levels of mRNAs encoding a family of cysteine-rich proteins of wool. *Journal of Investigative Dermatology, 102*(2), 178–185.
31. Powell, B. C., Arthur, J. R., & Nesci, A. (1995). Characterisation of a gene encoding a cysteine-rich keratin associated protein synthesised late in rabbit hair follicle differentiation. *Differentiation, 58*, 227–232.
32. Gong, H., et al. (2012). An updated nomenclature for keratin-associated proteins (KAPs). *International Journal of Biological Sciences, 8*(2), 258–264.
33. Rogers, M. A., & Schweizer, J. (2005). Human KAP genes, only the half of it? Extensive polymorphisms in hair keratin-associated protein genes. *Journal of Investigative Dermatology, 124*, vii–ix.
34. Rogers, M. A., et al. (2006). Human hair keratin-associated proteins (KAPs). *International Review of Cytology, 251*, 209–263.
35. Itenge-Mweza, T. O., et al. (2007). Polymorphism of the KAP1.1, KAP1.3 and KRT.1.2 genes in merino sheep. *Molecular and Cellular Probes, 21*(5–6), 338–342.
36. Gong, H., Zhou, H., & Hickford, J. G. (2010). Polymorphism of the ovine keratin-associated-protein 1-4 (KRTAP1-4) gene. *Molecular Biology Reports, 37*, 3377–3380.

37. Gong, H., et al. (2011). Identification of the ovine keratin-associated protein KAP1-2 gene (KRTAP1-2). *Experimental Dermatology, 20,* 815–819.
38. Gong, H., Zhou, H., & Hickford, J. G. (2011). Diversity of the glycine/tyrosine-rich keratin-associated protein 6 gene (KAP6) family in sheep. *Molecular Biology Reports, 38*(1), 31–35.
39. Gong, H., et al. (2011). Identification of the ovine KAP11-1 gene (KRTAP11-1) and genetic variation in its coding sequence. *Molecular Biology Reports, 38*(8), 5429–5433.
40. Gong, H., et al. (2011). Search for variation in the ovine KAP7-1 and KAP8-1 genes using PCR-SSCP. *DNA and Cell Biology, 31*(3), 367–370.
41. Shimomura, Y., et al. (2002). Polymorphisms in the human high sulfur hair keratin-associated protein 1, KAP1, gene family. *Journal of Biological Chemistry, 277*(47), 45493–45501.
42. Wu, D.-D., Irwin, D. M., & Zhang, Y.-P. (2008). Molecular evolution of the keratin associated protein gene family in mammals, role in the evolution of mammalian hair. *BMC Evolutionary Biology, 25*(8), 241–255.
43. Langbein, L., et al. (2007). Novel type I hair keratins K39 and K40 are the last to be expressed in differentiation of the hair: Completion of the human hair keratin catalogue. *Journal of Investigative Dermatology, 127,* 1532–1535.
44. Yu, Z., et al. (2011). Annotations of sheep keratin intermediate filament genes and their patterns of expression. *Experimental Dermatology, 20*(7), 582–588.
45. Langbein, L., et al. (2010). The keratins of the human beard hair medulla: The riddle in the middle. *Journal of Investigative Dermatology, 130*(1), 55–73.
46. Deb-Choudhury, S., et al. (2010). Electrophoretic mapping of highly homologous keratins: A novel marker peptide approach. *Electrophoresis, 31*(17), 2894–2902.
47. Langbein, L., et al. (2001). The catalog of human hair keratins. II. Expression of the six type II members in the hair follicle and the combined catalog of human type I and II keratins. *Journal of Biological Chemistry, 276*(37), 35123–35132.
48. Elleman, T. C. (1972). The amino acid sequence of protein SCMK-B2A from the high-sulphur fraction of wool keratin. *Biochemical Journal, 130*(3), 833–845.
49. Elleman, T. C. (1972). The amino acid sequence of protein SCMK-B2C from the high-sulphur fraction of wool keratin. *Biochemcal Journal, 128,* 1229–1239.
50. Elleman, T. C., & Dopheide, T. A. (1972). The sequence of SCMK-B2B, a high-sulfur protein from wool keratin. *Journal of Biological Chemistry, 247*(12), 3900–3909.
51. Powell, B. C., et al. (1983). Mammalian keratin gene families: Organisation of genes coding for the B2 high-sulphur proteins of sheep wool. *Nucleic Acids Research, 11*(16), 5327–5346.
52. Rogers, G. R., Hickford, J. G., & Bickerstaffe, R. (1994). Polymorphism in two genes for B2 high sulfur proteins of wool. *Animal Genetics, 25*(6), 407–415.
53. Rogers, M. A., et al. (2001). Characterization of a cluster of human high/ultrahigh sulfur keratin-associated protein genes embedded in the type I keratin gene domain on chromosome 17q12-21. *Journal of Biological Chemistry, 276*(22), 19440–19451.
54. Flanagan, L. M., Plowman, J. E., & Bryson, W. G. (2002). The high sulphur proteins of wool: Towards an understanding of sheep breed diversity. *Proteomics, 2*(9), 1240–1246.
55. Swart, L. S., & Haylett, T. (1973). Studies on the high-sulphur proteins of reduced merino wool. Amino acid sequence of protein SCMKB-IIIA3. *Biochemical Journal, 133*(4), 641–654.
56. Fujikawa, H., et al. (2012). Characterization of the human hair keratin-associated protein 2 (KRTAP2) gene family. *Journal of Investigative Dermatology, 132*(7), 1806–1813.
57. Frenkel, M. J., et al. (1989). The keratin BIIIB gene family: Isolation of cDNA clones and structure of a gene and a related pseudogene. *Genomics, 4,* 182–191.
58. Yu, Z., et al. (2009). Expression patterns of keratin intermediate filament and keratin associated protein genes in wool follicles. *Differentiation, 77*(3), 307–316.
59. Rogers, M. A., et al. (2002). Characterization of a first domain of human high glycine-tyrosine and high sulfur keratin-associated protein (KAP) genes on chromosome 21q22.1. *Journal of Biological Chemistry, 277*(50), 48993–49002.
60. Parry, D. A. D., et al. (2006). Human hair keratin-associated proteins: Sequence regularities and structural implications. *Journal of Structural Biology, 155*(2), 361–369.
61. Shimomura, Y., et al. (2010). Mutations in the keratin 85 (KRT85/hHb5) gene underlie pure hair and nail ectodermal dysplasia. *Journal of Investigative Dermatology, 130,* 892–895.

Evolution of Trichocyte Keratins

Leopold Eckhart and Florian Ehrlich

Contents

4.1	**General Principles of Keratin Evolution**	34
4.2	**The Evolutionary Origin of Keratins**	35
4.3	**The Exon-Intron Organization of Keratin Genes Changed at the Divergence of Type I and Type II Keratins and at the Origin of Type I Hair Keratins**	35
4.4	**Conservation and Diversification of Keratin Features**	37
4.4.1	Evolution of Primary Structure and Expression Patterns of Keratins	37
4.4.2	Conservation of the IF Domain	38
4.4.3	Evolution of the Head and Tail Domains	38
4.4.4	Keratins of Different Types of Epithelia	38
4.4.5	Regulation of Keratin Gene Expression	38
4.4.6	Evolution of Keratin Pairs	38
4.4.7	Evolution of Keratins with Unique Structures	39
4.5	**Origin of Hair Keratins**	39
4.6	**Diversification of Hair Keratins**	42
	References	43

Abstract

The evolution of keratins was closely linked to the evolution of epithelia and epithelial appendages such as hair. The characterization of keratins in model species and recent comparative genomics studies have led to a comprehensive scenario for the evolution of keratins including the following key events. The primordial keratin gene originated as a member of the ancient gene family encoding intermediate filament proteins. Gene duplication and changes in the exon-intron structure led to the origin of type I and type II keratins which evolved further by nucleotide sequence modifications that affected both the amino acid sequences of the encoded proteins and the gene expression patterns. The diversification of keratins facilitated the emergence of new and epithelium type-specific properties of the cytoskeleton. In a common ancestor of

L. Eckhart (✉) · F. Ehrlich
Research Division of Biology and Pathobiology of the Skin, Department of Dermatology, Medical University of Vienna, Vienna, Austria
e-mail: leopold.eckhart@meduniwien.ac.at; florian.ehrlich@meduniwien.ac.at

reptiles, birds, and mammals, a rise in the number of cysteine residues facilitated extensive disulfide bond-mediated cross-linking of keratins in claws. Subsequently, these cysteine-rich keratins were co-opted for an additional function in epidermal follicular structures that evolved into hair, one of the key events in the evolution of mammals. Further diversification of keratins occurred during the evolution of the complex multi-layered organisation of hair follicles. Thus, together with the evolution of other structural proteins, epithelial patterning mechanisms, and development programmes, the evolution of keratins underlied the evolution of the mammalian integument.

Keywords

Evolution · Molecular phylogenetics · Gene duplication · Sequence motif · Cysteine · Epithelium · Claw · Nail · Tongue · Hair

4.1 General Principles of Keratin Evolution

Humans and most other mammals have more than 50 keratins that differ in their structures and tissue expression patterns. This astonishing diversity of keratins is the product of evolution, that is, changes of genes over time. In mammal species, significant changes of keratins have occurred over the range of millions of years while the breeding of domestic species may select for certain keratin phenotypes on a much shorter time scale [1, 2].

The evolutionary diversification of keratins involved gene duplications and changes in sequence, which are well-characterized processes of gene evolution in general [3–5]. The prevalent mode of gene duplication in gene families is tandem duplication which results in two gene copies located adjacent to each other. The molecular mechanism of this type of duplication involves "unequal crossing over" and homologous DNA recombination [6]. Other modes of gene duplication are whole-genome duplication, duplication of chromosome segments, and retroposition (mRNA reverse-transcription and insertion of the cDNA copy into a chromosome). The latter gene duplications lead to a dispersed distribution of gene copies. The internal organisation (exons and introns) of the genes is typically maintained in all types of duplications except for retroposition which causes the loss of intronic sequences and requires the acquisition of novel promoter and enhancer sequences that drive expression of the retro-transposed gene copy. Recombination between genes can result in chimeric genes whereby most often individual exons from two genes are combined in a single gene (exon shuffling). Obviously, only gene mutations and recombinations in germ line cells have the potential to directly contribute to gene evolution whereas the effects of mutations in somatic cells are limited to individuals.

In the course of a gene duplication and during the transmission of the daughter genes through the subsequent generations, the two gene copies accumulate different mutations both in the coding sequence, thereby altering the structure of the protein and its interactions with other molecules in the cell, and in the gene elements that regulate expression, thereby altering the spatiotemperal distribution of the encoded protein in the developing and aging organism. Gene duplications may lead to an increase in the abundance of the encoded protein (expression of two genes instead of one), subfunctionalisation (each copy having a specific function whereas the ancestral protein fulfilled both functions), and neofunctionalisation (one copy assumes a new function while the other copy keeps its original role) [7].

Besides the immediate evolutionary processes affecting keratins, it is important to consider the evolution of keratins in the cellular and organismal context. The evolution of keratins required that crucial interactions of keratins with other molecules in the cell were maintained in all generations. Changes in the structure and expression sites of interacting proteins occurred concurrently with changes of keratins. At higher levels of system hierarchy, the evolution of keratins was influenced by and drove the evolution of epithelia and epithelial derivatives (such as hair) within species, and the diversification of species. One of the most striking examples of the integration of processes at different system levels is the co-evolution of cysteine-rich keratins and keratin-associated proteins which facilitated the hard cornification of keratinocytes, the emergence of hair and the evolution of mammals (see below).

4.2 The Evolutionary Origin of Keratins

Keratins belong to the superfamily of intermediate filament proteins. Accordingly, these proteins are characterized by the presence of an intermediate filament (IF) domain (for details, see other chapters of this book). Other proteins that were previously named "keratins" because of their presence in keratinized tissues are phylogenetically unrelated to keratin IF proteins and should be renamed to avoid confusion. In particular, the so-called "beta-keratins" are reptilian and avian skin proteins that lack an IF domain and have evolved from genes within the epidermal differentiation complex, a gene cluster without an evolutionary link to keratin genes [8]. Accordingly, the use of the term "beta-keratin", as opposed to "alpha-keratins" for true keratins with an intermediate filament domain, should be discontinued. The evolution of the "beta-keratins" which are now termed "corneous beta proteins" has been reviewed elsewhere [9].

Keratins have evolved from ancestral IF protein genes. The primordial IF protein and founding member of the IF superfamily was a lamin that was present in a common ancestor of all metazoans [10]. Lamins (type V IF proteins) have been conserved in Porifera (sponges), Ctenophora (comb jellies), Cnidaria (sea anemonies and jellyfish), Nematoda, Arthropoda, Mollusca, Echinodermata, Hemichordata and Chordata. Lamins have a prototypical IF domain, that consists of four alpha-helical segments (Chap. 3, Sect. 3.2) with end motifs that are also present in other IF proteins. Close to the carboxy-terminus of lamins, there is an immunoglobulin domain which is preceded by a nuclear localisation signal and followed by an isoprenylation motif [10]. The latter sequence elements lead to the localisation of lamins at the inner nuclear membrane.

The evolutionary loss of the nuclear localisation signal and of the isoprenylation motif led to the origin of the cytoplasmic IF (cIF) proteins. cIF proteins outside of chordates retained the immunoglobulin domain whereas cIFs lost this domain in chordates. At the nucleotide sequence level, the positions of introns within the IF-coding region were conserved between lamin and cIF genes, but additional introns appeared in the gene segment encoding the carboxy-terminus of cIFs [10].

The primitive cIFs have diversified (in an uncertain order of events) into type I and type II IF proteins (which together are referred to as keratins), type III IF proteins (vimentin, desmin, glial fibrillary acidic protein (GFAP), peripherin), type IV IF proteins (neurofilament proteins, synemin, syncoilin, α-internexin) and type VI IF proteins (nestin). In chordates, cIFs (including keratins) have lost a stretch of 42 amino acid residues in coil 1b [11].

4.3 The Exon-Intron Organization of Keratin Genes Changed at the Divergence of Type I and Type II Keratins and at the Origin of Type I Hair Keratins

Keratins K8 and K18 are considered the founding members of the type II and type I keratin families, respectively (Fig. 4.1) [12]. They are present in all chordates, and in many species the genes encoding them are located next to each other, indicating that they have emerged by tandem duplication of a single primordial keratin gene. In mammals *KRT8* and *KRT18* are neighboring genes (on human chromosome 12q13) with opposite orientation [12]. On the 3'-side of the *KRT8* (type II) gene, there is the cluster of type II keratin genes. This pattern strongly suggests that the type II keratin gene cluster has evolved by tandem gene duplications. All type II keratin genes with the exception of *KRT7* and *KRT86* have the same ori-

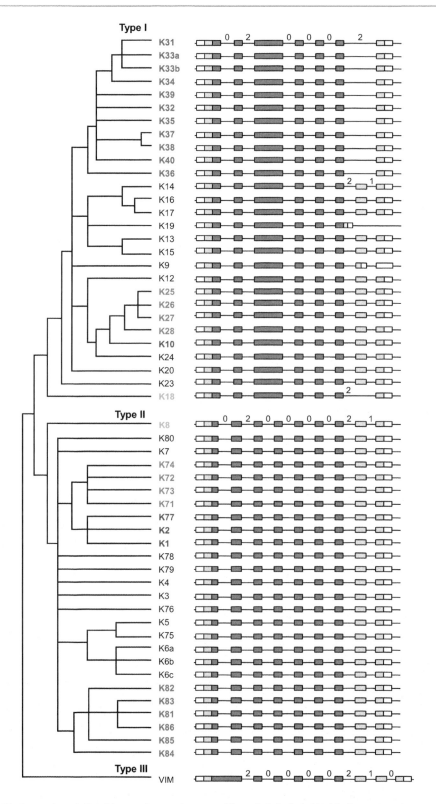

Fig. 4.1 Phylogenetic relationships and exon-intron structures of keratin genes. The phylogenetic tree of keratins (left) was modified from Vandebergh and Bossuyt [50]. Vimentin is included in this diagram to represent type III IF genes. Hair keratins are highlighted by red fonts, whereas orange and purple fonts indicate keratins of the inner root

entation as *KRT8*. Human type I keratin genes other than *KRT18* are located on chromosome 17q21. They are also arranged in the form of a cluster and all of these genes have the same orientation. Interestingly, the type I keratin gene cluster is interrupted by genes of the *keratin-associated protein* (*KRTAP*) family, indicating that the latter may have originated by modification of an ancestral type I keratin gene (discussed in [13]).

The assumption that *KRT8* and *KRT18* have originated by gene duplication raises the question as to which one of the two more closely resembles the single ancestral (primordial) keratin. In this regard it may be important to note that *intron 3 of KRT8* (and all other type II keratins) is homologous to intron 2 of type III IF genes whereas this intron is absent in *KRT18* (and all other type I keratins). Since this intron appears to have been present in the ancestor of type 1–3 IF genes [14], it can be concluded that the primordial *KRT18* gene has lost this intron after *KRT18* has originated by duplication from the more ancient primordial *KRT8* gene (Fig. 4.1).

Besides the aforementioned intron, all other introns within IF-coding region are located at homologous positions in type I and type II keratin genes. The introns in the 3′-terminal gene segment (downstream of the IF-coding region) are apparently not located at homologous positions but spliced in equivalent phases (positions within triplets). In most type I and all type II keratin genes, the last intron is spliced in phase 1 and the last-but-one intron is spliced in phase 2. By contrast, some type I keratin genes deviate from this pattern. *KRT9* has one coding and one non-coding exon at its 3′-end, and *KRT19* lacks homologs of two 3′-terminal exons. *KRT18* and all type I hair keratin genes are characterized by the presence of a single exon (spliced in phase 2) that encodes the tail domain (Fig. 4.1). In contrast to *KRT18* of mammals and other species within the clade Euteleostomi (bony fish and tetrapods), the *KRT18* gene of the elephant shark (*Callorhinchus milii*), a representative of the more "primitive" Chondrichthyes (cartilaginous fishes), has 2 exons, that together encode the tail domain. This pattern suggests that the presence of the 2 exons was the ancestral condition of *KRT18* and that this structure was inherited to other epithelial type I keratins already present in the common ancestor of Chondrichthyes and Osteichthyes [15]. It appears likely that the similar 3′-terminal exon-intron-structure of mammalian *KRT18* and type I hair keratin genes (Fig. 4.1) are the product of convergent evolution rather than the consequence of direct evolution of type I hair keratin genes from *KRT18*. However, further comparative studies of the keratin gene organisation in lung fish and amphibians are necessary to evaluate when and how the exon-intron structure of type I hair keratin genes has originated.

4.4 Conservation and Diversification of Keratin Features

4.4.1 Evolution of Primary Structure and Expression Patterns of Keratins

Keratins have evolved under selective constraints on their amino acid sequence and expression patterns. A comprehensive description of the diversification of keratins is beyond the scope of this chapter, but several principles will be summarized here. Of note, the currently available phylogenetic tree of keratins contains uncertain branching points and better resolution of these phylogenetic ambiguities is required to improve the scenarios of the evolution of keratins and domains within individual keratins.

Fig. 4.1 (continued) sheath and of the suprabasal epidermis. The primordial keratins K8 and K18 are highlighted by green fonts. In the comparative depiction of gene structures (right), exons are depicted as boxes. Dark and light shading indicated segments that encode the intermediate filament domain and the terminal (head and tail) domains, respectively. White parts of the boxes correspond to the non-coding regions of exons. Numbers above introns indicate the phase of splicing, i.e. the position of the intron relative to the reading frame (phase 0, between codons; phase 1, between the first and second nucleotide of a codon; phase 2, between the second and third nucleotide of a codon). *K* keratin, *VIM* vimentin

4.4.2 Conservation of the IF Domain

The IF domain has shown limited sequence variability throughout evolution, indicating that the conserved amino acid residues have essential roles in the function of keratins. Accordingly, sequence comparisons between homologous keratins of different species and different keratins (paralogs) within the same species allow to define critical residues and to predict the impact of mutations identified in human patients [16–20].

4.4.3 Evolution of the Head and Tail Domains

The sequences of head and tail domains show higher variability than that of the central rod domain. In type I and type II keratins of the epidermis these domains have independently evolved high contents of glycine and a regular spacing of aromatic residues (tyrosine and phenylalanine) which, according to the prevalent models, facilitate the formation of loops and intermolecular binding [21, 22]. By contrast, hair keratins have acquired high contents of cysteine residues in the terminal domains but also in the central rod domain [23], indicating that disulfide bonds distributed over the entire protein may link hair keratins to other proteins within mature hair.

One of the conserved sequence motifs outside of the IF domain is a carboxy-terminal motif (CTM) located in the tail domain of many but not all keratins [24]. A CTM of the consensus sequence VKTVETRDGEVI is present in K8 and type III IF proteins (vimentin, desmin, GFAP), indicating that this sequence element has originated in the common ancestor gene of type III and type II (and therefore also type I) IF proteins. Many type I and type II keratins have variants of the ancestral motif which have been denoted CTM-A and CTM-B [24]. CTM-A (EIRDGKVI) appears to be homologous to the second half of the ancestral CTM and is present in type I keratins, including hair keratins K36 and K39. CTM-B (VKFVST) is homologous to the amino-terminal half of the ancestral CTM and present in type II keratins, including hair keratin K84. Conservation of CTMs in a subset of keratins suggests that they are functional and under selective constraints in these keratins. However, the ancestral CTM was lost in other keratins, both of the epithelial and hair keratin types, indicating that CTMs are not strictly essential for keratin functions.

4.4.4 Keratins of Different Types of Epithelia

Keratins expressed in simple epithelia have more basal positions in the phylogenetic tree and appear to have homologs in phylogenetically more distant clades than keratins of stratified epithelia. This pattern suggests that keratins of simple epithelia have evolved early in evolution and keratins of stratified epithelia have evolved from copies of these ancestral keratins.

4.4.5 Regulation of Keratin Gene Expression

The evolution of keratins involved not only the diversification of amino acid sequences but also the emergence of regulatory mechanisms to specifically direct the expression of individual keratins to distinct types of epithelia [25, 26]. Various transcription factors such as AP1, AP2, Sp1, ets factors and others bind to the promoters and enhancers of epithelial keratin genes [27] and NF-kappa, Hoxc13, Foxn1 and others bind to the promoters of hair keratin genes [28–30]. However, many aspects of the regulation of keratin expression and the evolutionary history of the transcription factor binding sites remain to be defined.

4.4.6 Evolution of Keratin Pairs

The expansions of type I and type II keratins occurred in parallel and facilitated the evolution of epithelium-specific and skin appendage layer-specific pairs of heterodimerising keratins such as K1 and K10 (in the epidermis), K3 and K12 (in the corneal epithelium), and K4 and K13 (in the oral epithelium) [25, 26]. The evolution of any new pair of keratins, that are located on different chromosomes but co-expressed, is unlikely to be mediated by mutations that occur concurrently in 2 genes, but by selection for coordinated subfunctionalisation of keratin gene copies. These processes may share similarities to those proposed for the evolution of dimerising tubulins, the components of microtubules [31, 32]. At present, the fascinating phenomenon of co-evolution of keratin genes at two different loci awaits further investigations.

4.4.7 Evolution of Keratins with Unique Structures

Despite conservation of many structural principles of keratins, unique deviations from the canonical organization of keratins have evolved. Examples include, but are not limited to, the loss of the tail domain in K19 and the elongation of the tail domain of K78 in murine rodents [33].

4.5 Origin of Hair Keratins

The evolution of hair was a major event in the evolution of mammals. The hair follicle is a complex mini-organ and its origin depended on evolutionary innovations of genes that encode structural components of mature hair (such as hair keratins and keratin-associated proteins), components of epithelia that serve as scaffolds of the growing hair (such as trichohyalin of the inner root sheath) and regulators of gene expression and cell differentiation [34–36]. Hair keratins are heavily cross-linked via disulfide bonds whereas other keratins form only few or even none of these covalent protein connections [23]. From the comparison of amino acid sequences of keratins from phylogenetically diverse animals including fish, amphibians and amniotes, it can be concluded that keratins originally had a low cysteine content, and the high cysteine content of hair keratins arose by amino acid sequence changes at positions that originally contained other residues.

The recent availability of genome sequences from phylogenetically diverse vertebrates has allowed the identification of keratins most similar to hair keratins in non-mammalian species. These studies demonstrated that both the groups of type I and type II hair keratins have orthologs in reptiles and these orthologs have cysteine contents similar to those of mammalian hair keratins [23]. Specifically, the green anole lizard (*Anolis carolinensis*) was investigated as the first reptilian species with a fully sequenced genome. Two orthologs of mammalian type I hair keratins and 4 orthologs of type II hair keratins were identified in this lizard. The genes encoding hair keratin homologs show shared synteny with mammalian hair keratin genes, that is, they are flanked by homologous neighboring genes. Phylogenetic analyses confirmed a close relationship between mammalian hair keratins and reptilian hair keratin-like proteins. *In situ* localisation of representative type I and type II hair keratin homologs of the lizard showed that they were specifically expressed in the matrix of claws. This expression pattern was similar to the expression of mammalian hair keratins in the human nail matrix and in the matrix of claws and hooves of other mammals [37–39]. These data strongly suggest that a common ancestor of mammals and lizards already had hair keratin-like proteins that were expressed in the growth zone of claws. Accordingly, "hair keratins" were originally "claw keratins" before they were co-opted for an additional function in hair (Fig. 4.2). Consequently, the number of "hair keratins" increased during the evolution of hair and some of the hair keratins have adapted an expression specific to only one of the layers of the hair fiber (cuticle, cortex, medulla) and specific to stages of cell differentiation within these layers (for details, see the other chapters of this book) [40].

Although this evolutionary scenario is consistent with available phylogenetic and gene expression data, it does not describe the entire early evolution of hair keratins. First, mammalian and reptilian "hair keratins" have additional sites of expression that may be either ancestral or derived features. Some of the lizard claw keratins were also detected, though only at the mRNA level, in the tongue and in the scaled abdominal skin [23].

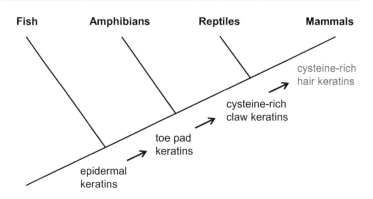

Fig. 4.2 The evolutionary origin of hair keratins. Based on the species distribution of hair keratin homologs and their tissue expression patterns (Eckhart et al. [23] and Vandebergh et al. [48]), a hypothesis on the evolutionary history of hair keratins was developed

In mammals, hair keratins are present not only in hair and claws (and homologous structures) but also in the filiform papillae of the tongue, in the scales on the tail of rodents [41, 42] and in the horn of cattle, sheep, antelopes and other ruminant artiodactyls [43]. While horns have certainly not been inherited from a common ancestor of reptiles and mammals, it is conceivable that both filiform papillae on the tongue and rigid scales were already present in an ancestor of all modern amniotes. It is likely that the first of these hair keratin-expressing structures (claws, filiform papillae of the tongue, hard epidermal scales) were the primordial sites of hair keratin expression and function. Indeed, hypotheses about filiform papillae of the tongue and scales as precursors of hair have been put forward [44]. Assuming that all these skin appendages evolved earlier than claws, hair keratin homologs may have had original functions in these structures before they contributed to the evolution of claws.

To address the question of the sequence of appearance of skin appendages during the evolution of tetrapods, it may be informative to perform comparative studies in amphibians, i.e. the phylogenetically basal tetrapods. Several taxa of frogs and caudates have claws (hence the name clawed frog for Xenopus) but amphibian claws are morphologically different from and probably not homologous to claws of amniotes [45]. The tongues of frogs have filiform papillae that are probably homologous to the papillae on the tongue of amniotes [46]. Scales are generally absent in amphibians but they may have been present in the last common ancestor of tetrapods [47].

Recently, comparative genomics has revealed potential orthologs of hair keratins in frogs. The cysteine content of these frog keratins (1–5 cysteine residues) is markedly lower than that of mammalian hair keratins (20 or more cysteine residues) [48, 49]. Interestingly, these hair-keratin-like proteins are expressed in the toe pads of tree frogs. This distribution supports the hypothesis that evolutionary precursors of hair keratins were originally expressed at the toes of early tetrapods and, by the increase of the cysteine content, acquired the feature of disulfide bond-mediated cross-linking that contributed to the rigidification of the digit tips and the origin of the amniote-type claws (Fig. 4.2). The expression of keratins in the tongues of the amphibians is not known, which is, in part, due to the fact that the model species Xenopus has secondarily lost a normal tongue.

Molecular phylogenetics does not clearly identify single other keratins or specific groups of other keratins that would be most similar type I and type II hair keratins, respectively [50] (Fig. 4.1). Together with considerations of the species distribution of keratin orthologs and of the arrangement of keratin genes, results of phylogenetics are compatible with the hypotheses that K8 and K18 are the most basal keratins and K23 and K80 represent other ancient phylogenetic lineages of keratins. In this scenario, the primordial hair keratins have originated either directly or indirectly (via other intermediate ancestral genes) by duplication and sequence modification of K23 (leading to type I hair keratins) and K80 (leading to type II hair keratins) (Fig. 4.3).

4 Evolution of Trichocyte Keratins

Type I keratins

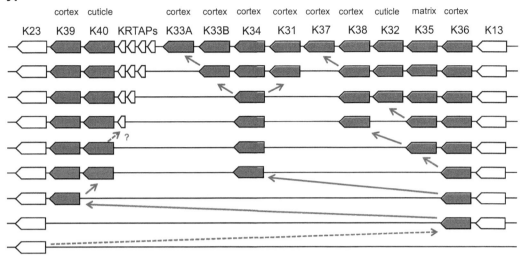

Type II keratins

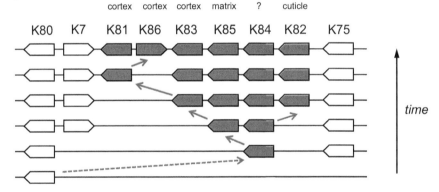

Fig. 4.3 Evolutionary scenarios for the diversification of hair keratins. The schematics show the arrangement of type I and type II keratin genes (arrows indicating the gene orientation, i.e. direction of transcription) in the human genome (top lines) and putative arrangements of genes in the evolutionary ancestors. The genes have diversified by duplications and subsequent accumulation of mutations. In this simplified model, the hypothetical parent gene of each gene duplication is depicted beneath the gene copy retaining more of the ancestral features (structure and expression pattern). Predominant sites of expression within hair follicles are indicated above the keratin genes. The possible origin of KRTAP genes from keratin genes is indicated by an arrow with a question mark. These diagrams illustrate examples of possible evolutionary scenarios

The evolution of hair follicles also involved the evolution of the inner root sheath (IRS), companion layer, and outer root sheath (ORS) which are essential for the growth of hair [25, 51]. Phylogenetic analyses show that hair keratins of both types are not closely related to IRS keratins [50] (Fig. 4.1). By contrast, type I (K25-K28) and type II (K71-K74) IRS keratin genes are most closely related to the respective keratins of the suprabasal layers of the epidermis (type I: K10; type II: K1, K2, K77) [52]. The keratins of the companion layer (type I: K16, K17; type II: K6, K75) are most closely related to keratins of the epidermal basal layer (type I: K14; type II: K5). Finally, keratins of the ORS include those of the epidermal basal layer (type I: K14; type II: K5) and others (type I: K15, K16, K17, K19; type II: K6) [40]. Thus, molecular phylogenetics of keratin genes indicates that the hair follicle has evolved by modification of epithelial differentiation programs, of which the hair fiber (characterized by cornification with maintenance of stable connections between corneocytes) is most different from the interfollicular epidermis (characterized by cornification with desquamation of corneocytes). The outer layers (ORS and companion layer) appear to be derived from the epidermal basal layer, and the IRS is equivalent to the suprabasal epidermis. IRS and epidermis likely share the feature of disintegration of cell-cell contacts at the end of terminal differentiation.

In summary of the early evolution of hair keratins, hair keratins are phylogenetically clearly distinct from keratins of other hair follicle compartments. Ancestors of mammalian hair keratins have existed before hair evolved and these ancestral hair keratin-like proteins were very likely expressed in claws before they were co-opted for functions in hair. Further studies of keratin gene expression in non-mammalian vertebrates and molecular phylogenetics will help to evaluate whether claws were the original sites of hair keratin expression [23].

4.6 Diversification of Hair Keratins

Modern mammals have multiple type I and type II hair keratins which have arisen by gene duplication and sequence modifications. A phylogenetic tree of mammalian hair keratins is indicated in Fig. 4.1, however, the bootstrap supports for several branches of this tree are weak and, therefore, the order of branching should be regarded uncertain. Among type I hair keratins there is strong support for monophyly of the group comprised of K31, K33a, K33b, and K34, all of which are expressed in the hair cortex, and a close relationship of K37 and K38 (Sect. 1.1, Table 1.1.1). Among type II hair keratins, K81, K83, and K86 are closely related and all of them are expressed in the hair cortex (Sect. 1.1, Table 1.1.2).

K36 and K84 are the most basal hair keratins in phylogenetic analyses. Interestingly, both are expressed in the filiform papillae of the tongue and in the nail but only at low levels in hair, lending support to the hypothesis that the primordial hair keratins had functions outside of hair. Although it is important to consider that K36 and K84 are not identical to the primordial cysteine-rich keratins of amniotes, it may be helpful to visualize the evolutionary diversification of hair keratins in a simplified manner by assuming that gene duplications result in two daughter genes, one of which remains functionally more similar to the previous generation whereas the other assumes a new role. This concept leads to a hypothetical scenario for the diversification of hair keratins that is depicted in Fig. 4.3. The schematic concept of keratin evolution remains to be extended to hair keratins without human orthologs. For example, sheep have a *KRT87* gene which is absent in humans [53]. Ultimately, a comprehensive model of hair keratin evolution will need to include species with extreme integumentary phenotypes such as cetaceans, i.e. whales and dolphins, which evolved from hairy terrestrial ancestors into hairless aquatic animals and lost multiple hair keratins [54].

Several issues in the evolution of hair keratins require further investigation. One of these topics is the contribution of individual keratins to the evolution and structure of different types of hair. Mammals have hair of very different physical properties and functions, and it has long been hypothesised that the predominant functions of hair have changed during mammalian evolution. A recent hypothesis emphasizes the ancient and conserved association of hair and sebaceous glands [55] while an alternative hypothesis suggests that primitive hair had a mechanosensory function similar to the whiskers of extant mammals and only later a pelage evolved and hair was used for thermoinsulation, protection against environmental insults and camouflage [56]. These evolutionary processes were associated with the diversification of hair follicles resulting, for example in mice, in vibrissae and four distinct types (guard, awl, auchene, and zigzag) of pelage hair [57]. Human *KRT37* is expressed in vellus hair but not in long scalp hair [58]. The closely related *KRT38* is expressed in a scattered manner in the cortex of human hair [58] and its ortholog is expressed in an asymmetrical manner, i.e. on one side of the hair cortex only, in the wool follicles of sheep [59]. Similarly, *KRT31*, *KRT35*, and *KRT85* of the hair fiber and *KRT27* and *Trichohyalin* of the IRS are expressed asymmetrically in sheep hair follicles to determine the deflection of the bulb and curvature of wool fibers [60]. Furthermore, *KRT31* and *KRT41* are expressed in a mutually exclusive manner on opposite sides within the cortex of hair in chimpanzees and gorillas, whereas *KRT41* is inactivated in humans and *KRT31* is expressed throughout the entire hair cortex in human hair [61]. The estimation that the pseudogenization of *KRT41* has occurred only about 200.000 years ago indicates that the evolution of hair keratins has affected the recent evolution of our species and is still ongoing.

References

1. Ng, C. S., et al. (2012). The chicken frizzle feather is due to an alpha-keratin (KRT75) mutation that causes a defective rachis. *PLoS Genetics, 8*, e1002748.
2. Gandolfi, B., et al. (2013). A splice variant in KRT71 is associated with curly coat phenotype of Selkirk Rex cats. *Scientific Reports, 3*, 2000.
3. Taylor, J. S., & Raes, J. (2004). Duplication and divergence: The evolution of new genes and old ideas. *Annual Review of Genetics, 38*, 615–643.
4. Innan, H. K. F. (2010). The evolution of gene duplications: classifying and distinguishing between models. *National Review of Genetics, 11*, 97–108.
5. Andersson, D. I., Jerlström-Hultqvist, J., & Näsvall, J. (2015). Evolution of new functions de novo and from preexisting genes. *Cold Spring Harbour Perspectives in Biology, 7*(pii: a017996).
6. Arguello, J. R., et al. (2007). Origination of chimeric genes through DNA-level recombination. *Genome Dynamics, 3*, 131–146.
7. Ohno, S. (1970). *Evolution by gene duplication*. Berlin: Springer.
8. Strasser, B., et al. (2014). Evolutionary origin and diversification of epidermal barrier proteins in amniotes. *Molecular Biology and Evolution, 31*, 3194–3205.
9. Alibardi, L. (2016). The process of cornification evolved from the initial keratinization in the epidermis and epidermal derivatives of vertebrates: A new synthesis and the case of sauropsids. *International Review of Cell and Molecular Biology, 327*, 263–319.
10. Peter, A., & Stick, R. (2015). Evolutionary aspects in intermediate filament proteins. *Current Opinion in Cell Biology, 32*, 48–55.
11. Riemer, D., Karabinos, A., & Weber, K. (1998). Analysis of eight cDNAs and six genes for intermediate filament (IF) proteins in the cephalochordate Branchiostoma reveals differences in the IF multigene families of lower chordates and the vertebrates. *Genetic Research, 211*, 361–373.
12. Zimek, A., & Weber, K. (2005). Terrestrial vertebrates have two keratin gene clusters; Striking differences in teleost fish. *European Journal of Cell Biology, 84*, 623–635.
13. Strasser, B., et al. (2015). Convergent evolution of cysteine-rich proteins in feathers and hair. *BMC Evolutionary Biology, 15*, 82.
14. Dodemont, H., Riemer, D., & Weber, K. (1990). Structure of an invertebrate gene encoding cytoplasmic intermediate filament (IF) proteins: Implications for the origin and the diversification of IF proteins. *EMBO Journal, 9*, 4083–4094.
15. Schaffeld, M., Höffling, S., & Jürgen, M. (2004). Sequence, evolution and tissue expression patterns of an epidermal type I keratin from the shark *Scyliorhinus stellaris*. *European Journal of Cell Biology, 83*, 359–368.
16. Lane, E. B., & McLean, W. H. (2004). Keratins and skin disorders. *The Journal of Pathology, 204*, 355–366.
17. Schweizer, J., et al. (2007). Hair follicle-specific keratins and their diseases. *Experimental Cell Research, 313*, 2010–2020.
18. Strnad, P., et al. (2011). Unique amino acid signatures that are evolutionarily conserved distinguish simple-type, epidermal and hair keratins. *Journal of Cell Science, 124*, 4221–4232.
19. Toivola, D. M., et al. (2015). Keratins in health and disease. *Current Opinion in Cell Biology, 32*, 73–81.
20. Coulombe, P. A. (2017). The molecular revolution in cutaneous biology: Keratin genes and their associated disease: Diversity, opportunities, and challenges. *The Journal of Investigative Dermatology, 137*, e67-e71.
21. Parry, D. A. D., & Steinert, P. M. (1999). Intermediate filaments: Molecular architecture, assembly, dynamics and polymorphism. *Quarterly Reviews of Biophysics, 32*, 99–187.
22. Badowski, C., et al. (2017). Modeling the structure of keratin 1 and 10 terminal domains and their misassembly in keratoderma. *Journal of Investigative Dermatology, 137*, 1914–1923.
23. Eckhart, L., et al. (2008). Identification of reptilian genes encoding hair keratin-like proteins suggests a new scenario for the evolutionary origin of hair. *Proceedings of the National Academy of Sciences of the United States of America, 105*, 18419–18423.
24. Eckhart, L., Jaeger, K., & Tschachler, E. (2009). The tail domains of keratins contain conserved amino acid sequence motifs. *Journal of Dermatological Science, 54*, 208–209.
25. Moll, R., Divo, M., & Langbein, L. (2008). The human keratins: Biology and pathology. *Histochemistry and Cell Biology, 129*, 705–733.

26. Bragulla, H. H., & Homberger, D. G. (2009). Structure and functions of keratin proteins in simple, stratified, keratinized and cornified epithelia. *Journal of Anatomy, 214*, 516–559.
27. Eckert, R. L., et al. (1997). The epidermis: Genes on – Genes off. *Journal of Investigative Dermatology, 109*, 501–509.
28. Jave-Suarez, L. F., et al. (2002). HOXC13 is involved in the regulation of human hair keratin gene expression. *The Journal of Biological Chemistry, 277*, 3718–3726.
29. Gilon, M., et al. (2008). Transcriptional activation of a subset of hair keratin genes by the NF-kappaB effector p65/RelA. *Differentiation, 76*, 518–530.
30. Potter, C. S., et al. (2011). The nude mutant gene Foxn1 is a HOXC13 regulatory target during hair follicle and nail differentiation. *Journal of Investigative Dermatology, 131*, 828–837.
31. Findeisen, P., et al. (2014). Six subgroups and extensive recent duplications characterize the evolution of the eukaryotic tubulin protein family. *Genome Biology and Evolution, 6*, 2274–2288.
32. Nielsen, M. G., Gadagkar, S. R., & Gutzwiller, L. (2010). Tubulin evolution in insects: Gene duplication and subfunctionalization provide specialized isoforms in a functionally constrained gene family. *BMC Evolutionary Biology, 10*, 113.
33. Langbein, L., et al. (2016). Localisation of keratin K78 in the basal layer and the first suprabasal layers of stratified epithelia completes the expression catalog of type II keratins and provides new insights into sequential keratin expression. *Cell and Tissue Research, 363*, 735–750.
34. Dhouailly, D. (2009). A new scenario for the evolutionary origin of hair, feather, and avian scales. *Journal of Anatomy, 214*, 587–606.
35. Alibardi, L. (2012). Perspectives on hair evolution based on some comparative studies on vertebrate cornification. *Journal of Experimental Zoolology Part B: Molecular Developmental Evolution, 318*, 325–343.
36. Wagner, G. P. (2014). *Homology, genes and evolutionary innovation*. Princeton: Princeton University Press.
37. Perrin, C., Langbein, L., & Schweizer, J. (2004). Expression of hair keratins in the adult nail unit: An immunohistochemical analysis of the onychogenesis in the proximal nail fold, matrix and nail bed. *British Journal of Dermatology, 151*, 362–371.
38. Bowden, P. E., Henderson, H., & Reilly, J. D. (2009). Defining the complex epithelia that comprise the canine claw with molecular markers of differentiation. *Veterinary Dermatology, 20*, 347–359.
39. Carter, R. A., et al. (2010). Novel keratins identified by quantitative proteomic analysis as the major cytoskeletal proteins of equine (Equus caballus) hoof lamellar tissue. *Journal of Animal Science, 88*, 3843–3855.
40. Langbein, L., & Schweizer, J. (2005). Keratins of the human hair follicle. *International Review of Cytology, 243*, 1–78.
41. Langbein, L., et al. (2001). The catalog of human hair keratins. II. Expression of the six type II members in the hair follicle and the combined catalog of human type I and II keratins. *Journal of Biological Chemistry, 276*, 35123–35132.
42. Gomez, C., et al. (2013). The interfollicular epidermis of adult mouse tail comprises two distinct cell lineages that are differentially regulated by Wnt, Edaradd, and Lrig1. *Stem Cell Reports, 1*, 19–27.
43. Solazzo, C., et al. (2013). Characterisation of novel α-keratin peptide markers for species identification in keratinous tissues using mass spectrometry. *Rapid Communications in Mass Spectrometry, 27*, 2685–2698.
44. Dhouailly, D., & Sun, T. T. (1989). The mammalian tongue filiform papillae: A theoretical model for primitive hairs. In D. Van Neste, J. M. Lachapelle, & J. L. Antoine (Eds.), *Trends in human hair growth and alopecia research* (pp. 29–34). Dordrecht: Springer.
45. Maddin, H. C., et al. (2009). The anatomy and development of the claws of Xenopus laevis (Lissamphibia: Anura) reveal alternate pathways of structural evolution in the integument of tetrapods. *Journal of Anatomy, 214*, 607–619.
46. Kleinteich, T., & Gorb, S. N. (2016). Frog tongue surface microstructures: Functional and evolutionary patterns. *Beilstein Journal of Nanotchnology, 7*, 893–903.
47. Alibardi, L. (2003). Adaptation to the land: The skin of reptiles in comparison to that of amphibians and endotherm amniotes. *Journal of Experimental Zoology Part B: Molecular and Developmental Evolution, 298*, 12–41.
48. Vandebergh, W., et al. (2013). Recurrent functional divergence of early tetrapod keratins in amphibian toe pads and mammalian hair. *Biology Letters, 9*, 20130051.
49. Suzuki, K. T., et al. (2017). Clustered Xenopus keratin genes: A genomic, transcriptomic, and proteomic analysis. *Developmental Biology, 426*, 384–392.
50. Vandebergh, W., & Bossuyt, F. (2012). Radiation and functional diversification of alpha keratins during early vertebrate evolution. *Molecular Biology and Evolution, 29*, 995–1004.
51. Mlitz, V., et al. (2014). Trichohyalin-like proteins have evolutionarily conserved roles in the morphogenesis of skin appendages. *Journal of Investigative Dermatology, 134*, 2685–2692.
52. Langbein, L., et al. (2013). New facets of keratin K77: Interspecies variations of expression and different intracellular location in embryonic and adult skin of humans and mice. *Cell and Tissue Research, 354*, 793–812.
53. Yu, Z., et al. (2011). Annotations of sheep keratin intermediate filament genes and their patterns of expression. *Experimental Dermatology, 20*, 582–588.
54. Nery, M. F., Arroyo, J. I., & Opazo, J. C. (2014). Increased rate of hair keratin gene loss in the cetacean lineage. *BMC Genomics, 15*, 869.
55. Dhouailly, D., et al. (2017). Getting to the root of scales, feather and hair: As deep as odontodes?

Experimental Dermatology. https://doi.org/ 10.1111/exd.13391.
56. Maderson, P. F. A. (2003). Mammalian skin evolution: A reevaluation. *Experimental Dermatology, 12*, 233–236.
57. Sundberg, J. P. (Ed.). (1994). *Handbook of mouse mutations with skin and hair abnormalities.* Boca Raton: CRC Press.
58. Langbein, L., et al. (1999). The catalog of human hair keratins. I. Expression of the nine type I members in the hair follicle. *Journal of Biological Chemistry, 274*, 19874–19884.
59. Plowman, J. E. (2007). The proteomics of keratin proteins. *Journal of Chromatography B, 849*, 181–189.
60. Yu, Z., et al. (2009). Expression patterns of keratin intermediate filament and keratin associated protein genes in wool follicles. *Differentiation, 77*(3), 307–316.
61. Winter, H., et al. (2001). Human type I hair keratin pseudogene phihHaA has functional orthologs in the chimpanzee and gorilla: Evidence for recent inactivation of the human gene after the Pan-Homo divergence. *Human Genetics, 108*, 37–42.

Evolution of Trichocyte Keratin Associated Proteins

5

Dong-Dong Wu and David M. Irwin

Contents

5.1	Introduction of Keratin Associated Protein (KRTAP)	48
5.2	Classification of KRTAP Gene Family Members	48
5.3	Molecular Evolution of Gene Families and Phenotypic Evolution	50
5.4	Molecular Evolution of the KRTAP Gene Family and Evolution of Hair in Mammals	51
5.5	Extraction and Expansion of KRTAP Gene Family in Some Special Mammals	53
5.6	Concerted Evolution of the KRTAP Gene Family	53
5.7	Keratin Associated Protein in Domestic Mammals	54
	References	55

Abstract

The major components of hair are keratins and keratin associated proteins (KRTAPs). KRTAPs form the interfilamentous matrix between intermediate filament bundles through extensive disulfide bond cross-linking with the numerous cysteine residues in hair keratins. A variable number of approximately 100–180 genes compose the KRTAP gene family in mammals. KRTAP gene family members present a typical pattern of concerted evolution, and its evolutionary features are consistent with the evolution of mammalian hair. KRATP genes might be more important in determining the structure of cashmere fibers in domestic mammals like sheep and goats. KRTAP gene variants thus should provide information for improved wool by sheep and goat breeding.

D.-D. Wu (✉)
State Key Laboratory of Genetic Resources and Evolution, Kunming Institute of Zoology, Chinese Academy of Sciences, Kunming, China

D. M. Irwin
Department of Laboratory Medicine and Pathobiology, University of Toronto, Toronto, Canada
e-mail: david.irwin@utoronto.ca

Keywords

Keratin associated proteins (krtap) · Hair · Phenotypic evolution · Molecular evolution · Concerted evolution · Gene clusters · Mammals

5.1 Introduction of Keratin Associated Protein (KRTAP)

The major components of hair are keratins and keratin associated proteins. Keratins provide cellular stability against stress by forming an intermediate filament cytoskeleton for epithelial cells, while the keratin associated proteins (KRTAP) form the interfilamentous matrix between the intermediate filament bundles through extensive disulfide bond cross-linking with the large numbers of cysteine residues found in hair keratins [1]. Progression of hair follicle differentiation is characterized by the sequential activation of distinct sets of hair-specific keratin and KRTAP genes. The structural integrity and tight regulation of both groups of these genes is essential for proper hair growth.

Keratins present structural and functional conservation when compared among different mammals. In stark contrasts, the keratin associated proteins display high divergence, which is consistent with the diversity of hair phenotypes among mammals [2]. Due to its high diversity, studies of the evolution of the KRTAP gene family should contribute to our understanding of the evolution of mammalian hair. In this chapter, we will introduce the KRTAP gene family, its evolutionary features, and its association with the evolutionary origin of hair.

5.2 Classification of KRTAP Gene Family Members

Due to the economic importance of wool, the KRTAP family was first, and best, studied in sheep. With the development of large-scale genome sequencing technology, the genomes of many vertebrates, particularly mammals, have been successfully assembled, which has facilitated the identification of many genes, particularly genes within large gene families, such as the KRTAP gene family. It has been concluded that KRTAP genes are found only in the genomes of mammals and not in the other vertebrate species. For example, a KRTAP gene cluster within human chromosome 17 was found in mammals, but not in the genome of chickens (Fig. 5.1). The pattern is consistent with all hairs and is unique to mammals, but has not been found in other species.

In mammals, the KRTAP gene family has a dynamic distribution, ranging from less than 100 to over 180 genes. For example, about 188 genes were identified in the mouse genome, 175 in the sloth, while 122 were found in humans, but only 35 in dolphins [2, 3]. As the sequence composition of the KRTAP genes have no homology with other existing genes, it is likely that the KRTAP genes originated de novo from non-genic regions. There are two major groups of KRTAP genes: high/ultrahigh cysteine (HS-KRTAP, Fig. 5.2) and high glycine-tyrosine (HGT-KRTAP, Fig. 5.2), that are considered to have independently originated based on their distinct amino acid compositions [1]. Some KRTAPs also have a high content of serine residues, but are proposed to have originated from an ancestor with the HS-KRTAP genes. HGT-KRTAP genes appear to evolve more dynamically than the HS-KRTAP genes, with positive selection contributing to this. The reason for this is unclear, but might be due to their functional differences.

KRTAP genes have been further divided into about 30 subfamilies, i.e. KRTAP1-KRTAP35. Six of the subfamilies, i.e. KRTAP6-8, 19-21, are classified as HGT-KRTAP with high glycine-tyrosine amino acid composition, while the other 29 KRTAP subfamilies are HS-KRTAP with high/ultrahigh cysteine. Cysteine is important for the formation of the large number of disulfide bonds formed by these proteins, thus, changes in cysteine composition can result in different interactions among KRTAPs and between keratins and KRATPs leading to combinatorial complexity and thereby the morphological differences in hair fiber strength, rigidity and flexibility found between species [4].

In humans, KRTAPs include 101 intact gene members, which are arranged in tandem and clustered on chromosomes 11p15.5, 11q13.4, 17q21.2, 21q22.1, and 21q22.3 (Fig. 5.3). KRTAP genes are distributed mainly to five genomic regions in the genomes of placental and marsupial species: 17q21.2 Cluster 1 contains genes from subfamilies KRTAP 1, 2, 3, 4, 9, 17, 16, and 29. 21q22.1 Cluster 2 contains genes from subfamilies 13, 24–27 and all glycine-tyrosine rich

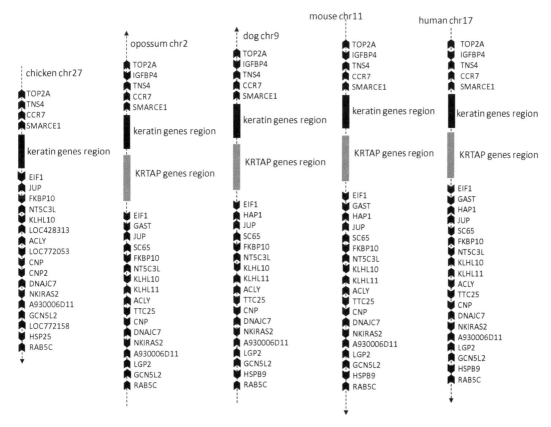

Fig. 5.1 Origin of the KRTAP gene cluster in mammals. Genomic neighborhoods surrounding the keratin genes in the genomes of the chicken and representative mammals. KTRAP genes are only found in mammalian genomes and one cluster originated near the keratin genes

---

HGT-KRTAP

>NP_853633.1| keratin-associated protein 6 -1 [Homo sapiens]
MCGSYYGNYYGTPGYGFCGYGGLGYGYGGLGCGYGSCCGCGFRRLG
CGYGYGSRSLCGYGYGCGSGSGYYY

HS-KRTAP

>NP_005544.4 | keratin-associated protein 5 -9 [Homo sapiens]
MGCCGCSGGCGSSCGGCDSSCGSCGSGCRGCGPSCCAPVYCCKPVC
CCVPACSCSSCGKRGCGSCGGSKGGCGSCGCSQCSCCKPCCCSSGC
GSSCCQCSCCKPYCSQCSCCKPCCSSSGRGSSCCQSSCCKPCCSSSG
CGSSCCQSSCCKPCCSQSRCCVPVCYQCKI

Fig. 5.2 Examples of human high glycine-tyrosine KTRAP (HGT-KRTAP) and high/ultrahigh cysteine KRTAP (HS-KRTAP) proteins. Human KRTAP6-1 has a high content of glycine and tyrosine amino acid residues while KRTAP5-9 has a very high content of cysteine residues in their protein sequences

Fig. 5.3 Distribution of KRTAP genes in the human genome. The order, orientation, and subfamily membership of human KRATP genes on the different chromosomes is displayed. Genes labeled in red genes are pseudogenes. (Modified from [2])

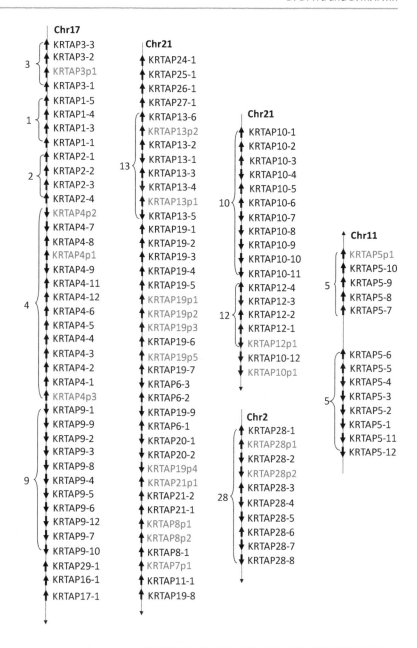

KRTAPs. 21q22.3 Cluster 3 possesses genes from subfamilies 10 and 12. Cluster 4 encodes genes of subfamily 28. Cluster 5 corresponds to genes of subfamily 5. In mice and rats, the new subfamilies 30 and 31 have been inserted into the genomic locations of subfamilies 4 and 9, respectively. In a similar manner, in the dog genome a new cluster has been generated on chromosome 31 that includes three genes, one from subfamily 10 and two from subfamily 12.

5.3 Molecular Evolution of Gene Families and Phenotypic Evolution

A gene family is a set of similar genes, formed by duplication from a single original gene, where each gene generally has similar biochemical functions. Gene duplications, changes in evolutionary rate and pseudogenization frequently occur within a gene family, with these processes contributing to functional innovation and adaptation of organisms. Therefore, to study the macular mechanisms underlying adaptive evolution, large scale studies have typically retrieved and examined genes within gene families. Studies of adaptive evolution have frequently examined gene families involved in the perception of the environment, such as the olfactory receptors [5–8], the vomeronasal receptors [9–11], and the sweet/umami and bitter receptors [12–15]. These studies have facilitated our understanding of the general evolutionary trends in gene families, genomic complexity and lineage-specific adaptation. Below, we use olfactory receptors and vomeronasal receptors as examples.

The olfactory receptor (OR) gene family is the largest gene family identified in vertebrate genomes, a gene family that follows the "birth-and-death" process, which is characterized by frequent gene duplications and losses [16]. The dynamic evolution of the OR gene family during the evolution of vertebrates likely reflects the functional requirement for differing olfactory abilities in different evolutionary lineages. For example, the number of gene losses in the primate lineage is much greater than that in other mammalian lineages, which is considered to be due to the acquisition of trichromatic vision by primates.

Studies of the vomeronasal receptor gene family have also found an association with adaptation. For example, the number of intact V1R genes is positively correlated with the morphological complexity of the VNO [10]. Intact V1R repertoire sizes vary at least 23-fold among mammals with functional VNOs and this size ratio represents the greatest among-species variation in the gene family size of all mammalian gene families [10]. The ratio of the number of intact V1R genes to intact V2R genes increased approximately 50 fold in the evolutionary transition from water to land [11]. Circumstantial evidence suggests that V1Rs tend to recognize airborne molecules, while V2Rs recognize water-soluble ligands. Comparison of the numbers of class II ORs to those of class I ORs, which are also suggested to bind to volatile and water-solvable molecules, respectively, shows a similar pattern of change during the evolutionary transition of vertebrates from an aquatic to a terrestrial environment.

Large-scale studies have demonstrated that the evolution of gene families can reveal properties concerning the evolution of specific phenotypes. Based on these earlier studies, we therefore believed that a study of the evolution of the KRTAP gene family should also reveal information on the origin and evolution of hair during the evolution of mammals.

5.4 Molecular Evolution of the KRTAP Gene Family and Evolution of Hair in Mammals

With the rapid development of next generation genome sequencing the genomes of a large number of vertebrate, particularly mammalian, genomes have been assembled de novo. For example, the Ensembl genome database contains many mammalian genomes (see Fig. 5.4) and these genomes have facilitated the exhaustive identification of the inventories of large gene families, such as KRTAPs, olfactory receptors, and vomeronasal receptors.

Many features of the evolution of KRTAP genes likely reflect patterns in hair evolution. For instance, KRTAP genes are unique to mammals, and not other animals, just like hair. In addition, the majority of the KRTAP gene subfamilies originated, and diverged, before the placental mammalian radiations. Accordingly, the mammalian ancestor should have had a high diversity in their KRTAP genes and therefore enjoyed a similar range and spectrum of hair characteristics that is seen in modern mammalian species. The rapid emergence of the KRTAP gene family correlates with the evolution of mammalian hair and the rapid emergence of plentiful hair might have contributed to the successful radiation of homothermal mammals by helping them retain body heat since hair acts as an insulator [17].

Variation in gene copy number is considered to be a pivotal factor underlying the complexity of functional traits controlled by large gene

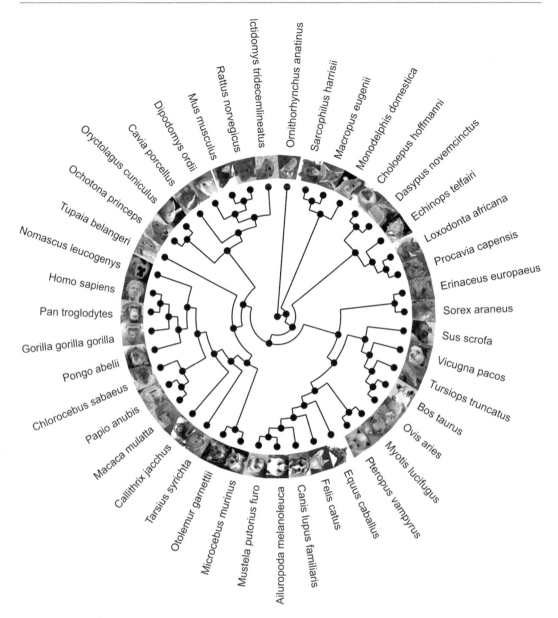

Fig. 5.4 Phylogenetic tree of mammals with de novo assembled genomes in the Ensembl database. Data was from Ensembl version 85. De novo assembled genomes were helpful for the retrieval of complete KRTAP gene repertoires in mammals and the study of their evolution

families. For example, rodents have a strong olfactory ability due to their large number of olfactory receptors in their genomes (~1000 genes). The numbers of intact V1R genes is also positively correlated with the morphological complexity of the VNO [10]. Variability in KRTAP gene numbers thus might reveal to some extent patterns in the evolution of hair. For example, KRTAP is unique to mammals, consistent with hair being unique to mammals, with specialization occurring within mammals. The numbers of KRTAP genes are highly divergent between species but homogenous within a species, suggesting that the diversity of hair observed among species is due to gene number content. In stark contrast to KRTAP, the number of hair keratin genes has changed little during the evolution of mammals (e.g. human contains 17 genes, mouse 15 genes and opossum 19 genes). Such a small gene repertoire thus limits its ability to shape this highly diverse feature among and within species, except with the emergence of KRTAP that modify the functions of these genes, with its large repertoire, yielding much greater functional diversity.

Genes with similar function arrayed tandemly in the genome can provide combinatorial complexity to biological diversity. For example, olfactory receptors, vomeronasal receptors, sweet/umami receptors and bitter receptors detect and distinguish the innumerable molecular diversity of odors and tastes in the environment by using a combination of receptors that have a huge diversity to recognize diverse ligands. In an analogous manner, with hair, the combination of KRTAP and keratin generates the high diversity of hair phenotypes that are observed between species, within species, even different locations on the skin of an individual [2].

Despite the similarity in the diversity of KRTAPs in the ancestor of mammals, the traits, and content, of hair in extant species has likely diverged significantly from their ancestral characteristics as the KRTAP sequences have diverged dramatically, both due to point mutations and recombination including gene conversion.

5.5 Extraction and Expansion of KRTAP Gene Family in Some Special Mammals

Humans have only recently lost their body hair, presumably because humans can obtain heat and can keep warm by using clothing. Despite a lack of hair, humans actually have a similar density of hair follicles as apes [18], which potentially explains why humans do not have a significantly smaller number of KRTAP genes. Perhaps, changes in the amount of human hair are due to a reduction in the expression of KRTAP genes. In contrast to humans, rodents have an expanded KRTAP gene family. Perhaps, mice and rats need more hair as they are adapted to the nocturnal environment [2]. A dramatic gene expansion of this family, with 175 members (50 in subfamily 9 and 37 genes in subfamily 20, respectively), is seen in the sloth (*Choloepus hoffmanni*), a nocturnal hairy mammal with long, coarse and shaggy fur that serves as a host for a number of different microorganisms [3]. In stark contrast, only nine intact KRTAP genes were found in the dolphin (*Tursiops truncatus*). This aquatic mammal is almost hairless, with only a few hairs (bristles) on the upper lip of the rostrum, which are shed soon after birth. The hairless pits on the rostrum of adults have specialized sensory functions. The epidermal surface also undergoes high proliferation and sloughing of epidermis cells, in order to maintain a smooth skin, a major advantage for swimming.

5.6 Concerted Evolution of the KRTAP Gene Family

There are two general types of gene families that differ with respect to their evolutionary patterns with birth-and-death processes and concerted evolution [16]. Under the birth-and-death process, after gene duplication some duplicates are maintained in the genome for a long time, which can either maintain the ancestral function or obtain a new function, while some duplicates will be lost by deleterious mutations [16]. The olfactory receptors, the vomeronasal receptors, and the sweet/umami and bitter receptors are examples of this type of gene family. Under concerted evolution, genes evolve in a concerted manner rather than independently due to gene conversion and/or unequal crossing over events, which spread mutations occurring in one gene across the entire gene family. Gene conversion is a nonreciprocal recombination where a segment of DNA of a recipient gene is copied from a donor gene. Unequal crossing over occurs between homologous chromosomes and results in nonreciprocal exchanges of material and chromosomes of unequal length. The two major mechanisms of concerted evolution induce similar results, e.g. increased similarity among genes. The only difference between the two is that unequal crossing changes the number of genes, but gene conversion cannot [16]. Consequently, concerted evolution will reduce the divergence among gene sequences.

The KRTAP gene family displays a typical concerted evolution pattern, with several interesting evolutionary features. For example, the numbers of genes in many of the KRTAP subfamilies is stable among mammals, e.g. subfamily 1, 2, 3. Most genes of a subfamily within a species are clustered together in phylogenetic trees, suggesting that sequence homogenisation is occurring within lineages but that divergence is occurring between species. In addition, these genes are usually tandemly clustered on the chromosomes. With this type of gene distribution, gene conversion and/or unequal crossing over should occur easily and frequently, and reduce the divergence among the sequences of the genes within a species. Frequent gene conversion/unequal crossing over events are detected among the different KRTAP genes. The high GC nucleotide content of KRTAP coding sequences is also consistent with concerted evolution of family, as gene conversion would bias sequences in favor of GC nucleotides [19, 20]. Biased gene conversion has been considered to result from gene conversion events that occur during the repair of double-strand break by recombination, with the repair process being biased for GC basepairs [19].

5.7 Keratin Associated Protein in Domestic Mammals

Wool is a textile fiber obtained from sheep and some other animals, including cashmere and mohair from goats, qiviut from muskoxen, angora from rabbits, and other types of wool from camelids. Wool fibre is a highly organized structure, whose main histological components include the cuticle, cortex (orthocortex, paracortex, and mesocortex), and medulla [21]. Wool fibre characteristics, such as diameter, crimps, and length, are essential parameters of the wool trait, as well as important indicators of the spinning efficiency of the wool. Almost all of the fibre is composed of the cortex, which consists of keratin intermediate filaments embedded in a matrix of keratin-associated proteins, which are cross-linked to the keratin intermediate filaments with inter-chain disulfide-bonding, to form the wool fiber after keratinization.

Considering the economic importance of wool, KRTAPs were originally and best studied in domestic animals such as sheep. The formation of KRTAP composites gives wool the special mechanical attributes of strength, inertness, and rigidity [21]. HGT-KRTAPs are predominantly present in the orthocortex of the wool fibre and are the first group of KRTAPs expressed after intermediate filament synthesis [22]. The content of HGT-KAPs in the fibre varies considerably both within and between species, ranging from less than 3% in human hair and Lincoln sheep wool, through 4–12% in merino wool, 18% in mouse hair, to 30–40% in echidna quill [23]. HGT-KAPs are present at much lower levels in the felting lustre wool mutant compared to normal wool [24], and in the felting lustre mutant wool follicles the HGT-KAP genes are down-regulated [25], suggesting HGT-KAPs have some association with wool crimp.

Identification of KRTAP genes that affect wool traits should lead to the development of gene-markers for improving wool quality. Three HGT-KAPs families have been investigated in sheep: KAP6, KAP7 and KAP8. Variation in a KRTAP6 gene has been reported to be associated with wool fibre diameter [26], and recently a 57-bp deletion in KRTAP6-1 has been found to be associated with variation in fibre diameter traits [27]. Five KRTAP6 genes have been reported in sheep. In contrast, there are three KRTAP6 genes in humans and a variable number of putative KRTAP6 genes identified in other species [3], suggesting that KAP6 is a diverse family that varies considerably in gene number content among species. For example, there appears to be an absence of any functional KRTAP6 gene in the hairless armadillo, and reports of nine KRTAP6 genes in rabbits and alpaca [3]. All these findings suggested that future investigation of KRTAP6 genes in sheep might shed further light on wool traits and their potential improvement through targeted breeding.

Skin Transcriptomes from sheep also revealed genes that might be important for the development of wool fiber. For example, Fan et al. [28] reported genes that are highly expressed genes in sheep skin, which was found to include NADH dehydrogenase subunit 5 (*NADH5*), cytochrome c oxidase subunit I (*COX1*), NADH-ubiquinone oxidoreductase chain 4 (*ND4*), keratin associated protein 9.2 (*KAP9.2*), keratin 27 (*Krt27*), high-sulfur keratin BIIIB4 protein, hair keratin cysteine rich protein (*Dmpk*), keratin-associated protein 1.4 (*KAP1.4*) and keratin-associated protein 3–2 (*KAP3.2*). When black sheep skin was examined keratin family members and keratin associated proteins displayed down regulation [28].

Dong et al. [29] successfully generated a de novo genome assembly for the goat. With this assembly, 49 keratin genes and 30 keratin-associated protein genes were annotated and 29 of the keratin genes and all 30 keratin-associated protein genes were detected for expression with FPKM >5 in two types of follicles [29]. Mammalian hair is a highly keratinized tissue produced by hair follicles within the skin. There are two kinds of hair follicles: the primary hair follicle produces the coarse coat hair in all mammals, and the secondary hair follicle can produce the cashmere or 'fine hair' in certain mammals, including goats and antelopes. Based on transcriptome data from primary and secondary hair follicles, it was found that two of the 29 keratin genes and 10 of the 30 keratin-associated protein genes were consistently differentially expressed between primary and secondary hair follicles in all sample sets, and all of these were expressed at a higher level in secondary than in primary follicles, suggesting that the keratin-associated protein genes may be more important in determining the structure of cashmere fibers. The two differentially expressed keratin genes (keratin 40 and 72) were type 1 and type 2, respectively. With keratin-associated proteins divided into three major groups, high sulfur, ultra-high sulfur and high glycine-tyrosine, the ten differentially expressed keratin-associated proteins all belong to the ultra-high sulfur group, suggesting that this group of proteins may be important for the formation of cashmere.

All of these reports suggest that variants on KRTAP genes are important markers for sheep and goat breeding for wool fibre.

References

1. Rogers, M. A., et al. (2006). Human hair keratin associated proteins (KAPs). *International Review of Cytology, 251*, 209–263.
2. Wu, D.-D., Irwin, D. M., & Zhang, Y.-P. (2008). Molecular evolution of the keratin associated protein gene family in mammals, role in the evolution of mammalian hair. *BMC Evolutionary Biology, 8*, 241.
3. Khan, I., et al. (2014). Mammalian keratin associated proteins (KRTAPs) subgenomes: Disentangling hair diversity and adaptation to terrestrial and aquatic environments. *BMC Genomics, 15*, 779.
4. Shimomura, Y., & Ito, M. (2015). Human hair keratin-associated proteins. *Journal of Investigative Dermatology Symposium Proceedings, 10*(3), 230–233.
5. Niimura, Y., & Nei, M. (2003). Evolution of olfactory receptor genes in the human genome. *Proceedings of the National Academy of Sciences of the United States of America, 100*(21), 12235–12240.
6. Niimura, Y., & Nei, M. (2005). Evolutionary dynamics of olfactory receptor genes in fishes and tetrapods. *Proceedings of the National Academy of Sciences, 102*(17), 6039–6044.
7. Niimura, Y., & Nei, M. (2006). Evolutionary dynamics of olfactory and other chemosensory receptor genes in vertebrates. *Journal of Human Genetics, 51*(6), 505–517.
8. Niimura, Y., & Nei, M. (2007). Extensive gains and losses of olfactory receptor genes in mammalian evolution. *PLoS One, 2*(8), e708.
9. Grus, W. E., Shi, P., & Zhang, J. (2007). Largest vertebrate vomeronasal type 1 receptor (V1R) gene repertoire in the semi-aquatic platypus. *Molecular Biology and Evolution, 24*, 2153–2157.
10. Grus, W. E., et al. (2005). Dramatic variation of the vomeronasal pheromone receptor gene repertoire among five orders of placental and marsupial mammals. *Proceedings of the National Academy of Sciences of the United States of America, 102*(16), 5767–5772.
11. Shi, P., & Zhang, J. (2007). Comparative genomic analysis identifies an evolutionary shift of vomeronasal receptor gene repertoires in the vertebrate transition from water to land. *Genome Research, 17*(2), 166–174.
12. Shi, J., et al. (2003). Divergence of the genes on human chromosome 21 between human and other hominoids and variation of substitution rates among transcription units. *Proceedings of the National Academy of Sciences, 100*(14), 8331–8336.
13. Shi, P., & Zhang, J. (2006). Contrasting modes of evolution between vertebrate sweet/umami receptor

genes and bitter receptor genes. *Molecular Biology and Evolution, 23*(2), 292–300.
14. Fischer, A., et al. (2005). Evolution of bitter taste receptors in humans and apes. *Molecular Biology and Evolution, 22*(3), 432–436.
15. Parry, C. M., Erkner, A., & le Coutre, J. (2004). Divergence of T2R chemosensory receptor families in humans, bonobos, and chimpanzees. *Proceedings of the National Academy of Sciences, 101*(41), 14830–14834.
16. Nei, M., & Rooney, A. P. (2005). Concerted and birth-and-death evolution of multigene families. *Annual Review of Genetics, 39*, 121–152.
17. Maderson, P. F. A. (2003). Mammalian skin evolution: A reevaluation. *Experimental Dermatology, 12*(3), 233–236.
18. Schwartz, G. G., & Rosenblum, L. A. (1981). Allometry of primate hair density and the evolution of human hairlessness. *American Journal of Physical Anthropology, 55*, 9–12.
19. Marais, G. (2003). Biased gene conversion: Implications for genome and sex evolution. *Trends in Genetics, 19*(6), 330–338.
20. Galtier, N. (2003). Gene conversion drives GC content evolution in mammalian histones. *Trends in Genetics, 19*(2), 65–68.
21. Plowman, J. E., et al. (2009). Protein expression in orthocortical and paracortical cells of merino wool fibers. *Journal of Agricultural and Food Chemistry, 57*(6), 2174–2180.
22. Rogers, G. E. (2006). Biology of the wool follicle: An excursion into a unique tissue interaction system waiting to be re-discovered. *Experimental Dermatology, 15*(12), 931–949.
23. Gillespie, J. M. (1990). The proteins of hair and other hard α-keratins. In R. D. Goldman & P. M. Steinert (Eds.), *Cellular and molecular biology of intermediate filaments* (pp. 95–128). New York: Springer.
24. Gillespie, J. M., & Darskus, R. L. (1971). Relation between the tyrosine content of various wools and their content of a class of proteins rich in tyrosine and glycine. *Australian Journal of Biological Sciences, 24*(4), 1189–1198.
25. Li, S. W., et al. (2009). Characterization of the structural and molecular defects in fibres and follicles of the merino felting lustre mutant. *Experimental Dermatology, 18*(2), 134–142.
26. Parsons, Y. M., Cooper, D. W., & Piper, L. R. (1994). Evidence of linkage between high-glycine-tyrosine keratin gene loci and wool fiber diameter in a merino half-sib family. *Animal Genetics, 25*(2), 105–108.
27. Zhou, H., et al. (2015). A 57-bp deletion in the ovine KAP6-1 gene affects wool fibre diameter. *Journal of Animal Breeding and Genetics, 132*(4), 301–307.
28. Fan, R., et al. (2013). Skin transcriptome profiles associated with coat color in sheep. *BMC Genomics, 14*, 389.
29. Dong, Y., et al. (2013). Sequencing and automated whole-genome optical mapping of the genome of a domestic goat (Capra hircus). *Nature Biotechnology, 31*(2), 135–141.

Structural Hierarchy of Trichocyte Keratin Intermediate Filaments

R. D. Bruce Fraser and David A. D. Parry

Contents

6.1	Introduction	58
6.2	Primary Structure	59
6.3	Secondary and Tertiary Structure	60
6.4	Modes of Molecular Assembly in the IF	61
6.5	Differences and Similarities Between the Reduced and Oxidised Structures of IF	64
6.6	Summary	67
	References	67

Abstract

Although trichocyte keratins (hair, wool, quill, claw) have been studied since the 1930s it is only over the last 30 years or so that major advances have been made in our understanding of the complex structural hierarchy of the filamentous component of this important filament-matrix composite. A variety of techniques, including amino acid sequence analysis, computer modelling, X-ray fibre diffraction and protein crystallography, various forms of electron microscopy, and crosslinking methods have now combined to reveal much of the structural detail. The heterodimeric structure of the keratin molecule is clear, as are the highly-specific modes by which these molecules aggregate to form functionally viable IF. The observation that hair keratin can adopt not one but two structurally-distinct conformations, one formed in the living cells at the base of the hair follicle in a reducing environment and the second in the fully differentiated hair in dead cells in an oxidized state, was unexpected but has major implications for the mechanism of hair growth. Insights have also been made into the mecha-

R. D. B. Fraser
Institute of Fundamental Sciences, Massey University, Palmerston North, New Zealand

Tewantin, QLD, Australia

D. A. D. Parry (✉)
Institute of Fundamental Sciences, Massey University, Palmerston North, New Zealand

Riddet Institute, Massey University, Palmerston North, New Zealand
e-mail: d.parry@massey.ac.nz

nism of the uppermost level of hair superstructure, relating to the assembly of the IF in the paracortical and orthocortical macrofibrils.

Keywords

Filament-matrix composite · Structural transition in trichocyte keratin · X-ray diffraction of keratin fibres · Molecular assembly · Disulfide bonds · DST crosslinks

Abbreviations

IF	intermediate filament
KAP	keratin-associated proteins
DST	disulfo-succinimidyl tartrate
ULF	unit-length-filament
HIM	helix initiation motif
HTM	helix termination motif
TEM	transmission electron microscopy
STEM	scanning transmission electron microscopy

6.1 Introduction

The trichocyte keratins are the major protein constituents in a wide variety of epidermal appendages that include wool, hair, quill, claw, hoof, horn and baleen. The ultrastructure of these materials, as visualized by transmission electron microscopy, has indicated a classical filament-matrix composite (Fig. 6.1: [1, 2]) in which 7–10 nm diameter filaments (IF), that provide the appendage with a high tensile strength are embedded in a matrix of keratin-associated proteins (KAPs). This matrix provides the appendage with its toughness, pliability, and resistance to attack by micro-organisms [3]. The IF and matrix components in trichocyte keratin each contain a large number of chains: for example, data obtained for human hair indicate that 17 IF chains are expressed [4] (Chapter 1.1, Table 1.1.1 & 1.1.2) together with 89 KAPs [5, 6] (Chapter 1.1, Table 1.2.3). In contrast, in the β-keratins that constitute the epidermal appendages of birds and reptiles (feathers,

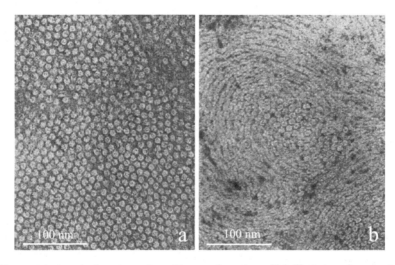

Fig. 6.1 (a) Electron micrograph of a cross-section of the paracortex of merino wool reduced and stained with osmium tetroxide, and further post-stained with lead and uranium salts. The intermediate filaments (light) are generally packed in a quasi-hexagonal manner in an osmiophilic matrix (dark). (b) A cross-section of the orthocortex, which illustrates a lower matrix content that in paracortex, and the packing of the IF in whorls. The inclination of the IF to the fibre axis increases with distance from the centre of the whorl. (Reprinted from J. Struct. Biol., 2003, Fraser et al. [22] with permission from Elsevier)

claws and scale), the filament and matrix components are combined in a single protein chain. In any one species, however, a wide variety of such chains have been identified, depending on the nature and function of the appendage [3, 7, 8].

The filaments in trichocyte keratin were initially referred to as microfibrils. However, it was subsequently observed that a major portion of the trichocyte keratin chains had a high degree of sequence homology with the proteins forming the class of structures known as intermediate filaments (see, for example, [9–11]). As a consequence of this similarity, the term microfibril lapsed in favour of intermediate filament (IF). Intermediate filament proteins that include vimentin, desmin, peripherin, neurofilament and nuclear lamin, as well as the trichocyte and epithelial keratins, exist in a wide variety of eukaryotic cells and each forms IF with diameters of about 7–10 nm [12–14]. The terminology "intermediate filament" arose initially from the observation that a third set of filaments existed in cells that were intermediate in diameter between the more commonly studied microtubules (25 nm) and microfilaments (7 nm). Modes of aggregation specific to each chain type, but which nonetheless have many features in common, lead to the formation *in vivo* of non-polar cytoskeletal filaments that provide the cell with its requisite mechanical properties, as well as other aspects related to function.

An additional level of superstructure – the macrofibril – is also evident in hair and wool. The macrofibrils in the paracortex are generally large, with irregular profiles (Fig. 6.1a), whereas those in the orthocortex are smaller and more circular in cross-section (Fig. 6.1b). The paracortex contains bundles of near-straight IF lying closely parallel to one another, and packed in a quasi-hexagonal manner. In the orthocortex, however, the IF have a spiral or whorl substructure with their angles of tilt increasing with radius measured from the centre of the whorl [15, 16]. The para- and orthocortex both contain similar or possibly identical IF proteins, but they differ significantly in both the type of KAPs present and their quantity [17–19]. Early work suggested that the KAPs were expressed later in development than the IF chains and at a point further up the hair follicle. While the bulk of the KAPs are indeed expressed subsequent to IF formation, there is now clear evidence that two KAPs in particular (KAP8, a high glycine-tyrosine protein, and KAP11, a high-sulfur protein) are expressed at high levels at the same time as the IF chains [5]. It would seem possible that these KAPs have a defining role in the assembly of the IF into the paracortical and orthocortical macrofibrils.

Supporting evidence comes from the work of Jones and Pope [20], who used developing human hair follicle cells to isolate trichocyte keratin structural components. Two types of assemblies of IF were observed (Chap. 11, Fig. 11.3), one being similar to that seen in the paracortex and the other to that in the orthocortex. The filament assemblies exhibited axial banding, thereby indicating that the IF were in axial register and that they formed a lattice with inter-IF bridges. It may be speculated that the latter involve the KAPs. Potential mechanisms by which macrofibril assembly proceeds *in vivo* have also been modelled [21–24] (Chap. 11).

6.2 Primary Structure

The amino acid sequences of sheep's wool (see, for example, [25, 26] and references therein) and human hair keratin IF protein chains (see the website http://interfil.org for a complete listing) have provided a wealth of information on the structure of the IF molecule and its modes of assembly into viable IF. Comparisons of these sequences have revealed that two homologous families of proteins are present. These have been designated Type I and Type II chains, by analogy with the nomenclature used for the sequences of the first sheep wool fragments characterized by Crewther and colleagues [27, 28]. The same Type I/Type II division of chains has also been observed for the epithelial keratins [12, 13, 29, 30]. In human hair (cortex, cuticle and medulla) a total of 11 Type I chains and six Type II chains have been identified [4]. These numbers are closely similar to those found in sheep wool, i.e. ten and seven respectively [26, 31]. Within the cortex alone, nine Type I chains and four Type II chains are expressed in human hair, compared with nine and five chains respectively for sheep wool.

The amino acid sequences of both Type I and Type II trichocyte (and epithelial) keratin chains display several highly conserved features. Each can be subdivided into three regions – an N-terminal (head) domain, an α-helix-rich, heptad repeat-containing central domain, and a C-terminal (tail) domain. The central domain contains about 310–315 residues and contains a heptad substructure of the form $(a\text{-}b\text{-}c\text{-}d\text{-}e\text{-}f\text{-}g)_n$. Residues in the a and d positions are generally apolar in character and are frequently leucine, valine or isoleucine residues. Overall, the heptad regions encompass about 284 residues (segments 1A, 1B and 2 are 35, 101 and 148 residues long, respectively). Short linkers, known as L1 (typically 11–14 residues) and L12 (16 and 17 residues respectively for Type I and Type II chains), connect segments 1A and 1B, and segments 1B and 2, respectively (Fig. 6.2). Highly conserved sequences occur in segment 1A and at the end of segment 2, and these are referred to as the helix initiation (HIM) and termination (HTM) motifs respectively.

Using a fast Fourier transform analysis it has been shown that segments 1B and 2 both display a non-random linear distribution of their acidic (aspartic and glutamic acid) and basic residues (arginine and lysine). In segment 1B the common periodicity is about 9.55 residues (1.42 nm), and in segment 2 it is close to 9.85 residues (1.46 nm). In each case the acidic and basic periods are approximately, but not exactly, 180° out of phase with respect to one another. Both segments are thus characterised by alternating bands of negative and positive charge.

The head and tail domains are characterized by high contents of cysteine residues that are involved in the formation of disulfide bonds as the hair undergoes terminal differentiation [32]. Many of these are believed to be intramolecular, whilst the remainder are formed with the cysteine-rich matrix proteins. Interestingly, however, there are also regions within the tail domain of some chains that display structural regularities based on an imperfect sequence repeat (see next section). Less sequence regularity has been observed in the head domains, though a nonapeptide quasi-repeat of the form (GGGFGYRSX) has been seen for Type II chains [33].

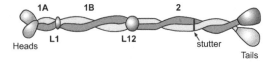

Fig. 6.2 Schematic diagram of the trichocyte keratin molecule. Each molecule is a parallel-chain, in-register heterodimer containing a Type I and a Type II chain. The rod domain, dominated by left-handed coiled-coil structure (Type I, light blue: Type II, cyan), suffers discontinuities in the underlying heptad (seven residue) pattern at linkers L1 and L12, as well as at a more central point in segment 2 (marked stutter). The N-terminal 35 residues in segment 2 has a hendecad sequence repeat, which implies that the two strands will lie nearly parallel to the axis of the molecule, rather than coiling around it in a left-handed manner. The acidic and the basic residues in segments 1B and 2 have near out-of-phase periodic dispositions (9.55 and 9.85 residues, respectively) that are important in the aggregation of molecules through salt bridges into viable IF through the A_{11}, A_{22} and A_{12} modes of interaction (see Fig. 6.4). The N-terminal domains (heads) are coloured green (Type I) and orange (Type II), and the C-terminal domains (tails) are coloured yellow (Type I) and red (Type II)

6.3 Secondary and Tertiary Structure

The heptad region, which also contains a preponderance of α-helix-favouring residues, has been shown to form a two-stranded α-helical coiled-coil rod structure about 46 nm long including the non-helical linker segments (Fig. 6.2). In addition, there is now strong evidence that the trichocyte (and epithelial) keratin molecule is a Type I/Type II heterodimer [10, 11, 34–37]. Further characterization has shown that the keratin chains lie parallel to one another (rather than antiparallel) and also are in axial register [38–40]. Interestingly, studies [41, 42] have shown that the trichocyte and epithelial keratin chains can heterodimerise. Furthermore, it has been shown that homodimers (Type I/Type I and Type II/Type II) can form but there is currently no evidence that these are functional structures, as distinct from transient intermediates. In addition, there is also now extensive evidence that promiscuity of chain association is not uncommon i.e. Type I chains can heterodimerise with several Type II chains, and vice versa (see, for example, [42, 43]).

The structural conclusions have been derived from both sequence analyses and from the fibre X-ray diffraction patterns recorded from highly oriented specimens of trichocyte keratin, mainly derived from porcupine quill. The patterns show two characteristic meridional reflections with spacings of about 0.515 and 0.1485 nm, and an equatorial maximum with a spacing of about 1 nm [44–46] as well as a strong near-equatorial layer line with an axial spacing of about 7 nm (Fig. 6.3: [47]). All of these features are characteristic of a coiled-coil rope structure with α-helical strands (Fig. 6.2 [47–49]). These observations were naturally associated with the oriented portion of the fibre structure i.e. the filaments, and, in particular, with the common stretch of heptad-containing sequence. Note that about three hendecad sequence repeats (each 11 residues long and closely similar to a heptad substructure) occur at the N-terminal end of segment 2, and this implies that the α-helical strands in the coiled-coil in this particular region will lie approximately parallel to one another rather than coiling in a left-handed manner [50, 51].

No fraction of the trichocyte keratin chain has yet been crystallized and studied by X-ray protein crystallography. Nonetheless, crystal structures are now available for chain fragments of several other IF, including vimentin [51–57], lamin [58] and epidermal keratin [59]. As each has an homologous rod domain, it is highly probable that the structural data pertaining to these other fragments will prove to be a close match to the structure that exists in the trichocyte keratins.

Sequence data in the head and tail domains of sheep trichocyte keratins indicate elements of regular secondary structure. For example, the sequence immediately C-terminal to the rod domain in Type II chains is predicted to form a β-sheet with four antiparallel strands linked by conserved β-turns [33]. The β-sheet has an apolar face that is suggestive of a pairing with a like sheet from a second (antiparallel) molecule via apolar interactions, thereby providing specific links between molecules within the IF. The tail domain of Type I chains displays a different type of sequence regularity. A proline-cysteine-X motif, where X is variable, is repeated up to ten times, with seven of them being contiguous [33]. Modelling suggests that this region may adopt a polyproline II left-handed helical conformation with three residues per turn, thereby placing the cysteine residues along one edge. The potential that this might interact with a like element of structure in a second (antiparallel) molecule, and thus form a series of disulfide bonds to stabilize the molecular assembly, is evident.

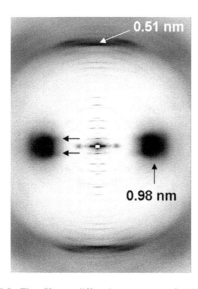

Fig. 6.3 The X-ray diffraction pattern of trichocyte keratin from porcupine quill is highly detailed and shows the equatorial and near-equatorial maxima at a lateral spacing of about 0.98 nm as well as a meridional reflection at a spacing of about 0.51 nm. The axial spacing of the near-equatorial layer lines, indicated by arrows, has a value of about 7 nm. (Reprinted from J. Struct. Biol., 2008, Parry et al. [87] with permission from Elsevier)

6.4 Modes of Molecular Assembly in the IF

From the observation that there was a regular linear disposition in the acidic and the basic residues, it was suggested by Crewther et al. [10] that three antiparallel arrangements of molecules were probable. Note that antiparallel modes were

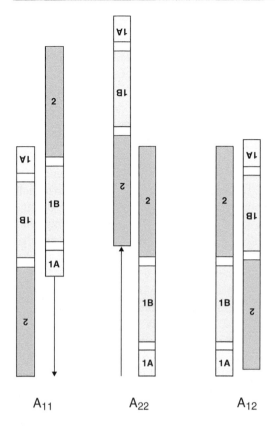

Fig. 6.4 The DST crosslink data are consistent with three modes of molecular interaction (A_{11}, A_{22} and A_{12}). A_{11} corresponds to an overlap of antiparallel 1B segments (blue), A_{22} to an overlap of antiparallel 2 segments (orange), and A_{12} to an overlap of antiparallel molecules. The axial displacements in each case are measured from the N-terminal end of segment 1A (yellow) to the C-terminal end of segment 2 (orange), and may be negative (A_{11}), positive (A_{22}), or either positive or negative (A_{12}). The heterodimer nature of the molecule is not indicated. Linkers L1 and L12 are represented by white rectangles between the 1A and 1B segments, and between the 1B and 2 segments respectively

proposed but, as will be shown below, characterization of induced crosslinks confirmed that these three modes were, indeed, fundamental to molecular aggregation in the trichocyte keratin IF. The slightly different periods in segments 1B and 2 indicated a higher preference for self-self interactions i.e. 1B-1B (A_{11}) and 2-2 (A_{22}) interactions than for mixed ones ie. 1B-2 interactions (A_{12}).

Experimental proof that supported these proposals came through the chemical crosslinking of spatially-adjacent lysine residues in intact IF, or sub-assemblies of them, with the periodate-cleavable bi-functional crosslinking reagent disulfo-succinimidyl-tartrate (DST) [40]. The crosslinking was performed under mild conditions that did not prevent subsequent assembly of the crosslinked proteins into functional IF. The crosslinked proteins were cleaved with cyanogen bromide and trypsin, and the peptides resolved by high-performance liquid chromatography (HPLC). By comparing the peptide peaks before and after crosslinking it is a simple matter to ascertain which of them have been shifted. These were then recovered for chemical characterization. By reacting these peptides with periodate, two peaks were produced, each of which can be sequenced, thereby allowing the individual lysine residues that had been adjoined to be identified. Using a least-squares analysis on the crosslink data the relative positions of the molecules can be determined (Fig. 6.5). This approach was used successfully for the trichocyte keratins in a reducing milieu to mimic the initial formation of the IF in the hair follicle, and in an oxidizing one to relate to the fully developed (disulfide-bonded) form of the hair as it emerges from the scalp [40, 61]. This work revealed that the intramolecular crosslinks were consistent with a parallel, in-register chain arrangement as predicted. In addition, all of the intermolecular crosslinks could be accounted for in terms of the A_{11}, A_{22} and A_{12} modes (Fig. 6.5). The A_{12} and A_{22} staggers were near identical for the reduced and oxidizing structures (i.e. about +2 h_{cc} and +187 h_{cc} respectively, where h_{cc} is the mean axial rise per residue in a coiled coil conformation), but the A_{11} stagger differed considerably from about -112 h_{cc} (reduced) to about -131 h_{cc} (oxidised). Trichocyte keratin

favoured over parallel ones on the basis of the observation that the phase difference between the common period for the acidic and basic residues, although close to 180°, was nonetheless displaced from it. These three modes were subsequently termed A_{11}, A_{22} and A_{12} (Fig. 6.4; [60]), and were defined as (a) an antiparallel overlap of a pair of 1B segments, (b) an antiparallel overlap of a pair of 2 segments, and (c) an antiparallel overlap of a pair of molecules. This theoretical result lacked experimental verification at the time that it was

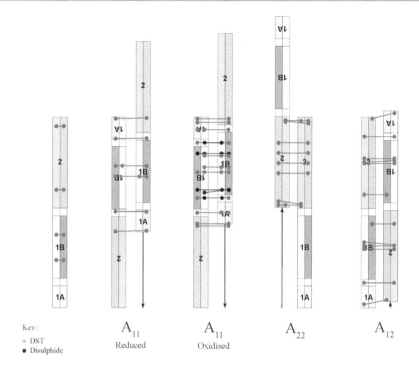

Fig. 6.5 Schematic representation of the DST crosslinks (red) characterized in (from left to right) a trichocyte keratin molecule, a pair of antiparallel molecules in the A_{11} mode (reduced) and then after molecular slippage has occurred (oxidized), a pair of antiparallel molecules in the A_{22} mode (reduced and oxidized), and a pair of antiparallel molecules in the A_{12} mode (reduced and oxidized). These data confirm that the molecule must be a heterodimer (the two chains are indicated by different shades of the same colour in each of the rod domain segments) and also that the chains lie in axial register and are parallel to one another (rather than being antiparallel). The disulfide bonds (black) occur only in the A_{11} oxidised structure and not in the A_{11} reduced form

Table 6.1 Axial staggers in the reduced and oxidised forms of trichocyte keratin IF

	Reduced[a]	Oxidised[a]	
	DST[b]	DST[b]	S-S[c]
A_{11}	−112.1 ± 2.6	−130.8 ± 1.6	−130.7 ± 0.1
A_{12}	−0.3 ± 2.4	2.6 ± 1.6	n.d.
A_{22}	185.9 ± 6.7	185.8 ± 1.6	187.8 ± 0.6
Molecular overlap	2.4 ± 8.6	−13.6 ± 3.6	−14.2 ± 0.6

n.d. not determinable
[a]measured in multiples of h_{cc}, the unit rise per residue in a coiled-coil conformation (0.1485 nm)
[b]determined using positions of DST crosslinks involving lysine residues
[c]determined using positions of cysteine residues

IF thus has two discrete structures depending on its stage of development (Table 6.1).

The molecular slippage of about 19 h_{cc} (about 2.82 nm) that occurs within the IF has several pronounced effects. Firstly, it results in the positions of the cysteine residues in a pair of antiparallel 1B segments in the A_{11} mode that are not in axial alignment in the reduced form to become closely aligned axially in the oxidized structure. The expectation is that these cysteine residues would then form intermolecular disulfide bonds that would stabilize the structure of the IF [62]. Secondly, the head and tail domains of the IF molecules, that lie predominantly on the surface of the intermediate filament, are arranged quite differently in the reduced and oxidized states (Fig. 6.4: [63]). In the reduced state, the terminal domains are arranged on a two-start left-handed helix. The heads and tails are also well separated from one another and in positions where interactions between them are likely to be minimal. In the oxidized state, however, the heads and tails are much closer to one another, and hence very much more likely to interact strongly with one another. The terminal domains lie on a one-start left-handed helix of pitch length 23.5 nm. Thirdly, modelling of the molecular slippage in the A_{11} mode [64] shows that this occurs along an hydropathic (apolar) stripe formed by a combination of both the Type I and Type II chains. It is thus a simple axial translation that does not involve a relative rotation of either of the two contributing segments. The molecular packing is non-optimal in the reduced case with near node-to-node interactions of the helical molecules. In contrast, in the oxidised structure the molecules are packed closer to one another in near optimal node-antinode mode. This change in molecular packing corresponds closely to the different radial dimensions of the IF observed in the reduced and oxidized states (Sect. 6.5).

Regarding the stabilization of the IF that occurs as a result of disulfide bond formation it is possible to perform a least-squares analysis similar to that used for the DST crosslink data, but using instead the positions of the cysteine residues as the input data [65]. There are, of course, far fewer data available than in the DST analyses that involved the crosslinking of lysine residues. However, the necessity that the cysteine residues must lie in almost exact axial register, if disulfide bonds are to form, does provide a very strong constraint on the relative axial displacements of the IF molecules, and leads to more precise estimates of both A_{11} and A_{22} (Table 6.1; [65]). The protofilament, defined as a pair of antiparallel molecular strands (four chains in section) with each molecular strand consisting of a linear array of similarly directed molecules with a small head-to-tail overlap [62, 66, 67], utilises the A_{11} and A_{22} modes of interaction. As the protofilaments thus contain all of the possible disulfide bonds between rod domain coiled-coil segments in the trichocyte keratin IF it follows that this sub-IF structure may indeed have a true physical existence.

The protofilament is thus a convenient subfibrillar element of structure in trichocyte keratin that simplifies the description of the IF postassembly. The packing of the two coiled-coil ropes has been shown [64] to be close to optimum [68] and in cross-section the protofilament will appear to be ribbon-like, rather than cylindrical. It is probable that the ribbon will have a slow twist and depending upon the section thickness will appear circular when viewed by TEM. It is of interest that ribbon-like sub filamentous structures of similar dimensions have been identified as components of the thick filaments in muscle [69].

It is not known whether the protofilament has any role in the actual mechanism of assembly. Details of trichocyte keratin assembly have not yet been determined experimentally but are believed to closely follow the mechanism established for other members of the IF family (see, for example, [38, 39, 70–73]). In these cases, the first step involves a lateral assembly of about eight tetramers. Each tetramer is a near half-staggered arrangement of antiparallel molecules in the A_{11} mode. The structure thus formed is about 50–60 nm long and has been termed a unit-length-filament (ULF). The second step involves axial aggregation of the ULF, exclusively through the A_{22} mode, to produce immature IF that are typically about 15–16 nm in diameter and several μm in length. The third step involves optimization of the molecular packing and a subsequent radial compaction to produce IF of diameter close to 10 nm.

6.5 Differences and Similarities Between the Reduced and Oxidised Structures of IF

The lateral dispositions of the constituent rod domains in the trichocyte keratin IF differ significantly between the oxidized and reduced structures. In the former case, studies [47, 49] used X-ray diffraction methods to show that the rod domains of the IF molecules lay on an annulus of radius 2.9 and 3.0 nm respectively for the dehydrated and hydrated forms. This is markedly smaller than the value (3.5 nm) deduced from the maximum in the radial density function obtained by Watts and colleagues [74] from their electron microscope observations of freeze-dried and vitrified specimens. It is clear that a radial compaction occurs between the reduced and oxidized states and that this happens at the same time as the axial rearrangement of molecules in the A_{11} mode.

STEM data has indicated that hard-keratin IF contain about 32 chains in cross-section [74, 75] and this value would be compatible with the post-IF formation of eight protofilaments. The challenge is, therefore, to find models compatible with both the STEM data and with the axial and lateral changes observed. Towards that end the fully keratinized (oxidized) trichocyte keratin IF has been studied extensively by high resolution X-ray diffraction by Fraser and colleagues [49, 66, 76, 77]. From the meridional and near-meridional regions of the diffraction patterns two possible surface lattices can be derived, the first involving seven protofilaments and the second eight. In both cases the protofilaments would assemble via the A_{12} mode. The dimensions of the surface lattices (or their mirror images) are defined by the axial projections of the a and b surface lattice vectors (z_a and z_b respectively) and the changes in azimuth involved (t_a and t_b respectively). The value of z_a corresponds to the axial distance between successive surface lattice points, and z_b to the axial stagger between adjacent protofilaments (Fig. 6.6). For the eight-protofilament model the calculated dimensions are: axial repeat = 47.0 nm, z_a = 7.42 nm, $t_a = -\pi/2$, z_b = 19.79 nm, $t_b = +\pi/4$, whereas those for the seven-protofilament model are: axial repeat = 47.0 nm, z_a = 7.42 nm, $t_a = -4\pi/7$, z_b = 19.79 nm, $t_b = +2\pi/7$.

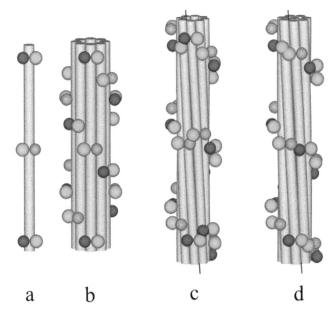

Fig. 6.6 Three-dimensional models of the reduced and oxidized IF structure in which the yellow cylinders represent the helical domains of the IF molecules. The green spheres represent the N-terminal domains of the Up molecular strands and the blue spheres to those of the Down molecular strands. The red spheres represent the C-terminal domains in the Up molecular strands and the orange spheres represent the C-terminal domains of the Down molecular strands. The crosslinking data have shown that the values of A_{12} and A_{22} are virtually unaltered between the reduced and oxidized states but that A_{11} changes markedly by about 2.82 nm. The resulting molecular slippage leads to several changes in the structure of the IF. (**a**) A single protofilament in the reduced form. (**b**) The terminal domains are well separated and arranged on a two-start left-handed helix in the reduced form, thereby giving rise to diagonal banding with a spacing of about 22 nm. In both the (8 + 0) and (7 + 1) models for the oxidized structure (**c** and **d** respectively), however, the terminal domains are in closer proximity and lie on a one-start helix of pitch length 23.5 nm. The positions of the cysteine residues in a pair of antiparallel 1B segments, that are not in axial register in the reduced form, are in almost perfect axial alignment in the oxidized form. Most of the cysteine residues would therefore be expected to form intermolecular stabilising disulfide bonds. (Reprinted from J. Struct. Biol, 2005, Fraser and Parry [63] with permission from Elsevier)

The former structure consists of a ring of eight protofilaments arranged on a constant radius, and is commonly referred to as the (8 + 0) model. The seven-protofilament model consists of the seven protofilaments arranged on a constant but slightly smaller radius than in the (8 + 0) structure. The concept that an eighth protofilament might also be centrally located was not considered by [76, 77], but it is now common practice to refer to this as the (7 + 1) model. It is important to note that the X-ray data on which these surface lattice structures are based pertain specifically to those collected for oxidized IF.

With these thoughts in mind it is now relevant to consider what the lateral structures of the IF might be in both the reduced and oxidized states. In the former situation X-ray microdiffraction studies on the hair follicle were undertaken by Er Rafik et al. [78]. These data indicated that the IF have a low-density core substructure (i.e. a ring structure with radius 3.5 nm) with an overall diameter of about 10 nm at a point just above the bulb region. Earlier cryo-electron microscopy on reduced IF had also indicated a ring structure with no evidence of a core structure [74]. These data are thus mutually consistent and support an (8 + 0) structure for reduced IF ([63]: axial repeat = 44.92 nm, z_a = 11.23 nm, t_a = $-\pi/2$, z_b = 16.85 nm, t_b = $+\pi/4$).

The optimum model for the oxidized structure of IF, however, has been less clear-cut, and two possibilities have been deemed worthy of consideration. The first model retains the (8 + 0) topology, but the constituent protofilaments pack closer to one another than in the "reduced" structure i.e. radial compaction occurs and the radius of the ring drops from 3.5 nm (deduced by Fraser and Parry [61] from the data of Watts et al. 2002 [74], see also Er Rafik et al. [78]) to 2.9–3.0 nm [47, 49]. The second model has (7 + 1) topology, with seven protofilaments lying on a ring of radius 2.9–3.0 nm with an eighth protofilament centrally-located.

Initially, the data available were not able to clearly differentiate one model from the other (Fig. 6.6). For example, early TEM studies of cross-sections of wool and hair showed both ring [79] and ring-core structures for oxidized trichocyte keratin IF [80, 81], and it was unclear which best represented the *in vivo* structure. Subsequent studies of electron stains, however, showed that this apparent anomaly was likely due to the characteristics of the particular electron stains used [16]. More recently, however, evidence has been accumulating on the relative merits of the (8 + 0) structure over that of the (7 + 1) arrangement. For example, the observed change in lateral dimensions of the IF between the reduced and oxidized states has now been shown to be consistent with an improved packing of the eight protofilaments in an (8 + 0) structure: in contrast, in a (7 + 1) arrangement, in which one of the protofilaments migrates from the ring to the centre of the IF during oxidation, sterically short contacts would occur between the protofilaments in the outer ring and that lying along the axis of the IF [82]. Furthermore, modelling of the intensity distribution in the 1 nm region of the equatorial X-ray diffraction pattern [82] favours the (8 + 0) model over the (7 + 1) structure, and this study reveals both the separation of the molecular strands with some accuracy and their ribbon-like arrangement. The physical dimensions defining the (8 + 0) structure for oxidized trichocyte keratin IF are as follows: axial repeat = 47.0 nm, z_a = 7.42 nm, t_a = $-\pi/2$, z_b = 19.79 nm, t_b = $+\pi/4$ [76, 77]. Interestingly, using the crosslink data the value of z_d (the extent of the dislocation of the surface lattice) is zero (within experimental error) for the (8 + 0) reduced structure, but is very large (about 17.82 nm) for the (8 + 0) oxidized model [61].

Studies of the sequence and amino acid composition of the trichocyte keratin head and tail domains have also now revealed conserved homologous subdomains (H1 and H2: [82] akin to those reported for the epidermal keratins [83]. As Steinert and Parry [84] have shown that H1 in the Type II epidermal chain was important in facilitating assembly at the two- to four-molecule level it is probable that the H subdomains in the trichocyte keratins will play a similar role.

As the evidence supports an (8 + 0) arrangement structure in oxidized trichocyte keratin IF it follows that the electron microscope observations that had revealed a central core must now be explained differently. One possibility [82] was that apolar, cysteine-rich head and/or tail domains occupy the core along the axis of the IF, thereby shielding them from water. This would necessarily result in a decrease in the efficacy of the reducing agents used prior to staining [85, 86], and lead to a stain-excluding core in the IF that had previously been attributed to the presence of a central protofilament.

6.6 Summary

Trichocyte keratin has a structural hierarchy of extraordinary complexity. Although hair and other epidermal appendages, such as quill and claw, have been studied for about 80 years it is only the recent acquisition of sequence data, allied to increasingly sophisticated experimental techniques, that have allowed much of the detail to be revealed. The intermediate filament trichocyte keratin chains all have cysteine-rich N- and C-terminal domains enclosing a central domain about 315-residues long containing α-helix-favouring residues with a heptad/hendecad substructure. Most of this region forms a left-handed two-stranded coiled-coil structure with two short breaks included (linkers L1 and L12). It also displays a regular linear disposition of its constituent acidic and the basic residues. This favours three specific modes of molecular aggregation (A_{12}, A_{11} and A_{22}). Uniquely, the axial stagger corresponding to A_{11} changes between the IF structure adopted at the base of the hair follicle in a reducing environment and the structure produced at the final stage of terminal differentiation in an oxidizing environment. The molecular slippage that occurs results in the axial alignment of the cysteine residues and the formation of disulfide bonds that stabilise the structure of the IF. Furthermore, X-ray, TEM and STEM data have allowed the surface lattice structures of both the reduced and oxidised IF structures to be deduced, and for the lateral compaction that occurs with increasing differentiation to be explained. The largest scale of the structural hierarchy pertains to the assembly of the IF into the paracortex- and orthocortex. The former is characterized by quasi-hexagonal packing of the constituent IF, and the latter by the spiral/whorl arrangement. Models for each have now been proposed.

References

1. Birbeck, M. S. C., & Mercer, E. H. (1957). The electron microscopy of the human hair follicle. Part1. Introduction and the hair cortex. *Journal of Biophysical and Biochemical Cytology, 3,* 203–214.
2. Rogers, G. E. (1959). Electron microscope studies of hair and wool. *Annals of the New York Academy of Sciences, 83,* 378–399.
3. Fraser, R. D. B., & Parry, D. A. D. (2014). Amino acid sequence homologies in the hard keratins of birds and reptiles, and their implications for molecular structure and physical properties. *Journal of Structural Biology, 188,* 213–224.
4. Schweizer, J., et al. (2007). Hair follicle-specific keratins and their diseases. *Experimental Cell Research, 313*(10), 2010–2020.
5. Rogers, M. A., et al. (2006). Human hair keratin-associated proteins (KAPs). *International Review of Cytology, 251,* 209–263.
6. Gong, H., et al. (2012). An updated nomenclature for keratin-associated proteins (KAPs). *International Journal of Biological Sciences, 8*(2), 258–264.
7. Fraser, R. D. B., & Parry, D. A. D. (2008). Molecular packing in the feather keratin filament. *Journal of Structural Biology, 162,* 1–13.
8. Fraser, R. D. B., & Parry, D. A. D. (2011). The structural basis of the filament-matrix texture in the avian/reptilian group of hard β-keratins. *Journal of Structural Biology, 173,* 391–405.
9. Dowling, L. M., Parry, D. A. D., & Sparrow, L. G. (1983). Structural homology between hard α-keratin and the intermediate filament proteins desmin and vimentin. *Bioscience Reports, 3,* 73–78.
10. Crewther, W. G., et al. (1983). Structure of intermediate filaments. *International Journal of Biological Macromolecules, 5,* 267–274.
11. Parry, D. A. D., & Fraser, R. D. B. (1985). Intermediate filament structure. 1. Analysis of IF protein sequence data. *International Journal of Biological Macromolecules, 7,* 203–213.
12. Parry, D. A. D., & Steinert, P. M. (1995). *Intermediate filament structure.* Heidelberg: Springer.
13. Parry, D. A. D., & Steinert, P. M. (1999). Intermediate filaments: Molecular architecture, assembly, dynamics and polymorphism. *Quarterly Reviews of Biophysics, 32*(2), 99–187.
14. Herrmann, H., & Aebi, U. (2004). Intermediate filament assembly: Molecular structure, assembly mechanism, and integration into functionally distinct intracellular scaffolds. *Annual Review of Biochemistry, 73,* 749–789.

15. Bryson, W. G., et al. (2000). High voltage microscopical imaging of macrofibril ultrastructure reveals the three-dimensional spatial arrangement of intermediate filaments in Romney wool cortical cells – A causative factor in fibre curvature. In *Proceedings of the 10th international wool textile research conference*.
16. Harland, D. P., et al. (2011). Arrangement of trichokeratin intermediate filaments and matrix in the cortex of Merino wool. *Journal of Structural Biology, 173*(1), 29–37.
17. Gillespie, J. M., & Reis, P. J. (1966). The dietary-regulated biosynthesis of high-sulphur wool proteins. *Biochemical Journal, 98*, 669–677.
18. Frenkel, M. J., Gillespie, J. M., & Reis, P. J. (1974). Factors influencing the biosynthesis of the tyrosine-rich proteins of wool. *Australian Journal of Biological Science, 27*, 31–38.
19. Powell, B. C., & Rogers, G. E. (1997). The role of keratin proteins and their genes in the growth, structure and properties of hair. In P. Jolles, H. Zahn, & E. Hocker (Eds.), *Formation and structure of human hair* (pp. 59–148). Basel: Birkhäuser Verlag.
20. Jones, L. N., & Pope, F. M. (1985). Isolation of intermediate filament assemblies from human hair follicle. *Journal of Cell Biology, 101*, 1569–1577.
21. Fraser, R. D. B., & Parry, D. A. D. (2003). Macrofibril assembly in trichocyte (hard α-) keratins. *Journal of Structural Biology, 142*(2), 319–325.
22. Fraser, R. D. B., Rogers, G. E., & Parry, D. A. D. (2003). Nucleation and growth of macrofibrils in trichocyte (hard-α) keratins. *Journal of Structural Biology, 143*, 85–93.
23. McKinnon, A. J. (2006). The self-assembly of keratin intermediate filaments into macrofibrils: Is this process mediated by a mesophase? *Current Applied Physics, 6*, 375–378.
24. McKinnon, A. J., & Harland, D. P. (2010). The role of liquid-crystalline structures in the morphogenesis of animal fibres. *International Journal of Trichology, 2*, 101–103.
25. Yu, Z., et al. (2009). Expression patterns of keratin intermediate filament and keratin associated protein genes in wool follicles. *Differentiation, 77*(3), 307–316.
26. Yu, Z., et al. (2011). Annotations of sheep keratin intermediate filament genes and their patterns of expression. *Experimental Dermatology, 20*(7), 582–588.
27. Crewther, W. G., Inglis, A. S., & McKern, N. M. (1978). Amino acid sequences of α-helical segments from S-carboxymethylkerateine-A. Complete sequence of a type-II segment. *Biochemical Journal, 173*, 365–371.
28. Gough, K. H., Inglis, A. S., & Crewther, W. G. (1978). Amino acid sequences of α-helical segments from S-carboxymethylkerateine-A. Complete sequence of a type-I segment. *Biochemical Journal, 173*, 373–385.
29. Hanukoglu, I., & Fuchs, E. V. (1982). The cDNA sequence of a human epidermal keratin: Divergence of sequence but conservation of structure among intermediate filament proteins. *Cell, 31*, 243–252.
30. Hanukoglu, I., & Fuchs, E. (1983). The cDNA sequence of a type II cytoskeletal keratin reveals constant and variable structural domains among keratins. *Cell, 33*(3), 915–924.
31. Clerens, S., et al. (2010). Developing the wool proteome. *Journal of Proteomics, 73*, 1722–1731.
32. Strnad, P., et al. (2011). Unique amino acid signatures that are evolutionarily conserved distinguish simple-type, epidermal and hair keratins. *Journal of Cell Science, 124*, 4221–4232.
33. Parry, D. A. D., & North, A. C. T. (1998). Hard α-keratin intermediate filament chains: Substructure of the N- and C-terminal domains and the predicted structure and function of the C-terminal domains of type I and type II chains. *Journal of Structural Biology, 122*(1–2), 67–75.
34. Parry, D. A. D., et al. (1977). Structure of α-keratin: Structural implication of the amino acid sequences of the type I and type II chain segments. *Journal of Molecular Biology, 113*, 449–454.
35. Steinert, P. M. (1990). The two-chain coiled-coil molecule of native epidermal keratin intermediate filaments is a type I-type II heterodimer. *Journal of Biological Chemistry, 265*, 8766–8774.
36. Hatzfeld, M., & Weber, K. (1990). The coiled coil of in vitro assembled keratin filaments is a heterodimer of type I and II keratins: Use of site-specific mutagenesis and recombinant protein expression. *Journal of Cell Biology, 110*, 1199–1210.
37. Coulombe, P. A., & Fuchs, E. (1990). Elucidating the early stages of keratin filament assembly. *Journal of Cell Biology, 111*, 153–169.
38. Steinert, P. M. (1991). Organization of coiled-coil molecules in native keratin 1/keratin 10 intermediate filaments: Evidence for alternating rows of antiparallel in-register and antiparallel molecules. *Journal of Structural Biology, 107*, 157–174.
39. Steinert, P. M. (1991). Analysis of the mechanism of assembly of mouse keratin 1/ keratin 10 intermediate filaments in vitro suggests that intermediate filaments are built from multiple oligomeric units rather than a unique tetrameric building block. *Journal of Structural Biology, 107*, 175–188.
40. Wang, H., et al. (2000). In vitro assembly and structure of trichocyte keratin intermediate filaments: A novel role for stabilization by disulfide bonding. *Journal of Cell Biology, 151*(7), 1459–1468.
41. Herrling, J., & Sparrow, L. G. (1991). Interactions of intermediate filament proteins from wool. *International Journal of Biological Macromolecules, 13*, 115–119.
42. Langbein, L., et al. (2010). The keratins of the human beard hair medulla: The riddle in the middle. *Journal of Investigative Dermatology, 130*(1), 55–73.
43. Smith, T. A., & Parry, D. A. D. (2007). Sequence analyses of type I and type II chains in human hair and epithelial keratin intermediate filaments: Promiscuous obligate heterodimers, type II template for molecule formation and a rationale for heterodimer formation. *Journal of Structural Biology, 158*(3), 344–357.

44. Astbury, W. T., & Woods, H. J. (1930). The X-ray interpretation of the structure and elastic properties of hair keratin. *Nature, 126*, 913–914.
45. Astbury, W. T., & Woods, H. J. (1933). X-ray studies on the structure of hair, wool and related fibres II. The molecular structure and elastic properties of hair keratin. *Philosophical Transactions of the Royal Society B: Biological Sciences, A 232*, 333–394.
46. MacArthur, I. (1943). Structure of α-keratin. *Nature, 152*, 38–41.
47. Fraser, R. D. B., MacRae, T. P., & Miller, A. (1965). X-ray diffraction patterns of α-fibrous proteins. *Journal of Molecular Biology, 14*, 432–442.
48. Fraser, R. D. B., MacRae, T. P., & Miller, A. (1964). The coiled-coil model of α-keratin structure. *Journal of Molecular Biology, 10*, 147–156.
49. Fraser, R. D. B., MacRae, T. P., & Suzuki, E. (1976). Structure of the α-keratin microfibril. *Journal of Molecular Biology, 108*, 435–452.
50. Parry, D. A. D. (2006). Hendecad repeat in segment 2A and linker L2 of intermediate filament chains implies the possibility of a right-handed coiled-coil structure. *Journal of Structural Biology, 155*, 370–374.
51. Nicolet, S. (2010). Atomic structure of vimentin coil 2. *Journal of Structural Biology, 170*, 369–376.
52. Strelkov, S. V., et al. (2002). Conserved segments 1A and 2B of the intermediate filament dimer: Their atomic structures and role in filament assembly. *EMBO Journal, 21*, 1255–1266.
53. Strelkov, S. V., et al. (2004). Crystal structure of the human lamin A coil 2B dimer: Implications for the head-to-tail association of nuclear lamins. *Journal of Molecular Biology, 343*, 1067–1080.
54. Meier, M., et al. (2009). Vimentin coil 1A – A molecular switch involved in the initiation of filament elongation. *Journal of Molecular Biology, 390*, 245–261.
55. Aziz, A., et al. (2012). The structure of vimentin linker 1 and rod 1b domains characterized by site-directed spin-labeling electron paramagnetic resonance (SDSL-EPR) and X-ray crystallography. *Journal of Biological Chemistry, 287*, 28349–28361.
56. Chernyatina, A. A., et al. (2012). Atomic structure of the vimentin central α-helical domain and its implications for intermediate filament assembly. *Proceedings of National Academy of Science USA, 109*, 13620–13625.
57. Chernyatina, A. A., Guzenko, D., & Strelkov, S. V. (2015). Intermediate filament structure: The bottom-up approach. *Current Opinion Cell Biology, 32*, 65–72.
58. Ruan, J., et al. (2012). Crystal structures of the coil 2B fragment and the globular tail domain of human lamin B1. *FEBS, 586*, 314–318.
59. Lee, C.-H., et al. (2012). Structural basis for heteromeric assembly and perinuclear organization of keratin filaments. *Nature Structural & Molecular Biology, 19*(7), 707–715.
60. Steinert, P. M., et al. (1993). Keratin intermediate filament structure: Crosslinking studies yield quantitative information on molecular dimensions and mechanism of assembly. *Journal of Molecular Biology, 230*, 436–452.
61. Fraser, R. D. B., & Parry, D. A. D. (2007). Structural changes in the trichocyte intermediate filaments accompanying the transition from the reduced to the oxidized form. *Journal of Structural Biology, 159*(1), 36–45.
62. Parry, D. A. D. (1996). Hard α-keratin intermediate filaments: An alternative interpretation of the low-angle equatorial X-ray diffraction pattern, and the axial disposition of putative disulphide bonds in the intra- and inter-protofilamentous networks. *International Journal of Biological Macromolecules, 19*(1), 45–50.
63. Fraser, R. D. B., & Parry, D. A. D. (2005). The three-dimensional structure of trichocyte (hard α-) keratin intermediate filaments: Features of the molecular packing deduced from the sites of induced crosslinks. *Journal of Structural Biology, 151*(2), 171–181.
64. Fraser, R. D. B., & Parry, D. A. D. (2014). Keratin intermediate filaments: Differences in the sequences of the type I and type II chains explain the origin of the stability of an enzyme-resistant four-chain fragment. *Journal of Structural Biology, 185*, 317–326.
65. Fraser, R. D. B., & Parry, D. A. D. (2012). The role of disulfide bond formation in the structural transition observed in the intermediate filaments of developing hair. *Journal of Structural Biology, 180*, 117–124.
66. Fraser, R. D. B., MacRae, T. P., & Parry, D. A. D. (1990). The three-dimensional structure of IF. In R. D. Goldman & P. M. Steinert (Eds.), *Cellular and molecular biology of intermediate filaments* (pp. 205–231). New York: Plenum Press.
67. Parry, D. A. D., Marekov, L. N., & Steinert, P. M. (2001). Subfilamentous protofibril structures in fibrous proteins: Cross-linking evidence for protofibrils in intermediate filaments. *Journal of Biological Chemistry, 276*, 39253–39258.
68. Rudall, K. M. (1956). Protein ribbons and sheets. In *Lectures on the scientific basis of medicine* (pp. 217–230). London: Athlone Press.
69. Reedy, M. K., & Perz-Edwards, R. J. (2016). Ribbons not subfilaments. *Biophysical Journal, 110*(Supplement 1), 13a.
70. Herrmann, H., et al. (1999). Characterisation of distinct early assembly units of different intermediate filament proteins. *Journal of Molecular Biology, 286*, 1403–1420.
71. Herrmann, H., et al. (2002). Characterisation of early assembly intermediates of recombinant human keratins. *Journal of Structural Biology, 137*, 82–96.
72. Mücke, N., et al. (2004). Molecular and biophysical characterization of assembly-starter units of human vimentin. *Journal of Molecular Biology, 340*, 97–114.
73. Parry, D. A. D., et al. (2007). Towards a molecular description of intermediate filament structure and assembly. *Experimental Cell Research, 313*(10), 2204–2216.
74. Watts, N. R., et al. (2002). Cryo-electron microscopy of trichocyte (hard α-keratin) intermediate filaments

reveals a low-density core. *Journal of Structural Biology, 137,* 109–118.
75. Jones, L. N., & Rivett, D. E. (1997). The role of 18-methyleicosanoic acid in the structure and formation of mammalian hair fibres. *Micron, 28*(6), 469–485.
76. Fraser, R. D. B., & MacRae, T. P. (1983). The structure of the α-keratin microfibril. *Bioscience Reports, 3,* 517–525.
77. Fraser, R. D. B., & MacRae, T. P. (1985). Intermediate filament structure. *Bioscience Reports, 5,* 573–579.
78. Er Rafik, M., et al. (2006). In vivo formation steps of the hard α-keratin intermediate filament along a hair follicle: Evidence for structural polymorphism. *Journal of Structural Biology, 154,* 79–88.
79. Rogers, G. E., & Filshie, B. K. (1962). Electron staining and fine structure of keratins. *International Congress of Electron Microscopy, 5,* O-2.
80. Filshie, B. K., & Rogers, G. E. (1961). The fine structure of α-keratin. *Journal of Molecular Biology, 3,* 784–786.
81. Millward, G. R. (1970). The substance of α-keratin microfibrils. *Journal of Ultrastructure Research, 31,* 349–355.
82. Fraser, R. D. B., & Parry, D. A. D. (2017). Intermediate filament structure in fully differentiated (oxidised) trichocyte keratin. *Journal of Structural Biology, 200,* 45–53.
83. Steinert, P. M., et al. (1985). Amino acid sequences of mouse and human epidermal type II keratins of 67,000 molecular weight provide a systematic basis for the structural and functional diversity of the end domains of keratin filament subunits. *Journal of Biological Chemistry, 260,* 7142–7149.
84. Steinert, P. M., & Parry, D. A. D. (1993). The conserved H1 domain of the type II keratin 1 chain plays an essential role in the alignment of nearest neighbour molecules in mouse and human keratin 1/keratin 10 intermediate filaments at the two- to four-molecule level of structure. *Journal of Biological Chemistry, 268,* 2878–2887.
85. Middlebrook, W. R., & Phillips, H. (1942). The action of sulphites on the cysteine disulphide linkages in wool. 3. The subdivision of the combined cysteine into four fractions differing in their reactivity towards sodium bisulphite. *Biochemical Journal, 36*(5–6), 428–437.
86. Lindley, H., & Cranston, R. W. (1974). The reactivity of the disulphide bonds of wool. *Biochemical Journal, 139,* 515–523.
87. Parry, D. A. D., Fraser, R. D. B., & Squire, J. M. (2008). Fifty years of coiled-coils and α-helical bundles: A close relationship between sequence and structure. *Journal of Structural Biology, 163,* 258–269.

Trichocyte Keratin-Associated Proteins (KAPs)

R. D. Bruce Fraser and David A. D. Parry

Contents

7.1	**Introduction**	72
7.2	**Primary Sequence Characteristics**	75
7.3	**Secondary Structure**	79
7.4	**Functional Roles of KAP Proteins**	81
7.4.1	KAP-KAP and KAP-IF Interactions	81
7.4.2	Mechanical Properties	82
7.4.3	Hair Diseases Related to KAP Proteins	84
7.5	**Summary**	84
	References	85

Abstract

The trichocyte (hard α-) keratins are epidermal appendages (hair, wool, hoof, horn, claw, baleen and quill) with a classic filament-matrix composite structure. In human hair, for example, keratin intermediate filaments (IF) of diameter 7.5 nm are embedded in a matrix formed from at least 89 different types of keratin-associated proteins (KAPs). The latter fall into three families, generally defined in terms of their cysteine residue or glycine plus tyrosine residue content. The KAPs, which infiltrate the space between the IF, are recognized as having especially important roles in the organisation of the IF into macrofibrils, in determining some of the most important physical attributes of the fully-keratinised hair fibre, including its hardness, toughness and pliability, and in linking IF to one another, either directly or indirectly, with a resultant increase in durability and resistance to degradation by microorganisms. Sequence data for many KAPs are now available, and repeating motifs of varying extent have been observed in a number of them. Little, however, is known about their three-dimensional structures, though modelling has indicated that some

R. D. B. Fraser
Institute of Fundamental Sciences, Massey University, Palmerston North, New Zealand

Tewantin, QLD, Australia

D. A. D. Parry (✉)
Institute of Fundamental Sciences, Massey University, Palmerston North, New Zealand

Riddet Institute, Massey University, Palmerston North, New Zealand
e-mail: d.parry@massey.ac.nz

local structural regularity is likely to exist. Current data suggest that the KAPs *in vivo* may adopt a variety of energetically-similar conformations stabilized predominantly by intramolecular disulfide bonds. The role of KAPs in hair diseases relates more to modulation in gene expression than to point mutations, in contrast to that observed for the IF proteins.

Keywords

Filament-matrix composite · Paracortex and orthocortex · Physical attributes of trichocyte keratin · Sequence repeats in matrix proteins · High-glycine-tyrosine proteins · High-sulfur proteins · Ultra-high-sulfur proteins · Matrix-related hair diseases

Abbreviations

IF	intermediate filaments
KAP	keratin-associated protein
HGT	high-glycine-tyrosine protein
HS	high-sulfur protein
UHS	ultra-high sulfur protein

7.1 Introduction

Studies of theprotein composition of mammalian epidermal appendages, such as hair and wool, have shown them to contain two classes of protein – a small number of low-sulfur (LS) proteins, derived from the intermediate filaments (IF), and a large number of keratin-associated proteins (KAPs), derived from the matrix (Fig. 7.1). Three classes of KAP have been identified: high sulfur proteins (HS), ultra high-sulfur-rich proteins (UHS) and glycine-tyrosine-rich proteins (HGT). Early studies of the composition and distribution of these three classes have been summarized by Gillespie [1] and Powell and Rogers [2]. Several important features emerged from these studies. Firstly, the IF (LS) proteins from different species and appendages were very similar and only contained a small number of components, in contrast to the large number of KAPs derived from the matrix. Secondly, there is a striking variation in both the content and the proportions of the KAPs in different appendages (Fig. 7.2). Another remarkable finding was that the contents and proportions of KAPs in a particular appendage were not fixed, but depended on nutritional status, as demonstrated in Fig. 7.3.

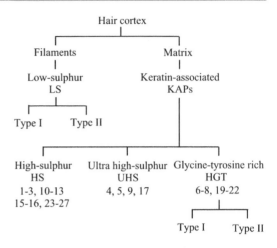

Fig. 7.1 Nomenclature used to describe the proteins present in the cortex oftrichocyte keratins, such as hair

Trichocyte keratin (hair) is a complex biological fibrous structure formed in the hair follicle. The bulk of each fibre consists of a cortex, sometimes with a central medulla formed from cell remnants and entrapped air. The cortex is surrounded by a cuticle formed from thin, overlapping flattened cells [3–5]. The IF component of the cortex is formed from the germative matrix cells that lie at the base of the hair follicle in the bulb region (Chap. 10). During development, the IF aggregate to form loose bundles. It now seems clear that some keratin-associated proteins (KAPs) are expressed simultaneously with the IF proteins and are involved with their ultra-structural organisation [6]. The KAPs form a matrix around and between each of the IF in the bundle. At terminal differentiation the cells die and the environment changes from a reducing one within the living cells to an oxidizing one. This allows both intra-molecular and inter-molecular disulfide bonds to be formed. Most of the cysteine resi-

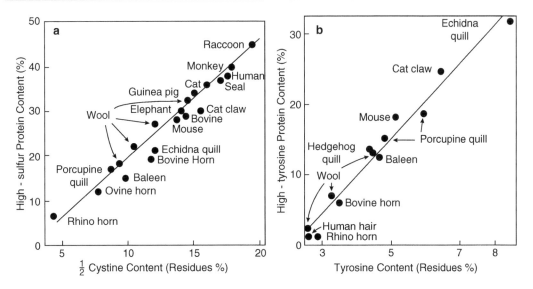

Fig. 7.2 Variability in composition of trichocyte (hard α-) keratins across species. (**a**) cysteine residues in sulfur-rich proteins, (**b**) tyrosine in glycine-tyrosine-rich proteins. (Reprinted from Ref. [1] with permission from Springer)

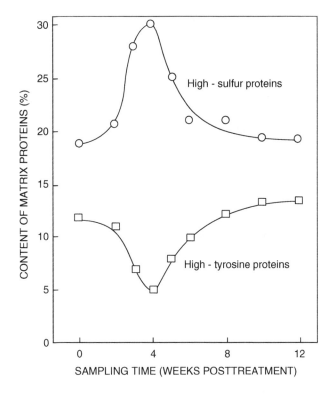

Fig. 7.3 Variability in the contents of HS and HGT KAPs in a single epidermal appendage (wool) after defleecing with epidermal growth factor. An inverse relationship between the levels of the HS and HGT proteins is not uncommon, and may be related to the para-ortho switch. (Reprinted from Ref. [1] with permission from Springer

dues in the cortical proteins are located in the KAPs and in the terminal domains of the IF chains, and the disulfide bonds formed will contribute significantly to the overall strength of the fully keratinised hair fibre.

Trichocyte keratin is a composite with a filament-matrix texture and many features of the structure and assembly of the filaments (IF) are understood (Chap.6). Much less is known about the structure of the KAPs that form the matrix.

Progress, however, is being made, largely as a consequence of two factors. Firstly, considerable quantities of sequence data have become available and, secondly, the expression pattern of the KAPs within the follicle has now been established. From the sequence data a number of families of KAP proteins have been recognised across several species. Human hair keratin, for instance, is believed to contain at least 89 different types of KAP chains *in toto* [6], and evidence has been presented that similar numbers of KAPs exist in other mammalian species [7].

The question then arises as to why there should be so many KAPs present *in vivo* in trichocyte keratins. An answer, in part, has been provided by Mercer [8]. He noted that keratins are produced by dying cells, and that it was likely that the rate of acceptance of genetic change would be much greater than for other structural proteins that remain in contact with living cells. The rate of point mutation in proteins (0.06 to 90 per hundred residues per 100 million years: [9]) is believed to have a direct correlation with the extent to which protein function demands a precise structure. Consequently, the rate at which mutations in KAPs are accepted is likely to be intrinsically higher than for IF proteins (or indeed proteins in general) because they take no further part in the biochemistry of the organism after they are laid down. This rationale may, in large part, help to explain the large number of KAP proteins that exist in hair and wool.

Initially, the KAPs that were isolated and characterized were defined in terms of their sequence composition as being high-sulfur proteins (HS), with cysteine contents up to 30 mol % and molecular weights in the range 5–55 kDa, ultra-high sulfur proteins (UHS) with cysteine contents in excess of 30 mol% and molecular weights typically in the range 10–25 kDa, and high-glycine-tyrosine proteins (HGT) with 35–60 mol% of glycine and tyrosine and molecular weights commonly in the range 5–10 kDa.

Several sub-groupings within the HS-proteins were described, (B2A, B2B, B2C, BIIIA, BIIIB). The B2 and BIII terminologies are now rarely used and, instead, each chain is specified as KAPa,b, where a refers to the family group and b to the member within that group, though a variant on this nomenclature has recently been proposed by Gong et al. [10]. The HS proteins include KAPs 1-3, 10-13, 15, 16 and 23-27, the UHS proteins encompass KAPs 4, 5, 9 and 17, and the HGT proteins constitute KAPs 6-8 and 19-22 [10, 11] (Chapter 1.1, Table 1.2.3). Some of these occur solely in the cortex of the hair fibre (KAPs 1-4, 6, 7, 9, 13, 15, 16, 20-23), but others (KAPs 5, 10, 12, 17, 19) are found only in the cuticle [11]. Yet others exist in both the cortex and the cuticle (KAPs 13, 23). The numbers of members in any one particular KAP family often show significant variations between species as, for example, in the KAP2 (11 in sheep and 5 in human), KAP10 (1 in sheep and 12 in human), KAP4 (27 in sheep and 15 in human) and KAP19 families (1 in sheep and 7 in human) [11]; (Yu, private communication). It seems probable that comparable variations will also be found in other species.

The first KAPs to be expressed at high levels in the human hair cortex are KAP8 (HGT) and KAP11 (HS), and these both first appear in the lower part of the hair follicle, including the bulb region (Fig. 7.4). It is thought that KAP8, in particular, may be involved in the aggregation of IF into loose bundles [12]. In contrast, the bulk of the KAPs, including KAPs 1-3 (HS), KAPs 4 and 9 (UHS), and KAPs 7, 19.1 and 19.2 (HGT), are strongly expressed later in the growth cycle and at a point much further up the hair follicle [6].

Structurally-distinct regions in the cortex of wool can be recognized in the form of the paracortical and orthocortical macrofibrils. The former are large and often irregular in outline (Fig. 6.1, Chap. 6). The IF in the paracortex pack in a quasi-hexagonal manner, the KAP content is higher than that in the fibre as a whole, and the percentages of the HGT and HS proteins present are lower and higher respectively than in the orthocortex. In contrast, the macrofibrils of the orthocortex are approximately circular in cross-section and have mean diameters in human hair of about 350 nm (maximum about 500 nm). The IF are packed in whorls or in a spiral arrangement, the KAP content is less than average across the hair fibre as whole, and the percentages of the HGT proteins and HS proteins are higher and lower respectively than in the paracortex. These observations indicate that both the quantity of the

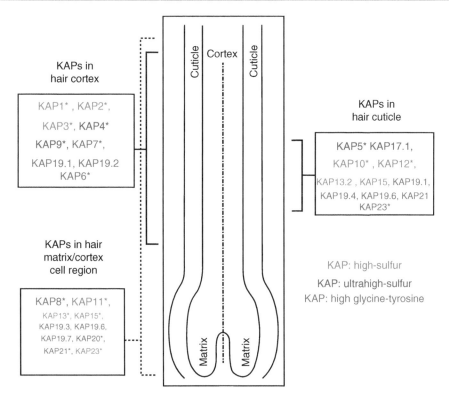

Fig. 7.4 KAP gene expression in human hair follicle. HS, UHS and HGT are coloured orange, red and dark blue respectively, and the asterisks relate to the families of proteins rather than to specific chains. The font size used is a reflection of the extent of mRNA expression (large font for strong expression, and smaller font for weak expression). Matrix refers to germative matrix cells. (Adapted from Rogers et al. [6])

KAPs and the types of KAP protein present directly influence the packing of the IF, and hence the physical properties of the hair fibre. It is also of interest to note that there is evidence that the orthocortex is formed slightly earlier and for a shorter period in the growth cycle than is the paracortex [13].

Evidence has been presented that the bilateral para/ortho-cortical organization of fine merino wool fibres relates directly to the crimp waves, with the paracortex on the inside and the orthocortex on the outside [14]. The separation into orthocortex and paracortex becomes less distinct as the fibre diameter increases and the crimp decreases, and in coarse fibres the orthocortical cells are grouped at the centre of the fibre with the paracortical cells distributed around the periphery [15]. Bilateral organisation has not been seen in human and other mammalian hair and it remains unclear as to the physical relevance of the observation in merino wool. There is also evidence that some KAPs in merino wool are expressed only in the paracortex (KAP4) or in the orthocortex (KAP6) [13, 16–18]. However, as KAP4 in rabbit was not confined to the paracortex [19] the pattern of evidence available thus far as to particular proteins being present in either the paracortex or the orthcortex (but not both) is both limited and inconsistent, and does not allow generalized conclusions to be drawn.

7.2 Primary Sequence Characteristics

Although some cortical cell KAPs contain no recognizable sequence periodicities, there are important exceptions. These include KAPs 1-4, 9, 13, and 16. Each of the chains in these protein families contains sequence regularities defined primarily in terms of pentapeptide motifs, based

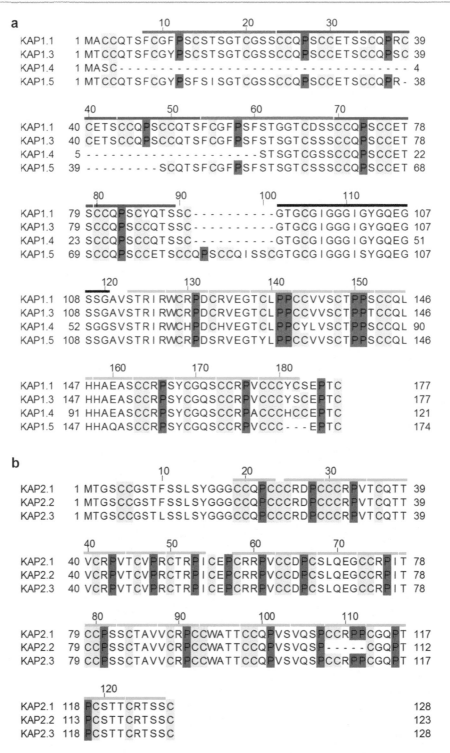

Fig. 7.5 Alignment of HS sequences from human hair. A comparison is shown of high-sulfur protein sequences in (**a**) the KAP1 and (**b**) the KAP2 families. In (**a**) the near-central glycine-rich linker domain is indicated by a black bar, the decapeptide repeats by blue bars, and the highly conserved 16-residue region containing a single residue and three poorly defined pentapeptides by red bars. The domain N-terminal to the linker thus consists of a pair of

on cysteine, proline and glycine residues [11, 20, 21]. The two most common pentapeptide repeats in human hair (referred to as A and B) are (C-C-X-P-X) and (C-C-X-S/T-S/T) respectively, with the di-cysteine being both unusual and of especial interest. Towards the centre of the KAP1 family of sequences is a highly conserved glycine-rich sequence 19 residues long (G-T-G-C-G-I-G-G-G-I-G-Y-G-Q-E-G-S-S-G) (Fig. 7.5a). The region N-terminal to this linker-like segment shows evidence of sequence duplication formed from a 16-residue region containing a single residue and three relatively poorly defined pentapeptide repeats, followed by three and two consecutive AB decapeptides respectively, each of which are almost entirely identical. In KAP1.4, however, this region comprises just one of the two repeats. The 60-residue region C-terminal to the linker contains 12 pentapeptides of the form (C-C-X-P-X) or (C-C-X-X-X), but these are much less well conserved in sequence and in organization than those in the N-terminal domain. In the KAP2 family the sequences are very highly conserved, and there is only one significant pentapeptide repeat of the form (C-C-X-P-X). This occurs 21 times consecutively in KAP2.2, and 22 times consecutively in both KAP2.1 and KAP2.3. However, a one-residue insertion and a two-residue deletion are required to maintain the phasing of the pentapeptide repeats. It needs to be emphasized that individual pentapeptides in this region often show significant variations from one another, and this includes the deletion of one complete pentapeptide in KAP2.2 (Fig. 7.5b). The N-terminal 18 residues in the KAP2 chains are rich in glycine and serine residues, and thus similar to the linker region in the KAP1 family. It is unlikely that the KAP2 family of proteins will adopt a fully regular structure, though numerous pentapeptide repeats do indicate a structure with limited order locally.

Other KAPs in human hair contain a predominance of just one particular repeat type (i.e. A or B). For example, KAP2.2, KAP3.1, KAP11.1 and KAP12.4 are A-rich (14A and 4B, 6A and 1B, 6A and 2B, 8A and 4B respectively) whereas KAP17.1 is B-rich (3A and 8B) [11]. Another KAP family member of human hair has a 20-residue period of the form BAAA (KAP16.1) whereas KAP5.1 has four consecutive 19-residue repeats, each of the form BABB with a one-residue deletion. KAP10.1 from cuticle has three contiguous copies of a 42-residue repeat constructed from seven A or B repeats plus a further unique one, separated by di-serine [11]. In essence, therefore, the KAP1 and KAP2 proteins in human hair are dominated by pentapeptide repeats, put together in various combinations, frequently imperfectly.

In sheep wool, the two pentapeptides are slightly different (C-C-Q/R-P-S/T and C/S/T-C-Q-P/T-S) and, in this instance, the KAP1 family contains a higher degree of sequence regularity than that seen in human hair. This is characterized by N- and C-terminal blocks, each about 75-residues long, that display a high degree of sequence homology with one another (Figure 3 in [11]). Each block consists of the two pentapeptide repeats that often alternate (but not always), thereby generating a quasi-decapeptide repeat. Several of the decapeptide repeats, too, are imperfect, as one- or two-residue deletions/insertions are not uncommon. The blocks are separated from one another by a glycine-rich central domain (45% glycine) with a length of 22-residues (in KAP1.3, for example, but similar composition characteristics are observed in other proteins of this family). KAP2 proteins also have some similarities to the KAP1 family but, in contrast, these proteins contain only a single block of similar length to those observed in the KAP1 family. The KAP2 block has a very high degree

Fig. 7.5 (continued) repeats (16-residue segment plus two or three contiguous decapeptides) in KAP1.1, 1.3 and 1.5, but only a single repeat occurs in KAP1.4. On the basis of homology, KAP1.5 contains both a decapeptide deletion and a decapeptide insertion. The domain C-terminal to the glycine-rich linker contains 12 contiguous pentapeptide repeats. (**b**) The KAP2 family of sequences is almost entirely conserved, though KAP2.2 lacks a complete pentapeptide. Overall, these proteins are characterized by 21 (KAP2.2) and 22 (KAP2.1, KAP2.3) contiguous pentapeptide repeats with a single one-residue insertion and a two-residue deletion. Pentapeptide repeats are indicated by green bars

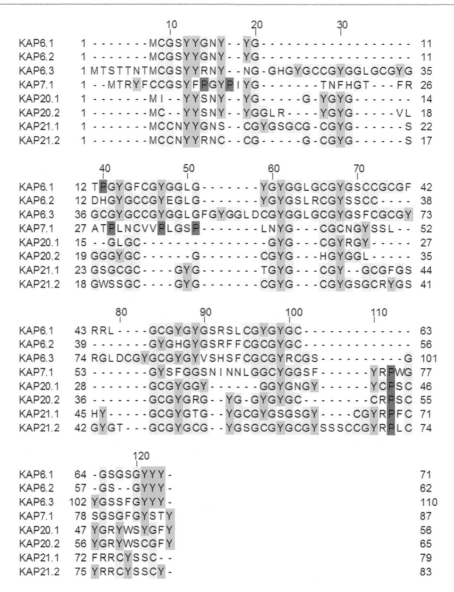

Fig. 7.6 Alignment of HGT sequences from human hair. Although there is little homology across the HGT chains as a whole it can be seen that the KAP21 family is largely characterized by (G-X)$_{26 \text{ or } 27}$, as are the KAP20 sequences by (G-X)$_{16}$. The latter, nonetheless, contain three one- or two-residues inserts that interrupt the phasing. The KAP6 proteins are different again. The sequence of KAP6.3 is characterised by a contiguous pair of 17-residue repeats of the form (H/C-G-Y-G-C-C-G-Y-G-G-L-G-F/C-G-Y-G-G), followed by a degenerate, approximately 25-residue repeat of the form (L-D-C-G-Y-G-X$_{0-3}$-C-G-Y-G-X$_{3-9}$-G-C-G-Y-R). KAPs 6.1 and 6.2 contain only a single repeat of each motif

of homology with both blocks in the KAP1 family [20]. In addition, the KAP2 proteins also contains a glycine-rich segment 21-residues long. It is, however, located at the N-terminal end of the chain instead of the near central location found in the KAP1 family of proteins. The fact that sequences of the human hair and sheep wool KAP1 and KAP2 families display significant species differences illustrates that the matrix in trichocyte keratin can tolerate a wide range of KAP1 and KAP2 proteins, provided that pentapeptide repeats are maintained.

None of the HGT proteins (KAPs 6-8 and 19-22) in human hair exhibit a pentapeptide structure, but sequence regularities of a different type have recently been recognized (Fig. 7.6: Fraser and Parry, unpublished). In the KAP21 family, for example, the sequences are dominated by a (G-X) motif repeated 26 or 27 times consecutively. Similarly, in the KAP20 sequences a (G-X) motif occurs 16 times, though in this case there are three one- or two-residue inserts that interrupt the phasing. The KAP6 proteins are different again. In the case of KAP6.3, the sequence is dominated by a contiguous pair of 17-residue repeats of the form (H/C-G-Y-G-C-C-G-Y-G-G-L-G-F/C-G-Y-G-G), immediately followed by two approximately 25-residue, degenerate repeats of the form (L-X-C-G-Y-G-X_{0-3}-C-G-Y-G-X_{3-9}-G-C-G-Y-G/R). KAP6.1 and KAP6.2 contain just one copy of each repeat. There are also some other examples of repeats in the HGT proteins. In sheep KAP6.1, for example, two contiguous 12-residue repeats of the form (G-Y-G-Y-G-S-R-S-L-C-G-S) have been reported, as have a pair of non-contiguous seven-residue motifs with sequence (Y-G-G-L-G-C-G) [13]. As a result of these observations it has become clear that the HGT proteins, as a group, are not devoid of sequence regularities as was initially believed.

7.3 Secondary Structure

It has not proved possible thus far to determine the three-dimensional structure of any KAP by X-ray protein crystallography or high-field NMR. Although the imperfect nature of the pentapeptide repeat mitigates against a completely regular conformation being adopted, the penta- and decapeptide motifs observed in the KAP1 and KAP2 families have been modelled [11, 20, 21]. In the latter study it was recognized that disulfide-stabilized pentapeptide rings were found naturally in crystals of some snake-venoms (see for example Betzel et al., [22]), illustrated in Fig. 7.7a. This observation was important since it established that the disulfide-stabilized pentapeptide ring structure postulated in the later study by Parry et al. [20] was stereochemically feasible.

The preferred conformation of the A repeat (C-C-X-P-X), amongst a number considered by Parry et al. [11], is that of a β-bend formed from residues 3, 4 and 5 in a reference pentapeptide and residue 1 in the pentapeptide C-terminal to it (Fig. 7.7b). In the first variant of this model the proline residue is in the *cis* form and the structure is based on the rotations about the mainchain single bonds observed from a study of homologous motifs recorded in the pdb database [11]. A disulfide bond between cysteine residue 2 in the reference pentapeptide and cysteine residue 1 in the pentapeptide C-terminal to it can be formed without difficulty. With a small variation of these dihedral angles, a stabilising hydrogen bond can also be formed between the C=O group of residue 3 in the reference pentapeptide and the NH group of residue 1 in the pentapeptide C-terminal to it. If the proline is in the *trans* form, however, it is still possible to form a disulfide bond as before, though the dihedral angles differ significantly from those found in the database search of homologous motifs (Fig. 7.7c). The hydrogen bond, however, forms naturally between the same groups as previously noted. It would seem that the probability of a disulfide bond being formed may be greater when the proline is in the *cis* rather than the *trans* form [11].

The fact that some limited sequence regularity exists, albeit imperfectly, strongly infers the probability that some local order will exist. The glycine-rich central region in the KAP1 family may be thought of as flexible, elastic-like linker capable of stretching out and becoming more ordered (but with a concomitant decrease in entropy) but, nonetheless, favouring a more compact conformation with a less ordered conformation (but with a corresponding increase in entropy). The effect of the compaction of the glycine-rich segments would be to shield as many of the apolar glycine residues as possible from the aqueous environment. In essence, the KAP1 proteins are thus conveniently envisaged as having elements of regular local order – the disulfide-bonded rings connected flexibly to one

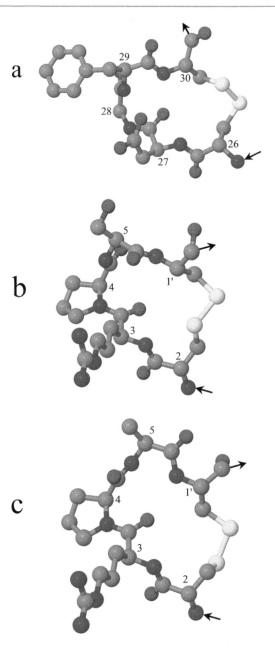

Fig. 7.7 Ball and stick diagrams of (**a**) a disulfide-stabilized pentapeptide ring present in snake venom [22] and possible counterparts (**b**, **c**) in HS KAPs [11]. In (**b**) the proline residue is in the *cis*-form and in (**c**) the proline residue is in the *trans*-form. The conformation of the pentapeptide repeat C2–X3–P4–X5–C1' is stabilised by a disulfide bond between C2 and C1' in consecutive pentapeptide repeats. The prime indicates a residue in the following pentapeptide. The structures are based on the average ϕ and ψ angles determined for homologous motifs recorded in the PDB database. With a small variation of the recorded dihedral angles a hydrogen bond can be formed in (**b**) between C=O in the residue following the firstcysteine residue and the N–H group in cys1'. In (**c**) the proline residue is present in the *trans* conformation and the hydrogen bond forms naturally. Atoms are coloured as follows: oxygen – red, nitrogen – blue, carbon – grey, sulfur – yellow

another, and thereby adopting a wide range of energetically-similar conformations. This scenario would seem to be more compatible with these proteins acting as a matrix and, consequently, having the physical attributes needed of a hair fibre, than it would if these proteins had a well-defined crystalline conformation of the type observed by X-ray crystallography for a wide range of globular proteins.

The HGT protein cortical sequences, that are rich in both glycine and tyrosine residues (KAPs 6, 8, 20, 21 and 22), have also been characterized, and initial indications were that these proteins would most likely adopt a random coil structure. Indeed, a degree of conformational randomness might well be important functionally. However, it is likely that the high occurrence of glycine residues with their very small sidechains will allow considerable conformational freedom to the HGT proteins, and it is believed that this will facilitate the formation of strong tyrosine-tyrosine interactions (see Sect.7.4.2).

In contrast to the idea of a completely random coil conformation, a number of the HGT proteins are now known to display sequence regularities indicative of regular secondary structure. For example, the KAP6.1 sequence in sheep wool, with its 12-residue repeat [13] characterized in large part by alternating glycine residues, suggests a β-chain conformation with one side composed totally of glycine residues. Silk sequences with this characteristic often form β-sheets with the sheets pairing via the glycine faces. It is not unlikely that something similar might occur here too. In human hair, both the KAP20 and 21 proteins contain multiple copies of a (G-X) repeat, which again would suggest a β-like conformation. In KAP20, the short interruptions present in the sequence might facilitate the formation of tight turns between β-strands, thereby giving rise to a small β-sheet with one face composed entirely of glycine residues. Assembly with a similar β-sheet in a second KAP20 molecule would not be unlikely. In the KAP21 family, however, the (G-X) motifs are continuous. Whether these allow the occurrence of tight turns too or whether the β-strand remains straight (and are thus very long) is a matter of conjecture. If the latter was to prove correct than a β-sheet could only be formed from the assembly of multiple molecules. The sequence repeats in the human KAP6 proteins also indicate that each repeat will adopt the same (or closely similar) secondary/tertiary structure.

It is of interest to note that the expression of one HGT protein (KAP8) occurs at about the same time as that of the IF proteins, thereby indicating that the HGT proteins may play a special part in the organisation of the IF into parallel bundles. It would be hard to imagine that this would occur if the HGT proteins were completely disordered. It is thus a working hypothesis that the matrix proteins as a whole may not have crystallographically-regular three-dimensional structures, but that they will contain regions with a well-defined secondary structure separated by regions with multiple degrees of freedom.

7.4 Functional Roles of KAP Proteins

7.4.1 KAP-KAP and KAP-IF Interactions

Fraser and MacRae [23], on the basis of swelling data, suggested that there was a physical limit to the separation of IF, irrespective of the degree of hydration of the trichocyte keratin involved. The implication of this conclusion was that IF-IF interactions, probably mediated by KAPs, were probable. The observation, in the sequences of the KAP1 family of sheep wool proteins, of two approximately 75-residue blocks dominated by pentapeptide repeats, and in human hair of a different but analogous organization of pentapeptide repeats, suggest a degree of regular secondary structure. The protein organisation was also suggestive that these might be capable of interacting with different IF, thereby limiting their separation. The fact also that these regions were separated from one another by a glycine-rich segment, that was capable of extension and flexibility but which nonetheless favoured a compact structure, was a further indication that such a role was not unlikely. Likewise, the presence of a similar single block in the KAP2 family could be taken as an indication of a similar (but clearly different) arrangement that might involve dimerization. There can be little doubt that the proteins of the KAP1 and KAP2 families, which are ancestrally-related, will have similarities in their structures and functions *in vivo*.

In the last few years some evidence has been presented that these ideas may have some foundation experimentally. Interactions identified to date are those between KRTAP2-1 (a HS protein equivalent to KAP2.1) with (a) itself, (b) the head of K86 and (c) K34 and K85 [24], but the role that these interactions play *in vivo* has yet to be determined. Another interaction characterized is that between KAP8.1 (a HGT protein) and the head domain of K85 [12]. In this instance, it has been proposed that KAP8.1 might play a specific role in the organization of the IF into the whorl or helicoidal structure characteristic of the orthocortex in hair. Although, in the cases noted above, neither the nature of the interactions nor the particular residues involved are known, it has become increasingly evident that specific linkages between IF and KAPs, and between KAPs themselves, do indeed play an important part in specifying the overall structure of the hair fibre as well as determining its physical properties.

7.4.2 Mechanical Properties

The common occurrence in biology of a filament-matrix composite is an indication of the special mechanical role that such a two-phase system plays *in vivo*. Other examples include the epidermal appendages of birds and reptiles with their end-to-end aggregation of β-rich central domains embedded in matrix formed from the remainder of the protein chains [25, 26], the end-to-end assembly of two-stranded coiled-coils in the egg cases of praying mantises, again surrounded by the remainder of the chains in a matrix structure [27], and the end-to-end aggregation of four-stranded coiled coils in the α-fibrous silks of the *Hymenoptera aculeata* (bees, wasp and ants) with their terminal domains forming a matrix [28]. In such structures, the filaments are thought to provide the required tensile strength of the fibre, whereas the matrix proteins are believed to determine its hardness, toughness and pliability [26]. In the case of the trichocyte keratins each of the numerous KAPs in the matrix is dominated by a relatively small number of amino acids. Sometimes these are in the form of quasi-repeats of sequence but, on other occasions, the dominance lies purely in the overall amino acid composition. Two features of particular importance in the sequences of the KAPs are those regions that are glycine-tyrosine-rich or cysteine-rich. While each would be predicted to lead to a specific mechanical attribute, the overall content of each feature would, in turn, be expected to modulate the physical properties appropriate to the function of hair *in vivo*.

The relationship between the mechanical properties of trichocyte appendages and their contents of matrix proteins has been discussed in detail by Fraser and MacRae [23] and a very clear correlation between the individual contents of the sulfur-rich and glycine-tyrosine-rich KAPs, reported by Bendit and Gillespie [29], is shown in Fig. 7.8. The HGT proteins are rich in glycine and tyrosine, both special amino acids but in quite different ways. Glycine, an apolar residue, has the smallest sidechain of any amino acid (a single hydrogen atom) and this will permit near complete freedom of rotation about the mainchain single bonds. The consequence of this is that the HGT protein chain will necessarily have a much higher degree of conformational flexibility than would a protein containing fewer glycine residues. It would be expected, therefore, that the HGT proteins would be capable of displaying a variety of energetically-equivalent conformations stabilized by apolar interactions [26].

Tyrosine residues, on the other hand, are large amino acids with an aromatic ring and a reactive hydroxyl group. They are similar to glycine, nonetheless, in being found predominantly in the interior of proteins rather than on surface regions exposed to water. There is considerable evidence

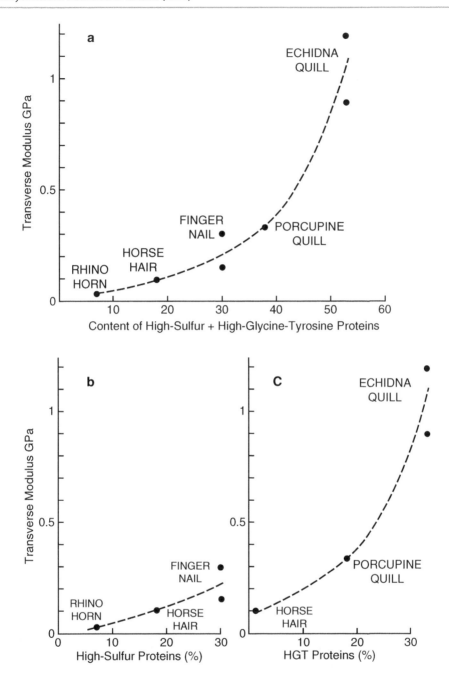

Fig. 7.8 The transverse compressional modulus of hard α-keratins in water, in gigaPascals (1 GPa = 10^{10} dyn. cm^{-2}), as a function of (**a**) the sum of high-sulfur (HS + UHS) and high glycine-tyrosine (HGT) protein contents, (**b**) the high-sulfur protein content at constant high-glycine-tyrosine protein content (~0%), and (**c**) the high-glycine-tyrosine protein content at constant high-sulfur protein content (~20%). Two values are shown for fingernail and echidna quill, reflecting different transverse mechanical properties in the radial and azimuthal directions in these keratins. (Reprinted from Bendit and Gillespie [29] with permission of Wiley)

that tyrosine residues interact strongly with one another, and that this occurs through the formation of hydrogen bonds between the centre of the aromatic ring and the tyrosine's hydroxyl group [30], or through ring stacking [31]. It follows that the mechanical implications of a HGT protein with its highly interactive tyrosine residues linked by conformationally-free glycine residues would be to both strengthen the material and provide it with pliability.

Consider, secondly, the role of cysteine-rich sequences. These will clearly result in the formation of many disulfide bonds in the fully differentiated (oxidized) hair fibre. The effect will result in an insoluble structure and one that is easily able to resist proteolysis. The swelling evidence suggests that the majority of the disulfide bonds are intramolecular (within the individual KAPs) but it is certain that some will be intermolecular (between the KAPs and the terminal domains of the IF proteins that form the filamentous component of the hair fibre) and therefore have a role in limiting the separation of IF to some finite extent.

7.4.3 Hair Diseases Related to KAP Proteins

Although a number of hair diseases have been recognized and studied in some detail, the expectation has been that the majority of these would relate to mutations in the IF proteins, rather than to the KAP proteins. The rationale for such thinking related to the observation that the IF occur in the highly organized portion of the hair fibre whereas the KAPs form the less highly structured portion of the structure i.e. the matrix. Also, because the family members of any particular KAP family display multiple small sequence changes, it seemed more likely that mutations in KAPs, as well as the polymorphic variations that occur in some KAP genes, would probably have a minor, possibly negligible, effect on structure and function. To a large extent, these thoughts have been verified from experience gained to date.

Nonetheless, there is limited evidence that KAP proteins can be the cause of significant hair diseases, though Rogers et al. [6] have suggested that the down regulation of KAP gene expression, especially in hereditary hair diseases, may prove to be much more important than the occurrence of point mutations. For example, in trichothiodystrophy there is a major decrease in the expression of a large group of UHS proteins, and hence of cysteine residue content in the hair fibre as a whole [6, 32, 33]. The decreased level of the KAP proteins expressed, and the consequential reduction in the number of cysteine residues available for crosslinking, led to a hair fibre that was both brittle and likely to fracture easily. A second hereditary hair disease that is believed to involve KAP proteins is hidrotic ectodermal dysplasia. This too resulted in a decreased level of KAP expression, and a concomitant change in amino acid composition characterized by less cysteine, proline and serine but more tyrosine and phenylalanine [34]. The cortical structure appeared looser than in healthy hair and there was some loss and/or damage to the cuticle. The number of data relating to KAP involvement in hair diseases remains very limited, however, and much more information will be required before generalisations can be made with any confidence.

7.5 Summary

In recent years, the acquisition of data relating to the KAPs, especially at the level of the amino acid sequence, and to the derivation of the expression patterns in thehair follicle, has been significant. Nonetheless, little is known at present about the three-dimensional structure adopted by any KAP. Structural analyses indicates that some KAP will almost certainly contain elements of local secondary structure, but the protein as a whole seems unlikely to have a fixed tertiary conformation. It can be argued that a matrix of highly ordered proteins would be most unlikely to bestow upon the hair fibre the physical properties for which the hair fibre has primarily evolved. It seems more probable, on the basis of the limited data currently available that the KAPs will adopt a variety of conformations dictated by the local molecular environment.

In contrast to the limited information on the structure of the KAPs, their functional roles have gradually become clearer, and evidence of IF-IF links mediated by KAPs has now been obtained. Furthermore, some understanding has been gained of the physical properties of keratin, such as its pliability and toughness, that are bestowed on it by matrix proteins rich in cysteine and in glycine and tyrosine residues. In addition, there are data indicative of the role played by KAPs in specifying the macrofibrillar organization of the cortex and, in the case of fine Merino wool, the differentiation into the paracortex and the orthocortex.

It is clear that many more data, possibly derived from the use of new techniques, will be required before the structures and functions of the KAPs are fully understood at the level of detail that is desired by those working in the field of trichocyte keratin.

References

1. Gillespie, J. M. (1990). The proteins of hair and other hard α-keratins. In R. D. Goldman & P. M. Steinert (Eds.), *Cellular and molecular biology of intermediate filaments* (pp. 95–128). New York: Plenum Press.
2. Powell, B. C., & Rogers, G. E. (1990). Hard keratin IF and associated proteins. In R. D. Goldman & P. M. Steinert (Eds.), *Cellular and molecular biology of intermediate filaments* (pp. 267–300). New York: Plenum Press.
3. Fraser, R. D. B., MacRae, T. P., & Rogers, G. E. (1972). Keratins: Their composition, structure and biosynthesis. In *The Bannerstone division of American lectures in living chemistry* (p. 320). Springfield: Charles C Thomas Publisher, Ltd.
4. Orwin, D. F. G. (1979). Cytological studies on keratin fibers. In D. A. D. Parry & L. K. Creamer (Eds.), *Fibrous proteins: Scientific, industrial and medical aspects* (pp. 271–297).
5. Orwin, D. F. G. (1979). The cytology and cytochemistry of the wool follicle. *International Review of Cytology, 60*, 331–374.
6. Rogers, M. A., et al. (2006). Human hair keratin-associated proteins (KAPs). *International Review of Cytology, 251*, 209–263.
7. Wu, D.-D., Irwin, D. M., & Zhang, Y.-P. (2008). Molecular evolution of the keratin associated protein gene family in mammals, role in the evolution of mammalian hair. *BMC Evolutionary Biology, 25*(8), 241–255.
8. Mercer, E. H. (1961). *Keratin and keratinization.* (1st ed., International series of monographs on pure and applied biology, Vol. 12, p. 316). Oxford: Pergamon Press.
9. McLaughlin, P. J., & Dayhoff, M. O. (1970). Eukaryotes versus prokaryotes: An estimate of evolutionary distance. *Science, 168*, 1469–1471.
10. Gong, H., et al. (2012). An updated nomenclature for keratin-associated proteins (KAPs). *International Journal of Biological Sciences, 8*(2), 258–264.
11. Parry, D. A. D., et al. (2006). Human hair keratin-associated proteins: Sequence regularities and structural implications. *Journal of Structural Biology, 155*(2), 361–369.
12. Matsunaga, R., et al. (2013). Bidirectional binding property of high glycine-tyrosine keratin-associated protein contributes to the mechanical strength and shape of hair. *Journal of Structural Biology, 183*(3), 484–494.
13. Powell, B. C., & Rogers, G. E. (1997). The role of keratin proteins and their genes in the growth, structure and properties of hair. In *Formation and structure of human hair* (pp. 59–148). Basel: Birkhäuser Verlag.
14. Horio, M., & Kondo, T. (1953). Crimping of wool fibers. *Textile Research Journal, 23*(6), 373–387.
15. Fraser, R. D. B., & MacRae, T. P. (1956). The distribution of ortho- and para-cortical cells in wool and mohair. *Textile Research Journal, 26*, 618–619.
16. Fratini, A., Powell, B. C., & Rogers, G. E. (1993). Sequence, expression, and evolutionary conservation of a gene encoding a glycine/tyrosine-rich keratin-associated protein of hair. *Journal of Biological Chemistry, 268*(6), 4511–4518.
17. Fratini, A., et al. (1994). Dietary cysteine regulates the levels of mRNAs encoding a family of cysteine-rich proteins of wool. *Journal of Investigative Dermatology, 102*(2), 178–185.
18. Yu, Z., et al. (2009). Expression patterns of keratin intermediate filament and keratin associated protein genes in wool follicles. *Differentiation, 77*(3), 307–316.
19. Powell, B. C., Arthur, J. R., & Nesci, A. (1995). Characterisation of a gene encoding a cysteine-rich keratin associated protein synthesised late in rabbit hair follicle differentiation. *Differentiation, 58*, 227–232.
20. Parry, D. A. D., Fraser, R. D. B., & MacRae, T. P. (1979). Repeating patterns of amino acid residues in the sequences of some high sulphur proteins from α-keratin. *International Journal of Biological Macromolecules, 1*, 17–22.
21. Fraser, R. D. B., et al. (1988). Disulphide bonding in α-keratin. *International Journal of Biological Macromolecules, 10*, 106–112.

22. Betzel, C., et al. (1991). The refined crystal structure of alpha-cobratoxin from Naja naja siamensis at 2.4-Å resolution. *Journal of Biological Chemistry, 266*, 21530–21536.
23. Fraser, R. D. B., & MacRae, T. P. (1980). Molecular structure and mechanical properties of keratins. In *The mechanical properties of biological materials*. SEB Symposium XXXXIV: Cambridge University Press.
24. Fujikawa, H., et al. (2012). Characterization of the human hair keratin-associated protein 2 (KRTAP2) gene family. *Journal of Investigative Dermatology, 132*(7), 1806–1813.
25. Filshie, B. K., & Rogers, G. E. (1962). An electron microscope study of the fine structure of feather keratin. *Journal of Cell Biology, 13*, 1–12.
26. Fraser, R. D. B., & Parry, D. A. D. (2014). Amino acid sequence homologies in the hard keratins of birds and reptiles, and their implications for molecular structure and physical properties. *Journal of Structural Biology, 188*, 213–224.
27. Bullough, P. A., & Tulloch, P. A. (1990). High-resolution spot-scan electron microscopy of microcrystals of an α-helical coiled-coil protein. *Journal of Molecular Biology, 215*, 161–173.
28. Fraser, R. D. B., & Parry, D. A. D. (2015). The molecular structure of the silk fibers from Hymenoptera aculeata (bees, wasps, ants). *Journal of Structural Biology, 192*, 528–538.
29. Bendit, E. G., & Gillespie, J. M. (1978). The probable role and location of high-glycine-tyrosine proteins in the structure of keratins. *Biopolymers, 17*, 2743–2745.
30. Levitt, M., & Perutz, M. F. (1988). Aromatic rings act as hydrogen-bond acceptors. *Journal of Molecular Biology, 201*, 751–754.
31. McGaughey, G. B., Gagne, M., & Rappe, A. K. (1998). π-Stacking interactions alive and well in proteins. *Journal of Biological Chemistry, 273*, 15458–15463.
32. Price, V. H., et al. (1980). Trichothiodystrophy. Sulfur-deficient brittle hair as a marker for a neuroectodermal symptom complex. *Archiv fur Dermatologische Forschung, 116*, 1375–1384.
33. Gillespie, J. M., & Marshall, R. C. (1983). A comparison of the proteins of normal and trichothiodystrophic human hair. *Journal of Investigative Dermatology, 80*, 195–202.
34. Gold, R. J., & Scriver, C. R. (1972). Properties of hair keratin in an autosomal dominant form of ectodermal dysplasia. *American Journal of Human Genetics, 24*, 549–561.

Part III

Hair Development

Introduction to Hair Development

Duane P. Harland

Contents

8.1 Finding Your Way Within the Anagen Follicle 89
8.1.1 Lateral Location – The Anagen Follicle's Ten Cell Lines 90
8.1.2 Longitudinal Location – Developmental Landmarks 90
8.2 Zone-by-Zone Synopsis of Hair Fibre Development 92
References 96

Abstract

The anagen phase of the hair follicle cycle is when the follicle is configured to grow hair. In short hairs (e.g., mouse underhairs and human eye lashes) anagen phase is short, but in the wool of sheep and in human scalp hair anagen is a prolonged state lasting for years. In this chapter we describe the morphological and biological divisions within the anagen follicle.

Keywords

Hair follicle · Morphology · Hair development · Inner root sheath · Developmental zones

D. P. Harland (✉)
AgResearch Ltd., Lincoln, New Zealand
e-mail: Duane.Harland@agresearch.co.nz

8.1 Finding Your Way Within the Anagen Follicle

Hair development is a biologically complex phenomenon because it combines active biological processes and chemically driven processes. Development occurs in a series of stages but the process and timing of development differs in each of the separate cell lines that make up the hair shaft and its supporting sheaths. Locating events within the follicle by cross-referencing of cell line and developmental stage is essential for understanding the process. This is never more important than when we are trying to match up data generated using different methods, and sometimes this is troublesome because different methods often visualise different features. For example, the last keratins expressed in the developing fibre do so above the top of the bulb [1] as demonstrated by light microscopy, and it might be useful to know how the onset of expression correlates with fibre mechanical hardness as demonstrated by atomic force microscopy [2]. Such an exercise is possible in this case only because both studies have adequately described the origin of their data in terms of lateral position across the follicles (cell line) and longitudinally with respect to key landmarks associated with particular stages of development. While lateral position is usually well described in most studies, longitudinal axis terminology is more variable.

8.1.1 Lateral Location – The Anagen Follicle's Ten Cell Lines

During development, a follicle can have up to ten concentric rings of cell lines (Table 8.1). Only some techniques (e.g., transmission electron microscopy or immunohistochemistry light microscopy) are able to clearly distinguish between all cell lines at all points along the follicle. Cell lines with a similar mode of development are often classified together into well-established tissue layers. Normally the fibre is distinguished from its surrounding root sheath, and when possible the root sheath is divided into the inner-root sheath and outer root sheath, dermal sheath and, in the bulb, the dermal papilla.

8.1.2 Longitudinal Location – Developmental Landmarks

While most researchers describe lateral positions within anagen follicles by cell line, the terminology to describe location along the proximal-distal axis has been more varied, and is still far from standardised. One tradition has been to use distance from the base of the follicle [3], but this approach is error prone because follicle shape is often subtly asymmetric, and because follicles vary in size; therefore the absolute distances at which events occur also varies. Most studies divide the follicle up either by specific landmarks, or by stages of the developmental process, or both. Morphological markers, at their least, usually include the bulb, suprabulbar region, isthmus and infundibulum [4] because these features

Table 8.1 Cell lines of an anagen hair follicle

Embryonic lineage	Tissue	Cell line	Features	Function
	Fibre	Medulla	Centre of fibre. Characterised by trichohyalin granules during formation and a wide range of randomly paired keratins. Collapses into chambers on fibre hardening/drying.	Important structural component in many mammalian hairs, increases fibre diameter.
		Cortex	Primary structural component of many fibres (wool, human hair) composed of elongated cells filled with ordered composite bundles (macrofibrils) of intermediate filaments and matrix proteins.	Thought to impart much of the mechanical properties to the fibre. Fibre curvature is controlled by across fibre differences in cortical cell lengths.
		Cuticle	Flattened cells that overlap around the cortex. Distinct cell shape changes and complex cytoskeletal arrangement during formation. Internal laminar amorphous keratin layers within mature fibre.	Protects the fibre from physical and chemical insult through high-density highly crosslinked exo-cuticle layers. Moderates water absorption via surface chemistry. The external morphology is moulded ridges (scales) which direct dirt and oil away from the skin.

(continued)

Table 8.1 (continued)

Embryonic lineage	Tissue	Cell line	Features	Function
Epidermal compartment	Inner root sheath (IRS)	Inner root sheath cuticle (IRSC)	Distinct cell shape changes and position of cellular junctions during development. Cytoskeletal organisation on fibre-side of cell with trichohyalin/intermediate filament network.	Supports the fibre during development. Moulds the fibre cuticle scales and develops specialised cell junctions that establish fibre surface chemistry.
		Huxley's layer	Accumulates a random filament network that intersects trichohyalin granules of increasing size. Develops a convoluted membrane with adjacent Huxley's cells containing abundant desmosomes.	Supports fibre during development. Plays a role in development of variation in fibre cross-sectional shape.
		Henle's layer	Similar to Huxley's but development is accelerated from the lower bulb until abrupt keratinisation at the top margin of the bulb. Hardened Henle's layer forms a continuous sheath around developing fibre above bulb level except for small windows.	Contains the developing post-bulb fibre (and remaining IRS layers) within a mechanically robust wrapper of circular profile.
	Outer root sheath (ORS)	Companion layer	It is typically not visible in light microscopy without specific staining. Cytoskeleton and cell junctions develop differently on each side of the cell, and notch-like invaginations connect companion to Henle's layer cells.	This narrow monolayer of cells used to be considered part of the outer root sheath. The companion layer appears to act as a shear plane between the outwardly growing fibre and IRS and the relatively immobile ORS.
		Outer root sheath (ORS)	Usually a monolayer around the bulb, expanded to a thicker structure above the bulb where cells are loosely cuboid. The ORS does not harden with the IRS or fibre. Cells are typically rich in glycogen. A pocket of stem cells adjacent to the arector pili muscle is quiescent during anagen.	Functions during anagen for the ORS are not well understood, but may include a skin immune role, a structural role in supporting the follicle shape and resorption of materials during the IRS breakdown.
Dermal compartment	Dermal sheath		Overlapping bands of collagen and fibroblast-like cells, also contains blood capillaries and neurons.	Likely role is likely structural, holding the follicle shape.
	Dermal papilla		This compartment in the lower bulb is not a cell line per se, but can contain many cell types (e.g., fibroblast-like cells, mast cells and blood capillaries).	Key role in the mediating active biological control of fibre growth and its organisation.

can be readily visualised at low magnification using a light microscope [5] and are common across hair types and species.

Morphological landmarks in longitudinal sections through follicles can be prone to artefacts produced by cutting direction, cutting angle and along-follicle changes in direction and straightness. Some key landmarks and potential artefacts are summarised in Table 8.2.

Other studies divide up the anagen follicle into zones based on developmental stages. Individual studies sometimes divide follicles into zones based on features associated with a particular phenomenon under investigation. For example, development of intermediate filament organisation in the cortex observed by X-ray diffraction [2, 6, 7] or successive waves of keratin gene expression [8, 9]. The authors normally pro-

Table 8.2 Common morphological features found in micrographs of anagen follicles along with potential artefacts associated with the markers

Feature	Location	Artefact risk
Sebaceous gland	Neck/infundibulum of follicle	None
Bulge (follicle cycling stem cell niche)	Below sebaceous glands at point where arrector pili muscle attaches.	Not clearly defined in all species–clear in rodent pelage follicles, but not in adult human scalp follicles or sheep wool follicles.
Adamson's fringe	Top of cloudy keratogenous zone, has a distinct inverted V shape.	The precise location of this feature is only visible if the section is perfectly aligned along the fibre axis. The fringe itself closely resembles an oblique cut through the various cell lines.
Keratogenous region	Cloudy or darkened region (depending on the microscopy method) within the developing fibre cortex, beginning just above the bulb.	Varies with section thickness and other factors; a good general indicator of location but with imprecise end points.
Hardening of Henle's layer	Occurs at the top of the bulb.	Very precise indicator. Is more clearly visible with staining or with Normarski or phase microscopy.
Top of the dermal papilla	Upper bulb	Location is very sensitive to misalignment of section to follicle axis, and differs between medulated and non-medulated fibres.
Widest point of the dermal papilla	Mid bulb	Affected moderately by alignment of section to follicle axis, but is the most precise landmark in the bulb region.
Neck of dermal papilla	Lower bulb, an excellent indicator that the section (at least at that point) aligns perfectly with the follicle axis.	None.

vide clearly defined micrographs of follicles that will allow clear positioning of events. More general developmental schemes are usually based on accumulated findings from many studies. An early example of this is seen in Edgar Mercer's book Keratin and Keratinization from 1961 (Fig. 8.1), and this was probably the inspiration for the scheme that we will use in this book.

The most comprehensive system combines both developmental stages with landmarks that are universal across hair types and species. Here we will use the ultrastructure morphology-based system developed by Don Orwin [11–15]. Orwin's scheme was based on sub-cellular features in both medullated and non-medullated fibres and is summarised in Table 8.3. It is Orwin's system that we will use in the following sections except that we suggest small changes to the boundary of zones to incorporate newer findings of biological processes (Fig. 8.2).

8.2 Zone-by-Zone Synopsis of Hair Fibre Development

Before going into detail about development within specific cell lines in Chap. 10, here is a summary of the overall process.

Zone A. All cells that make the hair, IRS and companion layer divide from a monolayer of outer-root-sheath (ORS) derived stem cells in the proximal bulb. These cells adhere to a basement membrane surrounding the neck and lower half of the follicle's dermal papilla. The progeny of these stem cells continue to divide but are initially undifferentiated. Messages from dermal papilla cells arranged into micro-niches are vital for the organization of cell layers.

Zone B begins when the streams of distally moving progeny cells display cell-line-specific differentiation. The first outward sign is the reshaping of the companion layer, inner root sheath cuticle, and fibre cuticle cell lines into monolayers which is associated with patterns of cell junctions and cell to cell alignment (probably involving primary cilia). In the cortex, significant cell reshaping and repositioning occurs. Consequently, the cortex at the distal end of Zone

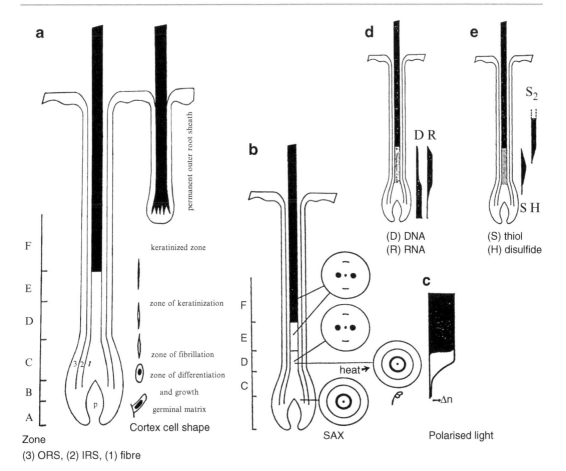

Fig. 8.1 Diagrams of features of the hair follicle from Mercer's book on keratinisation [10] which demonstrate that many of the basic concepts relating to hair development were established more than 50 years ago. (**a**). Mercer's developmental zones and changes in cortex cell morphology. (**b**). Changes in small angle x-ray diffraction pattern (SAX) along the follicle which relates to keratin filament organisation and orientation. Mercer notes a point after zone D in which heating the follicle no longer disrupts the organisation. (**c**). Degree of bifringence using polarisation microscopy increases rapidly during Zone C and D, but only the black-filled region survives heating. (**d**). DNA levels reduce and RNA levels increase in the upper bulb. (**e**). Thiols increase and then decrease as disulfide levels gradually increase above the bulb. (Reprinted from Keratin and Keratinization, 1961, Mercer [10] with permission from Elsevier)

B has fewer cells and these cells are highly elongated. In coloured hair follicles, melanocytes, their cell bodies attached to the upper half of the dermal papilla, release melanin granules into cells across the fibre cell lines.

Cell line specific trichokeratins and keratin associated products begin collecting rapidly from the beginning of Zone B. In medulla cells, a wide range of keratins are expressed but no keratin structures form, instead, vesicles and trichohyalin granules form. In the cortex, keratins assemble into highly organized bundles of intermediate filaments (IFs) called macrofibrils. In the cuticle, keratin builds up in the cytoplasm, and in the IRS layers, fibrillar networks and trichohyalin build up, most prominently in Henle's layer.

By the top of Zone B, IRS and fibre cell nuclei have lost transcriptional function and appear degraded and most cell shaping is complete and this is marked by high-energy metabolic events and the keratinization (hardening) of Henle's layer to form a restrictive tube within which all further development occurs.

Table 8.3 Summary of ultrastructurally-defined zones of the wool follicle

Zone	Histological synonyms[a]	Markers	Main biological processes
Zone A	Germative matrix	Mitosis and lack of clear cell line differentiation, distal boundary is a line (the critical level) angled proximally from the tip of the dermal papilla in non-medullated follicles or the widest part of papilla in medullated follicles [5]	Basement membrane-bound stem cells (mother cells) around dermal papilla neck continuously bud off daughter cells which divide further (possible transient amplifying cells).
Zone B	Hair matrix, elongation zone	Begins at first signs of cell line differentiation, in particular cell shape and alignment and first appearance of keratin in cortex.	Cell line differentiation especially in cell junctions, shape and position, cytoplasmic expression of keratin and trichohyalin. Nuclear function ceases in cortex, cuticle and IRS layers.
Zone C	Keratogenous	Begins where Henle's layer hardens.	Most cell shape/position changes completed, major keratin and KAP synthesis in fibre, increase in particular of KAP species. The gradual keratinisation process begins in the cortex and by the end of this zone, chemical processes have generally replaced active biological processes in fibre construction.
Zone D		Begins where cuticle keratin layer is continuous along membrane apposed to IRS cuticle (fibre isolated from IRS).	Keratin and KAP synthesis continue in cortex and cuticle while non-keratin cell components broken down.
Zone E	Adamson's Fringe (proximal)	Begins where remaining IRS layers harden.	Consolidation, keratinisation or hardening process associated with water loss, precipitation of keratin and internal changes to cortical IF structure.
Zone F	Isthmus	Begins where osmium staining is no longer effective in cortex (osmiophilia lost).	Fibre is in mature form. Probable extraction of material from IRS layers occurs and IRS takes on a porous appearance before fragmenting to release the fibre.
Zone G	Infundibulum	Begins where fibre is fully free of IRS material.	Sebaceous and sudoriferous (if present) glands coat fibre in material along with broken down remnants of IRS, the fibre emerges from the skin. The ORS merges with the skin epidermis and *stratum corneum*.

[a]Histological synonyms do not always match zone locations

Zone C is where final cuticle shape is defined, completing the process of cell reshaping that began in Zone B, with a transformation brought about by tilting of the cuticle monolayer into overlapping sheets. The inner root sheath cuticle becomes sculpted, and bulges into the fibre cuticle cells to form the fibre scale pattern. Within fibre cuticle cells there is a second wave of keratin-associated protein (KAP) expression and the exocuticle begins to form from irregular globular deposits. In the cortex, a second wave of keratin proteins express, macrofibril nucleation ceases, although macrofibril growth and infiltration by KAPs continues. There is a notable increase in cortex density over the distal parts of Zone C. All these events occur within a cornified tube formed by Henle's layer, which isolates the developing fibre from the outside except for small windows through which Huxley's cells contact companion layer cells. The companion layer forms an asymmetrical cytoskeleton the function of which is probably to allow it to interface between the distally moving IRS and the non-moving ORS cells.

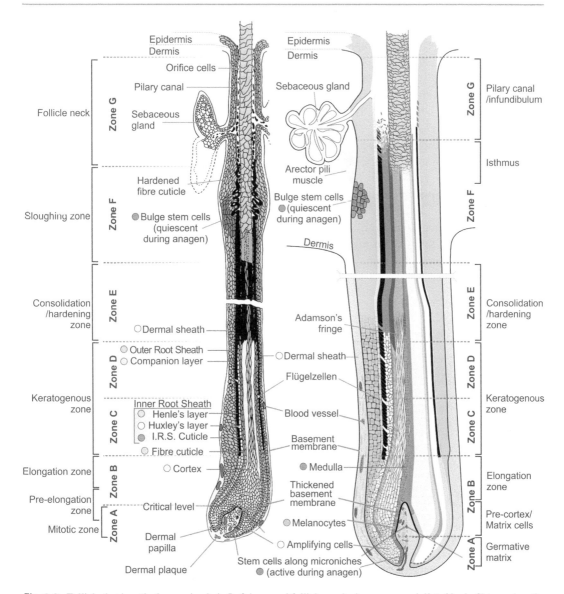

Fig. 8.2 Follicle that is actively growing hair. Left is a wool follicle producing a non-medullated body fibre, and on the right is a human scalp follicle growing a medullated hair

Zone D Consolidation of the presumed A-layer of the fibre exocuticle layer means that the fibre is effectively isolated from the IRS. While keratin structures continue to grow in cuticle and cortex cells, this zone is also associated with the targeted degradation of non-essential cell machinery, nuclear material and intercellular structures. The first signs of keratinization of the cortex are observed as a change in the appearance of the intermediate filaments and this rearrangement is due to the formation of disulfide bonds mediated by reactive oxygen species (possibly released from degenerating mitochondria).

Zone E. Keratinization of the fibre (which occurs in concert with cornification of the remaining IRS layers) involves considerable changes to the architecture of cortex IFs and numerous chemical cross-links are introduced between keratin and KAP proteins. The medulla cells undergo cornification and programmed collapse to form air spaces, and cell membranes with the fibre are converted into a cell membrane complex (CMC). There is significant shrinkage of the cortex (dehydration).

Zones A-E occur within the proximal 1.5 mm of the follicle, after which there is a variable length of no apparent change.

Zone F. The fibre becomes detached from the IRS and in a species-specific process, the IRS layers are partially reabsorbed through the companion layer and their remnants disintegrate into the pilary canal.

Zone G. Fragments of degraded IRS are mixed with sebaceous gland lipids and deposited on the surface of the newly emerging hair.

References

1. Langbein, L., et al. (2007). Novel type I hair keratins K39 and K40 are the last to be expressed in differentiation of the hair: Completion of the human hair keratin catalogue. *Journal of Investigative Dermatology, 127*, 1532–1535.
2. Bornschlögl, T., et al. (2016). Keratin network modifications lead to the mechanical stiffening of the hair follicle fiber. *Proceedings of the National Academy of Sciences, 113*(21), 5940–5945.
3. Chapman, R. E., & Gemmell, R. T. (1971). Stages in the formation and keratinization of the cortex of the wool fiber. *Journal of Ultrastructure Research, 36*(3–4), 342–354.
4. Schneider, M. R., Schmidt-Ullrich, R., & Paus, R. (2009). The hair follicle as a dynamic miniorgan. *Current Biology, 19*(3), R132–R142.
5. Auber, L. (1951). The anatomy of follicles producing wool-fibres, with special reference to keratinization. *Transactions of the Royal Society of Edinburgh, 62*, 191–254.
6. Baltenneck, F., et al. (2000). Study of the keratinization process in human hair follicle by X-ray microdiffraction. *Cellular and Molecular Biology, 46*(5), 1017–1024.
7. Rafik, M. E., et al. (2006). In vivo formation steps of the hard α-keratin intermediate filament along a hair follicle: Evidence for structural polymorphism. *Journal of Structural Biology, 154*(1), 79–88.
8. Langbein, L., et al. (2001). The catalog of human hair keratins. II. Expression of the six type II members in the hair follicle and the combined catalog of human type I and II keratins. *Journal of Biological Chemistry, 276*(37), 35123–35132.
9. Langbein, L., & Schweizer, J. (2005). Keratins of the human hair follicle. *International Review of Cytology, 243*, 1–78.
10. Mercer, E. H. (1961). *Keratin and keratinization* (1st ed., International series of monographs on pure and applied biology, Vol. 12, p. 316). Oxford: Pergamon Press.
11. Orwin, D. F. G., & Thomson, R. W. (1972). An ultrastructural study of the membranes of keratinizing wool follicle cells. *Journal of Cell Science, 11*(1), 205–219.
12. Orwin, D. F., & Woods, J. L. (1982). Number changes and development potential of wool follicle cells in the early stages of fiber differentiation. *Journal of Ultrastructure Research, 80*(3), 312–322.
13. Orwin, D. F. G. (1976). Acid phosphatase distribution in the wool follicle. I. Cortex and fiber cuticle. *Journal of Ultrastructure Research, 55*, 312–324.
14. Orwin, D. F. G. (1979). The cytology and cytochemistry of the wool follicle. *International Review of Cytology, 60*, 331–374.
15. Marshall, R. C., Orwin, D. F. G., & Gillespie, J. M. (1991). Structure and biochemistry of mammalian hard keratin. *Electron Microscope Reviews, 4*(1), 47–83.

Environment of the Anagen Follicle

9

Duane P. Harland

Contents

9.1	**The Skin**	97
9.2	**Follicle Groups**	98
9.3	**Peripheral Anagen Follicle**	98
9.4	**Dermal Sheath, Plaque and Papilla**	101
9.5	**Vasculature**	101
9.6	**Adipocytes**	103
9.7	**Nerves**	103
9.7.1	*Arrector pili* Muscle	105
9.8	**Follicle Stem-Cell Bulge**	106
	References	106

Abstract

Hair follicles are part of the skin. Almost universally, follicles are described as an epithelium-derived tubular down growth into the skin's dermis. Because follicles are complex structures, especially when in anagen phase and configured to actively grow fibres, it is easy to forget that they are part of a crowded environment within the skin. This chapter introduces some of the structures which surround the follicle as well as some of the peripheral parts of the follicle, including follicle groups, and the dermal sheath, vasculature, adipocytes, nerves and the *arrector pili* muscle.

Keywords

Hair follicle · Skin · Follicle groups · Hair follicle vasculature · Hair follicle lipids · Hair follicle arrector pili muscle · Hair follicle nerves

9.1 The Skin

Follicles are a particularly complex appendage of the skin. Like all squamous epithelia, skin has a dermal layer that makes up much of the skin's

D. P. Harland (✉)
AgResearch Ltd., Lincoln, New Zealand
e-mail: Duane.Harland@agresearch.co.nz

thickness over which there is a thin epidermis separated by a continuous basement membrane. The epidermal layer's primary function is to continuously generate cornified *stratum corneum*. The continuous production and shedding of these specialised cell remnants maintains a continuously renewed skin surface which provides mechanical protection and a barrier to water loss and invasion by microbes. A key concept is that epidermal and dermal compartments have different embryological origins and do not mix; cells do not routinely move either direction across the basement membrane. Below the dermis is generally a layer of connective or adipose tissue. In some species (e.g., human scalp) the bottoms of the follicles protrude through the dermis into the connective layer, while in others, (e.g., sheep) they are generally contained within the dermis. The follicle straddles the epidermal and dermal cell compartments, contains a basement membrane and both dermal and epidermal cells play critical roles in hair growth.

Published studies and books (including the one you are reading) often depict isolated follicles (e.g., Chap. 8, Fig. 8.2). This is a consequence of follicles being morphologically complex during anagen, and because they are composed of multiple cell lines, each with its own developmental biology. This can give the illusion that follicles are islands, isolated from the rest of the skin, but the reality is that skin is a very crowded environment. In addition to dense packing of follicles into groups, skin contains multiple glands, vasculature, nervous and immune cells as well as significant extracellular connective material (Fig. 9.1).

9.2 Follicle Groups

In the skin, one of the structures most commonly associated with a follicle is another follicle. This is because follicles usually form clusters, called follicle groups which are composed of one primary and a few secondary follicles (Fig. 9.2a). In many mammalian coats the primary and secondary follicles produce fibres that differ in structure and function. Once above the skin these functionally differing hairs give rise to the three-dimensional organisation of the coat, or pelage (Fig. 9.2b, c). It is not just the properties of individual hairs, but emergent properties of the pelage, and its modification over time (typically seasonal), that has resulted in the functional diversity in thermo-regulation and mechanical protection that has enabled mammals to colonize a wide range of habitats, including climatic extremes [2]. In wild mammals, different hair types normally emerge from follicles differing by the timing of their development in embryological skin [3, 4]. Primary follicles develop first in a hexagonal pattern across the skin and are identified in mature skin by their well-developed *arrector pili* muscle. Primary fibres in mature animals often produce robust guard hairs (sometimes filled with a central medulla composed of air-filled chambers).

Secondary follicles, as their name suggests develop after the primaries. In sheep, secondary follicles may result in fibres which emerge from a single or a shared epidermal orifice, but are always associated with a primary follicle [5] to which they are tethered by thin strands of *arrector pili* muscle, and thus form a follicle group [6, 7]. Something directly analogous appears to occur in the human scalp with distinct follicle groups (sometimes with hairs emerging from a common orifice) and interconnected by a local *arrector pili* network [8, 9]. Knowledge on human hair group biology is a developing but important area because dissociation between muscle and follicle appears to be correlated with follicle miniaturization brought on by androgens [10, 11]. In many mammals the secondary follicles produce very fine diameter fibres and have no sebaceous gland (e.g. sheep). However, it is not always the case. In human scalp hair, both primary and secondary follicles produce similar hairs and have sebaceous glands.

9.3 Peripheral Anagen Follicle

Enveloping the follicle is a sheath of dermal fibroblast-like cells and collagen that, at the base of the follicle, merges into the dermal papilla,

9 Environment of the Anagen Follicle

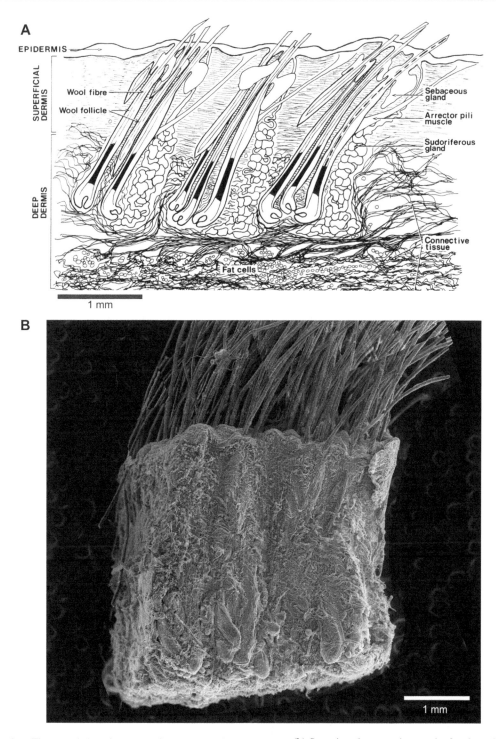

Fig. 9.1 The crowded environment of anagen hair follicles within the dermis and subcutaneous connective tissue of the skin demonstrated by sheep wool follicles (**a**) Schematic of the skin of a sheep showing three follicle groups. (**b**) Scanning electron micrograph of a glutaraldehyde and osmium fixed then critical point dried piece of sheep skin demonstrating the dense packing. Much of the material is collagen. (**a**: Reproduced from Orwin [1])

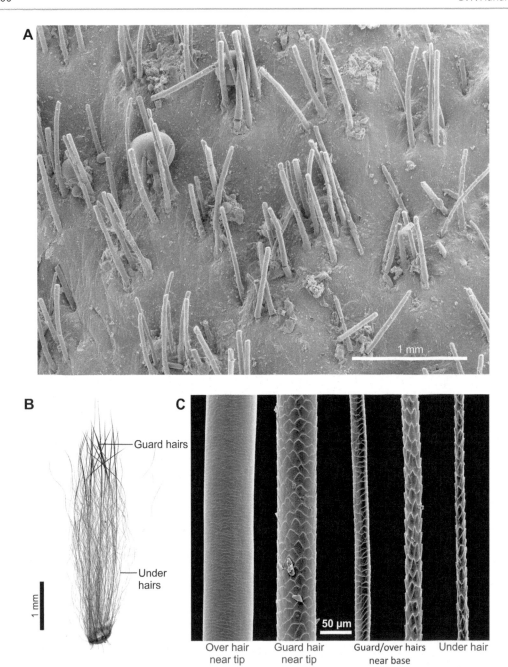

Fig. 9.2 Skin surface and hair types. (**a**) Scanning electron micrograph of a polymer replica of the surface of a sheep skin showing clusters of emerging hairs surrounded by debris. (**b**) Typical mammalian coat structure and fibre types. Robust guard hairs support a multitude of flexible under hairs. Sample is a discarded scab from a domestic cat (following inter-feline negotiations). (**c**) Scanning electron micrographs of cat hairs, showing that external morphology, cuticle scale pattern, shape and diameter between hair types and along single hairs vary greatly

forming the core of the follicle bulb. The epidermal parts, including apocrine-sweat and sebaceous glands, are separated from the dermal sheath and papilla by a basement membrane. Although the dermal papilla and dermal sheath do not contribute cells to the growing hair, interactions across the basement membrane are important for control of hair formation. If the epidermal cells of the bulb (sometimes called germative matrix cells) form the engine of hair assembly, the dermal papilla is more like the cockpit. While the dermal papilla is a specialized structure adapted to influence the hair assembly process, it is also under influence from local and systemic influences from outside the follicle. Chap. 10 goes into further detail about the follicle bulb and dermal papilla.

9.4 Dermal Sheath, Plaque and Papilla

The externally visible surface of a follicle is the dermal sheath, which is mostly composed of a dense network of connective material. Sometimes referred to as the connective tissue sheath, it is primarily composed of layers of variously oriented collagen fibrils in bands. The collagen structure of the dermal sheath suggests that it plays an important structural role in common with other similar cholesteric laminate structures in biology [12]. Intersecting, traversing and embedded with the sheath are fibroblasts and other peripheral structures such as blood vessels and neurons. The sheath is one of the least studied follicular structures despite a high level of apparent organisation. As the follicle's anagen phase becomes established, the dermal sheath of human hair thickens, eventually having distinct inner, middle and outer layers each with differing collagen organisation [13].

A thickened patch of dermal sheath at the proximal margin of the follicle bulb contains a pocket of fibroblasts (Fig. 9.3), and is called the dermal plaque. Directly distal of the plaque is the neck of the dermal papilla. The dermal sheath, plaque, and papilla are all of mesenchymal origin (i.e., derived from connective tissue associated stem-cells). During anagen, the dermal papilla is the only part of the dermal component that penetrates inside the bulb. Research on the dermal papilla has been dominated by its role in follicle morphogenesis and hair follicle cycling. Consequently, the approach and methods [14] are heavily aligned toward understanding the genetic messengers between the dermal papilla, the dermal sheath (DS) and the bulge-derived stem cells that are involved in generation of a new follicle during the telogen-anagen transition (Chap. 2).

The dermal components of a follicle do not contribute cells directly to a growing hair, being separated from the epidermally derived compartment by a basement membrane. However, the dermal papilla and sheath fibroblast cells differ considerably from their neighbours within the skin dermis. For example, dermal papilla and DS fibroblasts express specific enzymes and molecules, which can used to identify them [15]. The role of the dermal papilla is discussed further in the next chapter (Chap. 10).

9.5 Vasculature

Blood vessels, especially capillaries, surround most follicles, normally embedded within the dermal sheath and penetrating inside dermal papillae of follicles from a wide range of species including those of humans [16, 17], rodents [18, 19] and sheep [20–22] (Fig. 9.4). In small follicles, such as human vellus [16], the vessels are reduced to a few capillaries and do not penetrate the dermal papilla. However, vascular presence is not a necessity for hair fibre growth and is not, for example, found in the small follicles of neonatal rats [19].

Vasculature is seen as a conduit for systemic messenger molecules such as hormones [23]. How systemic messenger molecules such as hormones interact with the follicle is not necessarily simple and direct. For example, in hair follicles that are undergoing miniaturization during androgenic alopecia (age-related balding) there is a correlation to increased androgen levels in the blood, but rather than the hormones interacting directly with the developing hair fibre, research using cultured der-

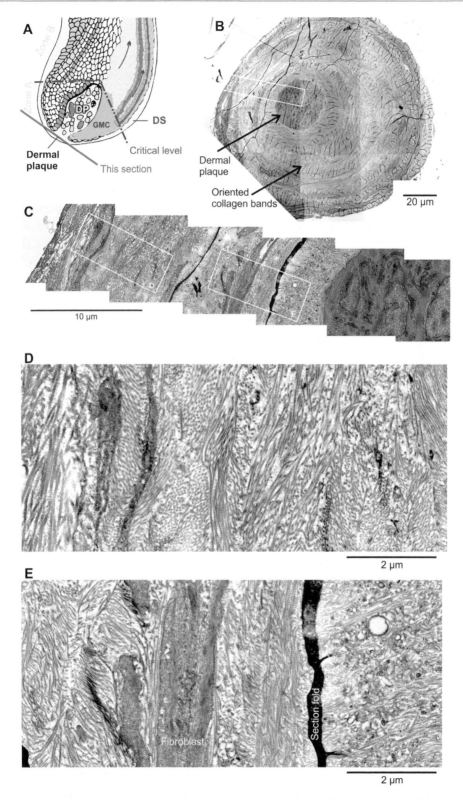

Fig. 9.3 The dermal plaque sits below the dermal papilla in wool follicles. (**a**) schematic indication of plaque location. (**b–e**) Transmission electron micrographs of transversely sectioned follicle showing dermal plaque cells surrounded by bands of collagen fibrils in various orientations

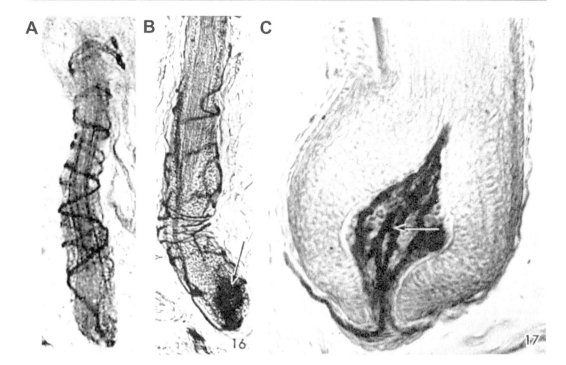

Fig. 9.4 Vasculature of the anagen follicle highlighted in dark staining based on acid phosphatase activity. (**a**) Blood vessels in a spiral pattern around a small merino sheep scrotum follicle. (**b**) Irregular vessels surround a larger merino wool follicle from the flank. In both (**a** and **b**) a vessel clearly leads to the follicle base. (**c**) Vessels within the dermal papilla of a merino lip-hair follicle. (Reprinted from Aust. J. Biol. Sci., 1968, Lyne and Hollis [27], with permission from CSIRO Publishing)

mal papilla cells indicates that the hormones stimulate increases in hormone receptor levels which must in turn affect the follicle [24]. It is assumed that most direct influence on the growing fibre in an anagen follicle will be via the dermal papilla rather than directly from the outside of the follicle. In the upper half of the follicle there are significant barriers to chemical penetration of the developing fibre due to the keratinized Henle's layer and possibly the membranes of the Companion layer [25].

Probably the primary role of the blood vasculature in follicles is to deliver a constant supply of nutrients and other metabolites sufficient to match the metabolic requirements of size of follicle, and therefore the size of the hair being produced. While there has been little work in this area, there is evidence from mouse follicles that vascularity is increased around follicles during anagen and that this is mediated by the outer root sheath [26]. Interference by over expression or blocking of this control mechanism during cell cycling affects follicle size.

9.6 Adipocytes

Dermal adipocytes surround the bulbs of follicles (Fig. 9.5). Differences in lipid crystallization observed in rat follicles during processing for TEM [19] suggests that the adipocytes that line up against the bulb basement membrane partition at least two different kinds of lipid composition or structural arrangement. The role that these follicle-attached adipocytes might play in the growth of hair is not understood, although they can sit adjacent to the basement membrane.

9.7 Nerves

Human scalp hair follicles [16], [30] and sheep body hair follicles [27] are surrounded by a loose network of nerves that converge into a well-organized ring of 5–12 palisade nerves just proximal of the duct of the sebaceous gland and just above the *arrector pili* muscle attachment point

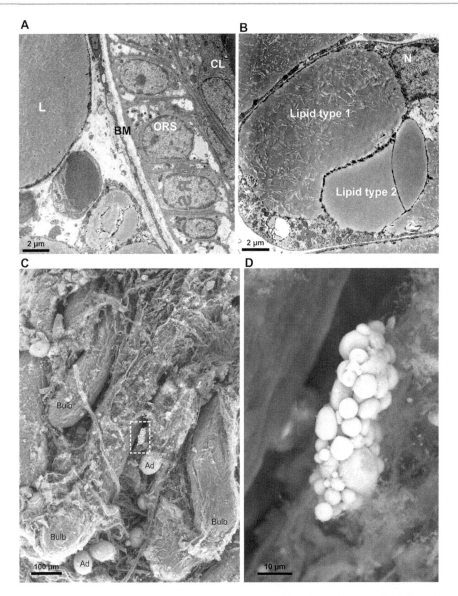

Fig. 9.5 Lipid structures associated with the external (dermal) surfaces of follicles. (**a**) Transmission electron micrographs of lipid body (L) containing cells associated with follicles in rat underfur. Adipocytes in this part of the dermal sheath almost touch the basement membrane (BM), bringing them physically close to the outer root sheath (ORS). CL, companion layer. (**b**) Part of an adipocyte showing multiple lipid vacuoles that each contain lipids that react very differently to chemical processing for microscopy. (**c**) Scanning electron micrograph of critical point dried sheep skin, treated with osmium to highlight lipids and imaged using mixed secondary electron and backscatter imaging. Lipid materials appear light. Ad = adipose cells, boxed feature (in D) is a complex of multiple lipid bodies close to the follicle dermal sheath. (Figures (**a**) and (**b**) reprinted from Hair Follicle, Differentiation Under the Electron Microscope, 2005, Morioka [19], with permission from Springer Nature; (**c**) and (**d**), with assistance of Veronika Novotna)

(Fig. 9.6) in the permanent part of the follicle which is retained during the follicle cycle (Chap. 2). This focal region contains both myelinated and unmyelinated neurons that are embedded in the dermal sheath [19]. For the most part, dense and organised nerves are a feature of the upper follicle of hairs and not associated with the bulb. An exception to this is in specialised sensory hairs (or vibrissae) such as facial whiskers. These hairs differ significantly from coat hair (which include human body and scalp hair) in both function and structure. Vibrissae are highly specialized active

9 Environment of the Anagen Follicle

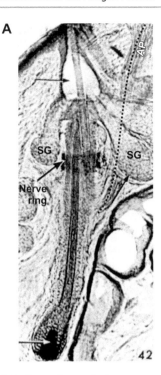
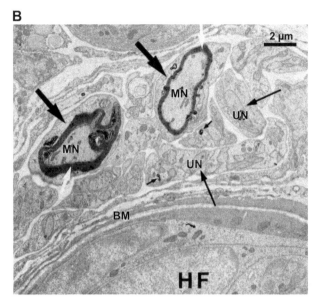

Fig. 9.6 Enervation of the anagen follicle. (**a**) Transmission light micrograph of a sheep wool follicle illustrating enervation. Unlike the loose network across much of the upper follicle, neurons form a well-organised palisade ring located between the sebaceous glands (SG) and where the arrector pili muscle (AP) attaches. (**b**) TEM micrograph of transverse section presumed upper follicle showing myelinated (MN) and unmyelinated (UN) neurons within the dermal sheath close to the follicle basement membrane (BM) in a rat follicle. Arrows point to Schwann cells encasing neuron. ((**a**) reprinted from Aust. J. Biol. Sci., 1968, Lyne and Hollis [27], with permission of CSIRO Publishing. (**b**) reprinted from Hair Follicle Differentiation Under the Electron Microscope, 2005, Morioka [19], with permission from Springer Nature)

mechano-sensory structures, and embryological, morphological and taxonomic studies have yet to convincingly determine the precise relationship between vibrissae and other mammalian hair, but all tend to agree that this split occurred very early in the evolution of hair [28, 29] and that vibrissae show specific adaptations for a nervous system role for active tactile sensing. Structurally, vibrissae follicles are surrounded by blood-filled sinus spaces containing mechano-receptive nerve endings, have specialized sebaceous glands, and often significant musculature that allow active intentional movement [27].

9.7.1 *Arrector pili* Muscle

The *arrector pili* muscle is a key external feature of follicles which is often simplistically depicted as an isolated strand of smooth muscle linking the upper follicle to the skin's upper dermis. However, sources cited by Chapman [7] indicate that German studies as early as 1885 [6] characterised the *arrector pili* muscles of domestic mammals and sheep as having a complex branched structure, with branches ending at each follicle in a follicle group. Early studies also indicated that in human skin, small branching fibres also led to the sebaceous glands [31]. The *arrector pili* attaches to the follicle slightly proximal of the upper stem cell bulge of the outer root sheath [32–34], and research on the prominent follicle bulges of mice indicate that during follicle development, the bulge expresses a specific muscle attachment factor (nephronectin) into the basement membrane [35]. At its other end, the muscle fans out just below the basal epidermis to connect via elastic fibres to a meshwork of collagen [7]. In sheep the multiple strands of the *arrector pili* muscle are interconnected by elastic and collagen fibres.

The main function of the *arrector pili* is to raise and lower hairs. It is the source of "goose bumps", also called *cutis anserina*. In many mammals, raising the stiff guard hairs that support the pelage can affect the coat's thermoregulation ability, and this effect is also used behaviourally by mammals to increase their apparent size during confrontations.

Studies have linked the branching nature of the *arrector pili* muscle in human scalp follicles [8, 9] to the poorly understood process of follicle miniaturisation which occurs as age-related balding in men and hair density thinning in women (androgenic alopecia). It appears that detachment of muscle strands from an individual follicle in a follicle group is correlated with that follicle's transformation over the follicle cycle into a miniaturized version resembling a vellus hair follicle, and that this is associated with androgens [10]. It is also worth noting that the attachment point of the *arrector pili* marks, at least crudely, the boundary between the distal part of the follicle that remains similar throughout the follicle cycle, or "permanent portion" and the proximal part that is extensively remodelled, or "transient portion" (Chap. 2).

9.8 Follicle Stem-Cell Bulge

The hair follicle bulge located in the upper (permanent) part of the follicle is an important focus of many studies of follicle cycling and development. The bulge, or sometimes "swelling" or "*wulst*", is a common feature of the outer-root sheath adjacent to the basement membrane of follicles from many species, having been long noted in human embryonic follicles [32], and in the embryonic and adult follicles of mice [36] and sheep [37]. The bulge is slightly distal to the attachment point of the *arrector pili* muscle [32–34]. In the curved follicles of sheep, the bulge can occur on either side of the follicle with respect to bulb deflection [37]. Morphologically, the bulge is particularly evident in developing human hair follicles [17, 32, 33] and in follicles of adult mice [36, 38], but, if visible at all, occurs only as a subtle swelling in human scalp hair follicles [39].

Stem cells within the bulge play a key role in replenishing the bulb stem cells that during anagen give rise to the fibre and its supporting structures [38, 40] (Chap. 2). However, during anagen's main fibre growth phase, the bulge stem cells are quiescent [41, 42], and there is currently no evidence that the bulge stem cells themselves play a direct role during hair formation during the anagen VI stage [43, 44].

What is less certain is the biology, and potential role of other cells in the outer root sheath, some of which may also be described as stem cells. Human scalp follicles (and sheep wool follicles) have an extended anagen VI phase. However, most of the research in this area has been carried out using mice, which have a very short anagen VI phase and a fibre with significantly different functional characteristics and morphology than any human hair. The relevance to scalp hair of results from studies examining the role of bulge stem cells using mouse vibrissae are even less clear because vibrissae follicles appear to have cycle phase that overlap to some extent [45].

It is important to realize that our understanding of the hair bulge, and the roles of the outer-root-sheath, and especially cell migration is at a relatively early stage, and research, especially in the stem-cell context, is progressing rapidly using human and mice models [34].

References

1. Orwin, D. F. G. (1989). Variations in wool follicle morphology. In G. E. Rogers, K. A. Ward, P. J. Reiss, & M. C. Marshall (Eds.), *The biology of wool and hair* (pp. 227–241). London/New York: Chapman & Hall.
2. Ryder, M. L. (1973). *Hair* (1st edn., Studies in biology, 58 p). The Institute of Biology (Eds.). London: Edward Arnold.
3. Rogers, G. E. (2006). Biology of the wool follicle: An excursion into a unique tissue interaction system waiting to be re-discovered. *Experimental Dermatology, 15*(12), 931–949.
4. Hardy, M. H. (1992). The secret life of the hair follicle. *TIGs, 8*(2), 55–61.

5. Hardy, M. H., & Lyne, A. G. (1956). The pre-natal development of wool follicles in merino sheep. *Australian Journal of Biological Sciences, 9*(3), 423–441.
6. Bonnet, R. (1885). Über dia muskulatur der haut und der knäueldrüsen. Bayerisches ärztliches intelligenzblatt.
7. Chapman, R. E. (1965). The ovine arrector pili musculature and crimp formation in wool. In A. G. Lyne & B. F. Short (Eds.), *Biology of the skin and hair growth* (pp. 201–232). Sydney: Angus and Robertson.
8. Poblet, E., Ortega, F., & Jiménez, F. (2002). The arrector pili muscle and the follicular unit of the scalp: A microscopic anatomy study. *Dermatologic Surgery, 28*(9), 800–803.
9. Song, W.-C., et al. (2006). A new model for the morphology of the arrector pili muscle in the follicular unit based on three-dimensional reconstruction. *Journal of Anatomy, 208*(5), 643–648.
10. Yazdabadi, A., et al. (2008). The Ludwig pattern of androgenetic alopecia is due to a hierarchy of androgen sensitivity within follicular units that leads to selective miniaturization and a reduction in the number of terminal hairs per follicular unit. *British Journal of Dermatology, 159*(6), 1300–1302.
11. Torkamani, N., et al. (2014). Destruction of the arrector pili muscle and fat infiltration in androgenic alopecia. *British Journal of Dermatology, 170*(6), 1291–1298.
12. Neville, A. C. (Ed.). (1993). *Biology of fibrous composites: Development beyond the cell membrane* (1st ed.). New York: Cambridge University Press. 214.
13. Ito, M., & Sato, Y. (1990). Dynamic ultrastructural changes of the connective tissue sheath of human hair follicles during hair cycle. *Archives of Dermatological Research, 282*, 434–441.
14. Ohyama, M., et al. (2010). The mesenchymal component of hair follicle neogenesis: Background, methods and molecular characterization. *Experimental Dermatology, 19*(2), 89–99.
15. Yang, C.-C., & Cotsarelis, G. (2010). Review of hair follicle dermal cells. *Journal of Dermatological Science, 57*(1), 2–11.
16. Montagna, W. (1962). *The structure and function of skin* (2nd ed.). New York: Academic.
17. Holbrook, K. A., et al. (1989). Morphogenesis of the hair follicle during the ontogeny of human skin. In G. E. Rogers, P. J. Reis, K. A. Ward, & R. C. Marshall (Eds.), *The biology of wool and hair* (pp. 15–36). London/New York: Chapman & Hall.
18. Parakkal, P. F. (1966). The fine structure of the dermal papilla of the guinea pig hair follicle. *Journal of Ultrastructure Research, 14*, 133–142.
19. Morioka, K. (2005). *Hair follicle, differentiation under the electron microscope – An atlas* (p. 152). Tokyo: Springer.
20. Ryder, M. L. (1956). The blood supply of the wool follicle. *W.I.R.A. Bulletin, 18*, 142–147.
21. Orwin, D. F. G. (1970). A polysaccharide-containing cell coat on keratinizing cells of the Romney wool follicle. *Australian Journal of Biological Science, 23*, 623–635.
22. Orwin, D. F. G. (1979). The cytology and cytochemistry of the wool follicle. *International Review of Cytology, 60*, 331–374.
23. Zouboulis, C. C., et al. (2007). Sexual hormones in human skin. *Hormone and Metabolism Research, 39*, 85–95.
24. Hibberts, N. A., Howell, A. E., & Randall, V. A. (1998). Balding hair follicle dermal papilla cells contain higher levels of androgen receptors than those from non balding scalp. *Journal of Endocrinology, 156*(1), 59–65.
25. Orwin, D. F. (1971). Cell differentiation in the lower outer sheath of the Romney wool follicle: A companion cell layer. *Australian Journal of Biological Science, 24*(5), 989–999.
26. Yano, K., Brown, L. F., & Detmar, M. (2001). Control of hair growth and follicle size by VEGF-mediated angiogenesis. *The Journal of Clinical Investigation, 107*(4), 409–417.
27. Lyne, A. G., & Hollis, D. E. (1968). The skin of the sheep: A comparison of body regions. *Australian Journal of Biological Science, 21*(3), 499–527.
28. Klauer, G. J., et al. (2001). Vibrissae-more than just hairs! *Journal of Morphology, 248*(3), 248–249.
29. Chernova, O. F. (2006). Evolutionary aspects of hair polymorphism. *Biology Bulletin, 33*(1), 43–52.
30. Montagna, W., & Parakkal, P. F. (1974). *The structure and function of skin* (3rd ed.). New York: Academic.
31. Fritsch, G. (1897). *Über die entstehung der rassenmerkmale des menschlichen kopfhaares.* KorrespBl. dtsch.Ges.Anthrop, 28.
32. Pinkus, H. (1958). Embryology of hair. In W. Montagna & R. A. Ellis (Eds.), *The biology of hair growth* (pp. 1–32). New York: Academic.
33. Akiyama, M., et al. (1995). Characterization of hair follicle bulge in human fetal skin: The human fetal bulge is a pool of undifferentiated keratinocytes. *Journal of Investigative Dermatology, 105*(6), 844–850.
34. Ohyama, M. (2007). Hair follicle bulge: A fascinating reservoir of epithelial stem cells. *Journal of Dermatological Science, 46*(2), 81–89.
35. Fujiwara, H., et al. (2011). The basement membrane of hair follicle stem cells is a muscle cell niche. *Cell, 144*(4), 577–589.
36. Paus, R., et al. (1999). A comprehensive guide for the recognition and classification of distinct stages of hair follicle morphogenesis. *Journal of Investigative Dermatology, 113*, 523–532.
37. Auber, L. (1951). The anatomy of follicles producing wool-fibres, with special reference to keratinization. *Transactions of the Royal Society of Edinburgh, 62*, 191–254.

38. Cotsarelis, G., Sun, T., & Lavker, R. M. (1990). Label-retaining cells reside in the bulge area of pilosebaceous unit: Implications for follicular stem cells, hair cycle, and skin carcinogenesis. *Cell, 61*, 1329–1337.
39. Ohyama, M., et al. (2006). Characterization and isolation of stem cell-enriched human hair follicle bulge cells. *Journal of Clinical Investigation, 116*(1), 249–260.
40. Alonso, L., & Fuchs, E. (2003). Stem cells of the skin epithelium. *Proceedings of the National Academy of Sciences of the United States of America, 100*(SUPPL. 1), 11830–11835.
41. Ito, M., et al. (2004). Hair follicle stem cells in the lower bulge form the secondary germ, a biochemically distinct but functionally equivalent progenitor cell population, at the termination of catagen. *Differentiation, 72*(9–10), 548–557.
42. Lyle, S., et al. (1998). The C8/144B monoclonal antibody recognizes cytokeratin 15 and defines the location of human hair follicle stem cells. *Journal of Cell Science, 111*(21), 3179–3188.
43. Morris, R. J., & Potten, C. S. (1999). Highly persistent label-retaining cells in the hair follicles of mice and their fate following induction of anagen. *Journal of Investigative Dermatology, 112*, 470–475.
44. Cotsarelis, G. (2006). Epithelial stem cells: A folliculocentric view. *Journal of Investigative Dermatology, 126*(7), 1459–1468.
45. Oshima, H., et al. (2001). Morphogenesis and renewal of hair follicles from adult multipotent stem cells. *Cell, 104*(2), 233–245.

Development of Hair Fibres

Duane P. Harland and Jeffrey E. Plowman

Contents

10.1	**Germative Matrix**	111
10.1.1	Zone A Contains Stem Cells Attached to the Dermal Papilla Basement Membrane	111
10.1.2	Replenishment of the Zone A Stem Cells	111
10.1.3	Cell Line Fate	113
10.1.4	The Dermal Papilla Pulls the Strings	113
10.1.5	Defining the Extent of Zone A	115
10.2	**Melanocytes and Hair Colouration**	116
10.3	**Development of the Cortex**	117
10.3.1	Extensive Cell Reshaping and Repositioning Occur in Zone B	117
10.3.2	Cortical Keratin Proto-Macrofibrils Seed in Zone B	120
10.3.3	Macrofibrils Grow by Coalescing, While Many New Keratins and KAPs Appear in Zone C	120
10.3.4	Macrofibril Keratinization Is an Extended Process of Intermediate Filament Reshaping and Disulfide Bond Formation	122
10.3.5	Cell Junctions Are Critical for Cell Reshaping in the Upper Bulb	125
10.3.6	Cell Junctions, Extracellular Matrix and Plasma Membranes Are Remodelled into the Cell Membrane Complex	126
10.3.7	Nuclear and Other Cytoplasmic Changes	127
10.3.8	Mitochondrial Degradation Is Peculiar in the Cortex	130

D. P. Harland (✉) · J. E. Plowman
AgResearch Ltd., Lincoln, New Zealand
e-mail: Duane.Harland@agresearch.co.nz;
Jeff.Plowman@agresearch.co.nz

10.4	**Development of the Medulla**	131
10.4.1	Early Differentiating Medulla Cells Migrate Up the Dermal Papilla	131
10.4.2	The Keratin Lottery Approach to Medulla Development	131
10.4.3	Some Interactions Occur Between Medulla and Cortex	133
10.4.4	Terminal Differentiation of the Medulla	133
10.5	**Development of the Cuticle and IRS Cuticle**	135
10.5.1	Cuticle Layers and Scale Edges Are Defined by Cell Shaping in the Bulb	135
10.5.2	Complex Patterns of Cell Junctions and Cytoplasmic Fibril Networks Develop	135
10.5.3	Fibre Cuticle Has a Specific Pattern of Early Keratin Expression	138
10.5.4	Cuticle Shape Finalisation and Keratin Structures Formation	139
10.5.5	Hardening of the Cuticle	140
10.5.6	The Cuticle Cell Membrane Complex and Development of Fibre Surface	140
10.6	**Development of the Inner Root Sheath**	142
10.6.1	Huxley's and Henle's Layers	142
10.6.2	IRS, Especially Huxley's Layer, Have a Role in Along-Fibre Shape Changes	145
10.7	**Development of the Companion Layer**	146
10.8	**Emerging Hair**	147
	References	149

Abstract

The growth of hairs occurs during the anagen phase of the follicle cycle. Hair growth begins with basement membrane-bound stem cells (mother cells) around the dermal papilla neck which continuously bud off daughter cells which further divide as a transient amplifying population. Division ceases as cell line differentiation begins, which entails changes in cell junctions, cell shape and position, and cell-line specific cytoplasmic expression of keratin and trichohyalin. As the differentiating cells migrate up the bulb, nuclear function ceases in cortex, cuticle and inner root sheath (IRS) layers. Past the top of the bulb, cell shape/position changes cease, and there is a period of keratin and keratin-associated protein (KAP) synthesis in fibre cell lines, with increases, in particular of KAP species. A gradual keratinization process begins in the cortex at this point and then non-keratin cell components are increasingly broken down. Terminal cornification, or hardening, is associated with water loss and precipitation of keratin. In the upper follicle, the hair, now in its mature form, detaches from the IRS, which is then extracted of material and becomes fragmented to release the fibre. Finally, the sebaceous and sudoriferous (if present) glands coat the fibre in lipid-rich material and the fibre emerges from the skin. This chapter follows the origin of the hair growth in the lower bulb and traces the development of the various cell lines.

Keywords

Hair follicle · Anagen · Hair shaft development · Developmental zones · Inner root sheath development

10.1 Germative Matrix

Hairs are made of the remnants of cells that have been sacrificially modified to create the specialised structures of the cuticle, cortex and medulla. The inner root sheath (IRS) and Companion layer that grow alongside the hair, supporting its development are also composed of material derived from dead cells. All these cells originate during the follicle's active fibre growth stage (anagen) from a monolayer of mitotically active stem cells located at the base of the follicle bulb. Further cell division occurs in a region close to the stem cells, but stops as cell line differentiation begins. This proliferative zone is referred to in many studies as the "germative matrix" or Zone A.

10.1.1 Zone A Contains Stem Cells Attached to the Dermal Papilla Basement Membrane

The proliferative cells of Zone A are of two kinds that are analogous to the proliferative cells found in the basal layer of the skin epithelium [1], which contains stem cells along the basement membrane that divides epidermis from dermis. In the epithelium these stem cells give rise to transient amplifying cells (TACs), which divide a number of times before eventually differentiating into specialized epithelial cells. What happens in the base of the follicle is analogous to skin, but with additional complexity. The Zone A stem cell layer is of epithelial origin [2], and are contiguous with the outer-root-sheath cells. The stem cells form a monolayer surrounding the lower part of the dermal papilla, that is, around the dermal papilla neck and extending no further than the widest part (Fig. 10.1).

Zone A stem cells differ ultrastructurally from the cells that divide from them, being columnar in morphology and this is similar across a wide range of follicles in different species including human [3], mouse [4] and sheep [5]. In wool follicles [6] and in human beard hair follicles [7], cells of the monolayer are attached to the basement membrane by hemidesmosomes (as is the case in the skin and other squamous epithelia [8]). While it appears that this attachment of the stem cells to the basement membrane is universal, there is some controversy. In human scalp hair it has been reported that hemidesmosomes are present, but only in a degraded form [8–10], and they have not been observed in early anagen mouse vibrissae follicles [11]. If this is the case, then it would suggest a significant gap in our understanding of the fine details of the relationship and division of functions that might occur within the Zone A stem cell population. However, it is also possible that these observations were made in the Zone B part (distal half, or apex) of the dermal papilla where degraded hemidesmosomes can also be observed in wool (Woods and Harland, unpublished).

Along the margin of the distal half of the dermal papilla, above the widest part, there is a significant thickening of the basement membrane.

10.1.2 Replenishment of the Zone A Stem Cells

Being an extension of the outer root sheath, Zone A stem cells originated from the stem-cell population in the "bulge" region during follicle remodelling at the onset of anagen [12]. Whether Zone A stem cells are replenished during the

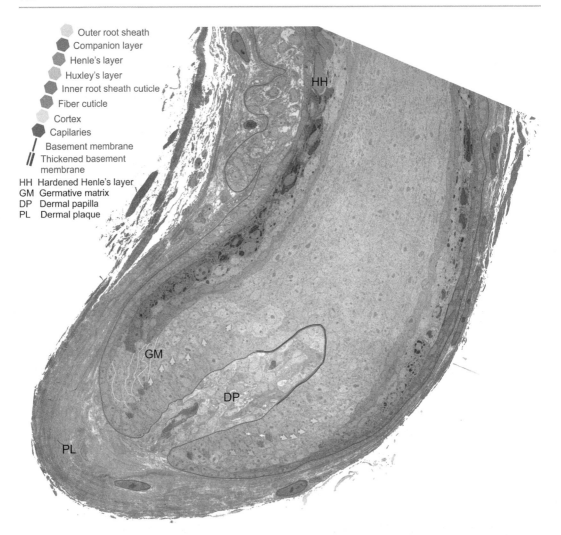

Fig. 10.1 Transmission electron micrograph of a wool follicle bulb showing main features and with differentiated epidermal cell lines overlaid in different colours. Coloured arrows point from basal stem cells aligned along dermal papilla neck to indicate movement of daughter cells. Cells dividing from each of the stem cells are probably fated to particular cell lines, but can divide again before the first signs of differentiation begins. (original micrograph courtesy of Joy L. Woods)

main growth phase of anagen (stage VI) from the outer root sheath cells (either by movement or division) is unknown. It is an important question for understanding the control mechanism of hair differentiation at a cellular level. There is limited evidence from wool that some cells in the outer root sheath actively divide during long-term anagen [13, 14], but no specific evidence that these are moving to the bulb during anagen VI.

The volume of the dermal papilla is positively correlated with follicle bulb volume and fibre volume/diameter in sheep wool and human scalp hairs and rat vibrissae [15–19]. However, it is not known if this is cause or effect, and the dynamics of the dermal papilla in terms of cell migration [20, 21], cell division [22, 23], and changes in extracellular matrix [19]. It is generally considered that the dermal papilla grows during early anagen (stages I–V), remains constant during anagen VI, and then decreases via cell migration out of the DP during the onset of telogen [24].

10.1.3 Cell Line Fate

In the epidermis, stem cells on the basement membrane (that separates dermis from epidermis) divide to produce TACs which divide a few more times before differentiating into the single keratinocyte cell line that develops into the stratum corneum. In the anagen follicle the situation is more complex, because instead of one cell line, there are seven. This raises a question – are cells already fated to a particular cell line when they divide from the hair stem cells along the dermal papilla neck, or is the fate decided by the position of individual daughter cells from the transient amplifying population as they leave Zone A? For example, are TACs that happen to be close to the periphery after migrating through the zone switched at that point to become Companion cells? The traditional view inherited from the development of stratum corneum has been that these TACs are genetically undifferentiated and homogeneous – simply a pool of blank slates which will later become fated to one cell line or another, but this view is changing.

In terms of cell-line differentiation, the cells of Zone A are particularly notable for their complete lack of specificity for all epithelial and hair-specific keratin antibodies in both human and sheep [25, 26, 92]. At this point the two tasks of these cells appear to be mitosis and migration. Studies with fluorescent probes in mice follicles and radio-labelled thymidine in human scalp follicles (Fig. 10.2) indicate that individual stem cells produce only a single cell line which then migrates along a particular pathway through the bulb. That is, each cell in Zone A is fated despite the lack of keratogenesis [27–29]. Single cell RNA analysis methods used to identify cell lineage differences within the TAC population and along the dermal papilla suggest that the dermal papilla is stratified into micro-niches which match up to groups of stem cells that then produce cells fated to particular cell lines or cells that will transform into one or another of cells with a similar phenotype such as the inner root sheath cell lines [30]. This latter plasticity of at least some of the cells is borne out by ultrastructural observations that appear to show dividing TACs can move to different cell lines [6].

Irrespective of where the cell type fate of Zone A cells is decided, the signalling pathways and the patterns of gene expression that underpin the onset of development of each cell line is not well understood [25]. However, from research on mice follicles, there is a rapidly increasing list of molecules that are expressed during anagen, either uniformly [31, 32], or more heterogeneously [33–35] within the wider bulb region. Transcriptomic libraries of specific cell-lines are at an early stage of development but can potentially indicate the expression levels for a very large number of genes and small non-coding RNA, which may act as messengers [36], and such tools are likely to reveal the control pathways more clearly when it is applied to a range of follicle types.

10.1.4 The Dermal Papilla Pulls the Strings

The dermal papilla does not directly contribute cells to the growing hair, being separated from the bulb cells of Zone A by a basement membrane. Nevertheless, interaction between the dermal papilla and the epidermal compartment of the lower bulb is seen as critical for maintenance of function during hair production [37–41].

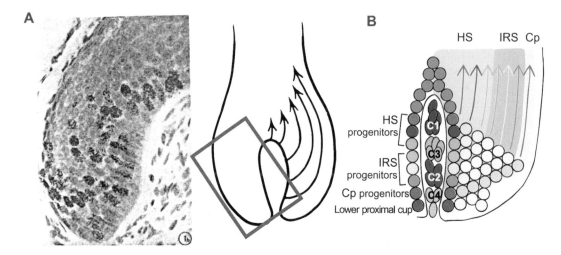

Fig. 10.2 Transient amplifying cells within the germative matrix bud off specific stem cells attached to the dermal papilla neck and then divide again while migrating into specific cell lines. (**a**) Migration has been observed using radio-labelling in humans. (**b**), Summary diagram from work on mice follicles of the spatial organisation of the dermal papilla micro-niches and Zone A cells. ((**a**) reprinted from Advances in Biology of Skin Vol IX, Hair Growth, 1967, Epstein and Maibach [29], and (**b**) reprinted from Cell, 2017, Yang, Adam [30], both with permission from Elsevier)

The micro-niches stratified within the dermal papilla that appear to program the specific cell line initially become organised at the earliest stages of anagen [30]. It might be tempting to imagine that once these niches, and the resulting pattern of fibre cell lines, has formed during the early stages of anagen, that the pattern is set for that part of the cycle. However, the system must also be flexible because control over hair growth at the level of Zone A involves not only maintenance of the order of cell lines, but also adjustments to the extent of production for each cell line. A simple example of a change in differentiation pattern that occurs in human scalp hair is along-fibre appearance and disappearance of medullation. Non-human hairs often contain more sophisticated changes in hair morphology and colouration during a single anagen phase [42] (Fig. 10.3a). Most work in this area has focused on rodent underfur, sometimes called zig-zags [43], and the guard hairs of rabbits [16, 44]. Along-fibre change in diameter and cross-sectional shape occur within follicles that do not change diameter or shape in line with the fibre inside. What does change is the distribution pattern of the cell lines, in particular the number and distribution of Huxley's cells [16, 44, 45], with rabbit guard hairs providing some of the most striking examples (Fig. 10.3b). Changes in cuticle scale pattern may also be associated with changes in fibre growth rate [46].

We know very little about the fine details of the organisation of the dermal papilla. The dermal papilla is a heterogeneous cell population during anagen [48, 49], containing fibroblasts alongside mast cells (in sheep but possibly not in humans) and often with capillaries. Dermal papillae fibroblasts often have extensive Golgi apparatus and rough endoplasmic reticulum [50, 51], and fibroblasts often have variable staining in transmission electron micrographs suggesting differing cytotologies (Fig. 10.4). Dermal papilla fibroblasts are well connected to one another by gap junctions, important for direct intercellular exchange of small molecules [52–54] and have pseudopodia that appear to make direct contact with the basement membrane [51]. Cadherin-based (adherens) junctions anchor together neighbouring cells [4, 50, 55]. The dermal papilla of wool follicles also contains mast cells that produce plasma-membrane bound granules [6, 56].

10 Development of Hair Fibres

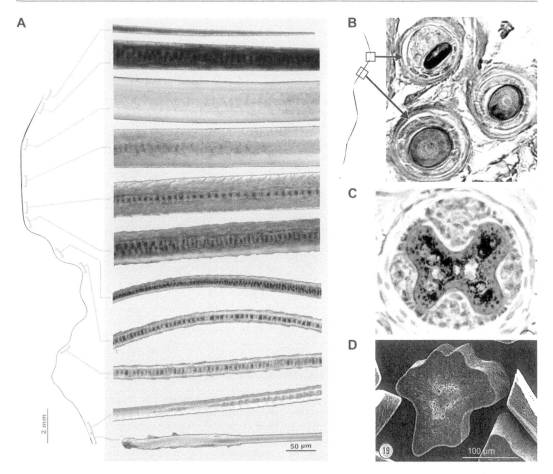

Fig. 10.3 Examples of along-fibre changes to hair morphology during anagen and examples of non-circular fibre cross-sections. (**a**) Australian Brush-tailed Possum guard hair illustrating changes to diameter, pigmentation, medullation pattern and cuticle structure. (**b**) Rat under hair, showing zig-zag regular morphology. Corresponding cross sections in mid follicle (Zone D/E) have a narrowing of hair between zig and zag, corresponding with an increase in Huxley layer cells. (**c**) More complex arrangement around a developing rabbit guard hair. Despite the "x" profile of the hair, the profile of the follicle remains circular due to changes in Huxley layer cells. (**d**) Cross section of whisker from the centre of a human moustache, showing irregular shape or "steak-boning". ((**b**) and (**c**) reprinted from Biology of the Skin and Hair Growth, 1965, Priestley and Rudall [45] and Straile [16], respectively (**d**) reprinted from Scanning Microscopy, 1990, Hess et al. [47])

The role these cells play, if any in hair growth is unclear. Surrounding all dermal papilla cells is an extracellular matrix (ECM) that is rich in basement membrane proteins and proteoglycans [9]. Proteoglycans are believed to have a major role in chemical signalling between cells, by binding and regulating the activities of secreted proteins.

All these cell membrane specializations infer a significant level of cell-to-cell organization and communication within the dermal papilla. During anagen stage IV, dermal papilla cells are organ-ised and functionally specialised, but that activity is not proliferative [21, 57].

10.1.5 Defining the Extent of Zone A

The extent of the proliferative region (Zone A) forms a rough cone or hemispheric shape centred on the dermal papilla. On longitudinal micrographs of follicles, the boundary is often drawn as a simple line called the Critical Level,

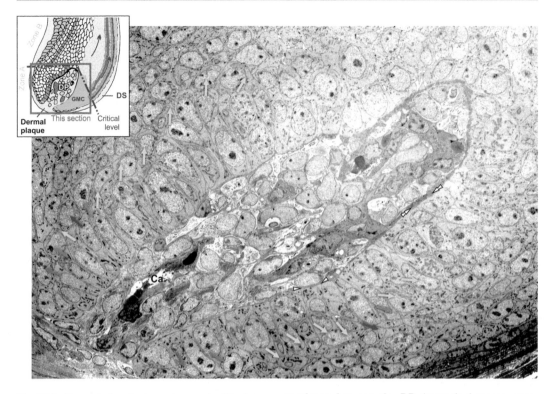

Fig. 10.4 Transmission electron micrograph of the lower wool follicle bulb centred on the dermal papilla (DP), showing its heterogeneous cellular composition. The neck of the DP and its connection with the dermal sheath is at the lower left of the micrograph, the DP apex is near the upper right corner. Stem cells are attached to the basement membrane between the DP (example between arrow heads). The basement membrane is thicker at the distal end of the DP (example between double arrowheads). This follicle is not melanised and does not produce a medulla. Coloured arrows indicate probable cell lines

or Auber's Line after histological work on medullated wool [15] in which the concept was first introduced. The location of the line with respect to the dermal papilla and its angle varies with the type of follicle. This line intersects the widest extent of the dermal papilla in sheep and human scalp follicles which produce a medullated fibre. In fibres without medulla, the line is generally drawn from the top of the dermal papilla. However, differentiation in the outer layers of the growing follicle (the Companion layer, and IRS) begins earlier than in the centrally located cell lines (e.g., cortex). Therefore the line is typically angled toward the follicle base [58], forming a kind of roughly defined conical cap [29].

The extent of the pre-differentiation region can be even better defined using gene expression methods. A clear marker for the germative cells is a total lack of any keratin expression [25, 26].

Here, we define Zone A as the region in which cells, often mitotically active, have not yet begun to differentiate into their cell-line specific phenotypes (based on shape, cell organelle organization or protein expression).

10.2 Melanocytes and Hair Colouration

In all mammal species examined, pigment-based hair colour originates from melanocyte cells interspersing the Zone B stem cells along the basement membrane around the upper part of the dermal papilla [59–62]. These cells resemble melanocytes found along the skin epithelium basement membrane, and within them melanin is produced and packaged into specialized melanosome organelles [63, 64] in a process that has close links to that of lysosome development [65].

Two types of melanin are produced, yellow pheomelanin and black eumelanin, and the resultant melanosomes differ slightly in morphology, eumelanosomes being elliptical and pheomelanisomes being variable, often spherical, and containing a vesiculoglobular matrix. In the mouse pelage follicle bulb, melanocytes are only active during anagen [66], and a melanocyte can switch which pigment is produced, but not produce more than one type of melanin at one time. Although melanin switching, and most aspects of the regulation of melanogenisis, has been primarily studied in mice, the general conclusions are probably applicable to all hair. Melanin production and switching appears to be regulated by signals from the dermal papilla in a process involving a host of genes, transcription factors and other small molecules [60]. For example, in mice, β-catenin in the dermal papilla appears to be a key molecule involved in melanin type switching [67]. However, β-catenin also clearly illustrates the difficulties in untangling signalling mechanisms in the dermal papilla because this molecule is also intimately involved in multiple pathways including those regulating keratinocyte activity during anagen, possibly regulating catagen onset and is essential for enabling a follicle to regenerate from stem cells in early anagen [68].

While melanin type is one factor that affects hair colour, the distribution of pigment granules within the hair shaft is also important. Extending from each melanocyte are numerous dendrites that, tentacle-like, reach out between the upwardly moving streams of cortex cells (Fig. 10.5). Mature melanosomes are transported to the dendrite extremities along a microtubule cytoskeleton [69] where they are transferred to target keratinocytes by a mechanism that probably involves partial phagocytosis of dendrite tips [70]. The analogy with tentacles continues beyond mere shape, because melanocyte dendrites are not static structures, but, assuming they behave in a similar way to melanocytes in the epidermis, move from cortex-cell to cortex cell, injecting melanosomes into different keratinocytes [71, 72]. Therefore, in addition to controlling melanin type and production, regulation of melanocyte dendricity may be an important determinant of hair colour.

How melanin is distributed within fibres differs between hairs of different species. Many species, for example rabbits, have guard hairs with horizontal stripes and have elaborate medullae which are decorated with melanin (Fig. 10.3); dendricity must be a partial controller of patterning [73, 74].

10.3 Development of the Cortex

10.3.1 Extensive Cell Reshaping and Repositioning Occur in Zone B

Hairs grow at a rate which is determined by the summation of a variety of factors. One of these will be the proliferative rate of the germative matrix from stem cells and TACs dividing. However, another major influence on growth rate is fibre extension which is mediated by cell reshaping cell migration, and this occurs within the upper follicle bulb. Wool fibre growth rate has been studied under various conditions by measuring growth period of weeks following clipping, or more accurately in single fibres from movement of radioactive cysteine (S^{35}) over a period of hours or days [75]. There is breed based variation in average growth rate, and also the capacity of follicles to maintain a constant rate in the face of environmental perturbation. Corredale sheep vibrissae hairs grow at about 500 micrometres per day, while merino wool growth is relatively less, but more robust in the face of environmental changes, falling in the range of 300–400 μm/day [76–78].

Much of this growth is powered by fibre extension of the cortex, which much like the other cell lines, undergoes a dramatic transformation as it moves through Zone B. This begins for the cortex just below the widest part of the bulb and ends at the top of the bulb, where Henle's layer hardens. Changes to the cortex cells during this transformation are similar in anagen follicles from all species so far examined. Cells change in shape from approximately spherical, to highly elongated and aligned with the long axis of the follicle (Fig. 10.6). Associated with this shape change

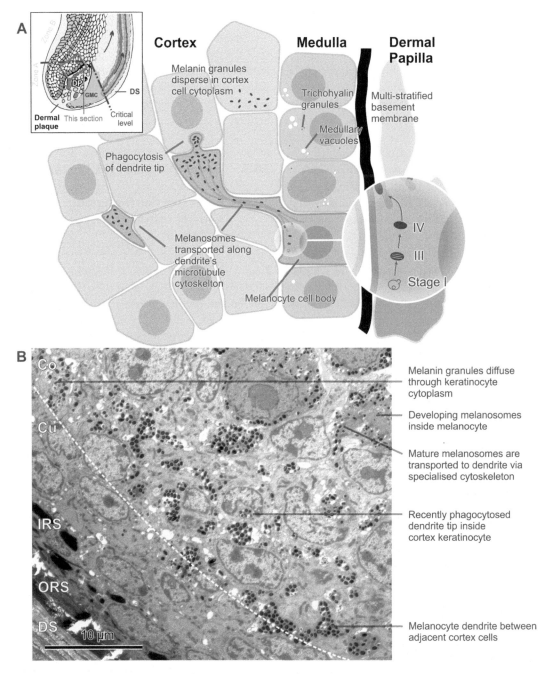

Fig. 10.5 Melanogenesis. (**a**) Summary of the process by which melanin granules are produced within melanocyte bodies attached to the dermal papilla basement membrane and are transported via tentacle-like dendrites to be injected into cortex cells that are streaming past. (**b**) Melanogenesis and melanin distribution in very low Zone B, adjacent to the dermal papilla, of a coloured wool follicle. Upper left corner of image is distal. *DS* Dermal sheath, *ORS* Outer-root sheath, *IRS* Inner root sheath, *Cu* Cuticle

is a decrease in the number of intercellular spaces, cell membranes become more closely apposed, and increases in numbers of cell junction complexes indicate increased cell adhesion and intercellular communication [3, 52, 56, 79].

10 Development of Hair Fibres

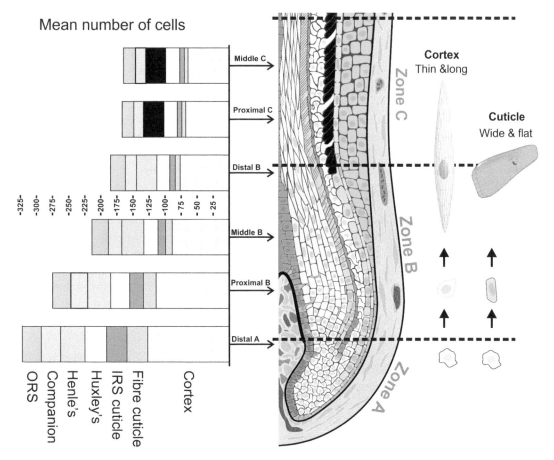

Fig. 10.6 Changes in cell number during early differentiation in wool follicles counted from transverse sections using transmission electron microscopy. Data were tabulated from a study by Orwin and Woods [6] plotted against a schematic of the lower follicle and cell shape changes. Stacked plot illustrates the changes in numbers of follicle cells per zone. IRS, fibre cuticle and cortex show dramatic reductions. Changes in cuticle cell number are probably mostly a result of cell morphology changes and some cell slippage, while cortex cell changes are likely caused by considerable cell slippage

Cell morphology changes and relative cell migration in Zone B, taken together, are probably one of the key determinants of final fibre diameter and shape, including curvature [80]. In addition to changes in individual cell morphology (from round to highly elongate or spicule-like), the number of cells (observed in wool follicle cross-sections) decreases significantly from proximal to distal Zone B [6]. This change in cell number, which parallels lesser changes in other cell lines through Zone B, appears to be a result of significant intercellular movement. It has been established in wool that these changes in cell number are not a result of cell death. The mechanisms that underpin, regulate and guide relative cell movement and cell reshaping are unknown, although are expected to involve cell junctions and the actin/myosin cytoskeleton (Fig. 10.6). Co-localized non-muscle myosin and actin are found throughout the cytoplasm of Zone B cells of rat pelage and vibrissae follicles [81], although to a lesser extent than in the developing cuticle. Changes in cell shape may also be associated with microtubules, which have been observed in Zone B wool cortex cells, and are often aligned parallel to the axis of the follicle [50, 82].

The extent of cell-to-cell movement and extension within the cortex is probably responsible for observations from radio-labelling studies that the inner root sheath appears to move more rapidly through the bulb than does the developing fibre [29, 83].

At the top of the bulb, Henle's layer undergoes sudden terminal cornification. The Henle's cells at that point are plate-like and are locked by desmosomes. After hardening, Henle's layer forms a restrictive sleeve that mechanically encases the other inner root sheath layers and the hair shaft. As cortex cells move into Zone C they appear to maintain their final shape and position [5, 6].

10.3.2 Cortical Keratin Proto-Macrofibrils Seed in Zone B

Macrofibrils of the mature hair shaft cortex are composed of a composite of trichokeratin intermediate filaments (IFs) and keratin associated proteins (KAPs), and it is in the proximal region of Zone B that the first synthesis and assembly of premature macrofibrils is observed.

From the onset of differentiation at the threshold of Zone B, cortex cells are characterised by a high concentration of cytoplasmic ribosomes, numerous mitochondria, but relatively small amounts of Golgi complexes, endoplasmic reticulum and vacuoles [3, 5]. This cytoplasmic environment suggests that protein production occurs directly into the cytoplasm. Keratin protein expression begins with Type I K35 and Type II K85, and then shortly (within a few cell lengths) is joined by Type I K31 [25]. K35 production ceases somewhere near the top of the bulb, but K31 continues.

The first ultrastructural indicator of keratin production does not occur until a few cell lengths after this onset of keratin protein expression. Visualised by transmission electron microscopy, the first signs of keratin ultrastructure are normally found in the cortex about level with the top of the dermal papilla. Intermediate filaments do not appear as a network of filaments, as is seen in the inner root sheath, but are always found as tightly aligned bundles. These bundles of IFs are attached at one end to a desmosome and are typically oriented along cell membranes, or they appear as isolated un-attached proto-macrofibrils in the cytoplasm (Fig. 10.7), this being the case in wool [84], mouse [79], and human scalp hair [3].

While desmosome-associated IFs are a well-known feature of keratinizing epidermal cells [85, 86], isolated bundles of IFs are only common in the cortex cell line.

How these initial bundles form is discussed in more detail in Chap. 11. These initial bundles are probably formed of heterodimers of K31 and K85 (K35's role is less clear) and they probably begin forming before the appearance of the first matrix-forming KAPs. We use the word "probably" because the precise relationship between the onset of expression, cytoplasmic protein concentration and the onset of ultrastructure is not well understood (Chap. 7). What we do know is that the first KAPs begin to express in the mid-upper bulb, above the level of the dermal papilla and this occurs after the first expression of keratin and probably about the same time as the first keratin structures (proto-macrofibrils) appear or shortly after this point.

What role KAPs play in the initial stages of proto-macrofibril assembly (either associated with desmosomes or free within the cytoplasm) is unclear. Some studies suggest keratin-KAP interactions are essential for seeding of proto-macrofibrils from desmosomes [88] or for free floating proto-macrofibrils [89]. However, observations of proto-macrofibrils from below the region of KAP expression and also those in which KAPs appear to be infiltrating matrix free macrofibrils (Fig. 10.8) suggests a more complex relationship [90, 91].

10.3.3 Macrofibrils Grow by Coalescing, While Many New Keratins and KAPs Appear in Zone C

After their initial appearance, proto-macrofibrils appear to grow by coalescing with neighbours (Fig. 10.9). In the upper bulb the Type I keratin K38 expresses, but does so in a patchy manner across the human cortex [25] and is associated with a subset of orthocortex cells on one side of the cortex [92], meaning it may be associated with a particular type of cell-type/macrofibril architecture.

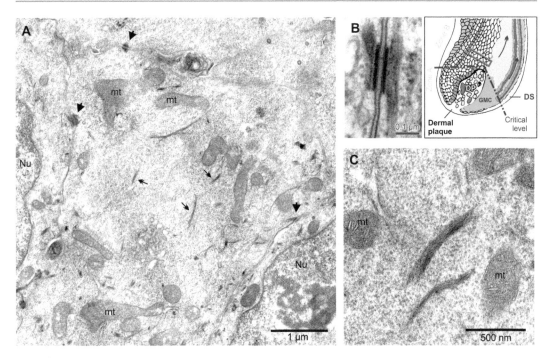

Fig. 10.7 The first signs of keratin ultrastructure in the cortex (approximate location inset top right). (**a**) Transmission electron micrograph of cytoplasm of a cortex cell just above the top of the dermal papilla in a wool follicle. The two forms of initial keratin are indicated by arrows – single head arrows for desmosomes and smaller double-head arrows for cytoplasmic bundles (or tactoids). Nu and mt indicate nuclei and mitochondria respectively. (**b**) A desmosome with attached keratin bundle. (**c**) Examples of two keratin bundles spontaneously formed in the cytoplasm. ((**b**) reprinted from J. Text. Eng., 2016, McKinnon et al. [87], with permission from The Textile Machinery Society of Japan)

The top of the bulb where Zone B ends, and the hardening of Henle's layer marks the start of Zone C, appears to be a significant threshold, not only for cell reshaping, but also for the diversity of keratins and KAPs that are expressed, probably by post-transcriptional gene expression [95] because the cortex cell nuclei are by this point non-functional [96, 97], discussed further below. It has also been suggested, but not extensively verified, that macrofibril nucleation appears to cease early in Zone C and continued macrofibril growth, which extends into Zone D, appears to be entirely a result of increases in the size of existing macrofibrils and coalescence of apposing macrofibrils [98].

Around the region of the Zone B/C transition, in human hair [25, 26] and wool [92] there is a flush of new keratins expressed (Fig. 10.10). In human scalp and beard hair this starts with expression of proteins K33a, K33b, K36, K37 (in vellus hairs only), K81, K83 and K86, with K34 being triggered slightly after the others. The expression of all of these proteins continues throughout Zone C and into Zone D.

It is not only new keratins that appear in Zone C, but also a plethora of new KAPs. In human hair this makes up to 19 high sulfur (HSP), 24 ultra-high sulfur (UHSP), and 14 high-glycine tyrosine (HGTP) KAPs in the cortex (22, 34, and 9 respectively in sheep). This increase in new KAP species starts near the beginning of Zone C with the expression of proteins from 8 new KAP families [100] (Fig. 10.10). These include HSP KAP families 1, 2, and 3 and HGTP family 6, as well as additional proteins from the HGTP 19 family not expressed in the earlier wave of KAPs. There are also UHSP proteins from families KAP4 and 9. These new proteins join the KAPs that continue their expression from the first wave that started in Zone B.

It is presumed that these mixtures of KAPs are the molecular components that continue to

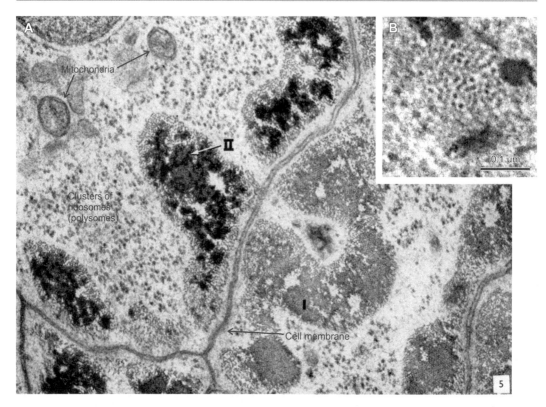

Fig. 10.8 (a) Transmission electron micrographs of cortex cell from Zone C of wool follicle showing infiltration of growing macrofibrils by matrix material (KAPs). Label I and II indicate macrofibrils in adjacent cells in which matrix material appears to have different levels of electron density from heavy metal staining. Gaps within the macrofibril appear to be devoid of cytoplasmic components (e.g., ribosomes). (b) Example of a cross-section of an early forming keratin bundle from proximal Zone B, showing filaments with spaces. ((a) reprinted from J. Ultrastruct. Res., 1971, Chapman and Gemmell [90], with permission from Elsevier; (b) reprinted from J. Text. Eng. McKinnon, Harland [87], permission from The Textile Machinery Society of Japan)

infiltrate the growing macrofibrils to form the inter-filament matrix material, a process which continues throughout Zone C. The key message is that with all these new keratins and KAPs, there are a large number of possible protein-protein associations that might occur between keratins and keratins, keratins and KAPs, and KAPs and KAPs.

10.3.4 Macrofibril Keratinization Is an Extended Process of Intermediate Filament Reshaping and Disulfide Bond Formation

The cornification of the cortex, or keratinization process as it is typically referred, occurs in several stages which begin just distal of the bulb (mid Zone C) and are completed within Zone E (just above the Keratogenous region). This gradual change is in contrast to typical epithelial cornification, which is seen in the inner root sheath cell lines and which involves a relatively sudden transition involving primarily enzymatically-mediated ε-(N-γ-glutamyl)-lysine bonds between loosely networked IFs and keratin-associated proteins (KAPs) such as loricrin or trichohyalin [85].

Outward signs of the keratinization process at the ultrastructural level in Zone C and D are subtle, but it has long been known that there is a change in the robustness of the intermediate filament (IF) alignment just distal of the top of the bulb. Small angle x-ray diffraction patterns of the upper bulb show a distinct pattern related to inter-

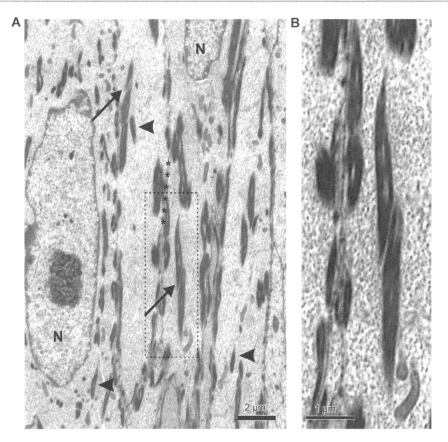

Fig. 10.9 Fusion/coalescence of developing macrofibrils is a feature though Zones B, Zone C and Zone D of the wool follicle. (**a**) This image is a longitudinal section from proximal Zone C of a fine wool follicle. Arrows indicate fusing macrofibrils being deformed at their interface. Arrowheads indicate well-formed proto-macrofibrils of a shape expected of paracortex-type architecture. N = nuclei and * = example of a plasma membrane between cells to which developing macrofibrils are attached or associated. (**b**) is an enlargement of the area in a dashed box. (Reprinted from J. Struct. Biol., 2011, McKinnon and Harland [93], itself modifed from J. Ultrastruct. Res., 1982, Woods and Orwin [94], with permission from Elsevier)

mediate filament and microfibril development. When subjected to high temperatures the pattern disappears, but this is not the case just distal of the bulb in the mid Keratogenous region [103, 104] (See Chap. 8, Fig. 8.1). Recent studies have used x-ray microbeam diffraction methods which allows more precise localisation of changes and have indicated that loose association of IFs observed in the upper bulb and just distal of the bulb (low Zone C) becomes successively tighter and more crystalline as Zone E is approached [105]. Studies using x-ray microbeam diffraction divide this region into overlapping zones, suggesting a mixed population of filaments in multiple states (Fig. 10.11). In the third zone (400–800 μm from the follicle base in human scalp hairs) IFs are thought to aggregate laterally into a tighter conformation, or a hexagonal paracrystalline array. This corresponds to a rapid increase in material hardness as measured by atomic force microscopy [106], and is about the same region where changes in filament appearance using transmission electron microscopy, and proteomic changes are first observed [91].

These changes are mediated by three intertwined processes. First there is the accumulation of oriented IFs surrounded by a matrix of KAPs within macrofibrils, which leads to large scale alignment of IFs within the fibre cortex, beginning in the mid bulb (see Chap. 11). Secondly

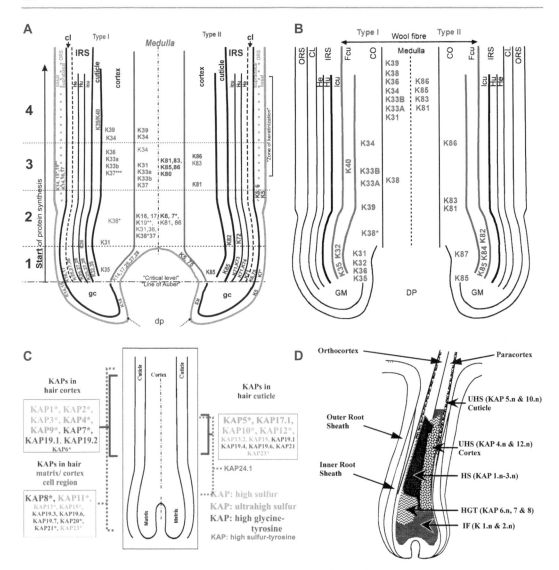

Fig. 10.10 Schematic summaries of expression order for keratins in (**a**) human beard hair and (**b**) medullated wool. Similarly, summaries of KAP expression in (**c**) human and (**d**) sheep. Abbreviations: *cl* Companion layer, *He* Henle's layer, *Hu* Huxley's layer, *Icu* IRS cuticle, *Fcu* fibre cuticle, *CO* Cortex; *gc* or *GM* Germative compartment/matrix, *dp* Dermal papilla. ((**a**) reprinted from J. Invest. Derm., 2010, Langbein et al. [26], with permission from Elsevier; (**b**) reprinted from Exp. Derm., 2011, Yu et al. [99] with permission from Blackwell Publishing; (**c**) reprinted from Int. Rev. Cytol., 2006, Rogers et al. [100], with permission from Elsevier, to include data from Rogers et al. [101]; (**d**) reprinted from Annals of the New York Academy of Sciences, 1991, Powell et al. [102], with permission from Blackwell Publishing)

there is a significant reorganization of internal IF structure including a radial contraction from Ca. 10 to Ca. 7.5 nm diameter, this being the final axial architecture of the filament (see Chap. 6). This second process is powered by the formation of disulfide bonds between keratin dimers and tetramers within the IF, and the third process is also a process of disulfide bond formation between IF proteins and KAPs and within and between KAPs. This latter process, like many aspects of KAP function is not well understood.

At the beginning of Zone E, which is generally marked by a pattern known as Adamson's fringe in light microscopy of longitudinally cut human

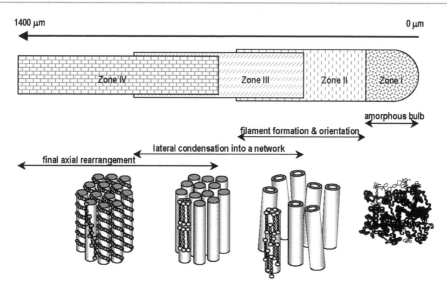

Fig. 10.11 Summary of changes in IF network based on analysis of x-ray data from human hair follicles by Rafik and colleagues. How these zones based on diffraction pattern changes relate to ultrastructural differences has not be clearly established, but Zone III and IV probably correspond to Zones D and E. A change in IF architecture from hollow to filled defines Zone III and final changes hypothesized to relate to KAP cross-link formation defines Zone IV. (Reprinted from J. Struct. Biol., 2006, Rafik et al. [105], with permission from Elsevier)

scalp hair follicles, there is a dramatic change in the appearance of the cortex along with the cuticle and inner root sheath. This is the finalisation of fibre cornification, and at the beginning of Zone E the cortical macrofibrils are at their maximum size and the surrounding cytoplasm has been degraded to the point that by mid-Zone E few recognizable components remain [5, 107–109]. The terminal hardening process which is the finale of keratinization is still not fully understood, but during the course of its movement through Zone E, macrofibrils become closely apposed, a term referred to by some authors as "consolidation". In wool, this results in an approximate 25% reduction in diameter which is also a process of dehydration [15, 90, 98].

The remaining cytoplasm and nuclear material is concentrated in between the macrofibrils. If and how this remnant material is modified (for example by crosslinking enzymes like transglutaminase) is not entirely resolved. It is possible that chemical crosslinking enzymes may play a role, being present in the cortex [110]. Nevertheless, these remnants take on several forms which are largely dependent on the ability of local macrofibrils to fuse laterally into extended masses. In cells containing macrofibrils in which lateral fusion between neighbours is not possible due to incompatible angles of peripheral IFs [89, 93, 111] (See Chap. 11), some of the remaining cytoplasm remains trapped between adjacent macrofibrils in the form of intermacrofibrillar material. The remaining cytoplasm, and typically most of the cytoplasm of cells in which macrofibrils have undergone large-scale lateral fusion, is forced into cytoplasmic remnants that have a stellate shape in cross-section.

10.3.5 Cell Junctions Are Critical for Cell Reshaping in the Upper Bulb

Desmosomes are predominantly found in tissues that are subjected to strong physical stress, such as the epidermis, and are linked intracellularly via plaque proteins (desmoplakins and plakoglobin) to bundles of intermediate filaments [112–114]. They are well documented in wool and human hair follicle cells both ultrastructurally and cytochemically [79, 84, 90, 98, 115–117]. As cortex cells move through Zone B in wool follicles, the number of desmosomes per cortex cell perimeter increases, then markedly decreases at the Zone B to Zone C transition [84]. Desmosomes in epithelial cells are associated with changes in cell shape and cell-type specific positioning [118] and it is interesting that the increase in desmosome numbers in the cortex coincides with what is probably the highest levels of cell inter-positioning and cell morphological change. Throughout Zone B, three desmosome proteins, desmoglein 2, 3 and 4, are expressed. One of them, desmoglein 4, a transmembrane desmosomal cadherin, begins expressing in Zone B and is specific to follicles and the more differentiated layers of the epidermis, being expressed in the stratum granulosum. The importance of desmosomes in the developing cortex is highlighted by knockout mutants of desmoglein 4 (human, mouse and rat) which experience hypotrichosis [117]. While the intermediate filament anchoring function of desmosomes is normally considered within the context of a cytoskeleton, in the hair cortex their role may be expanded to also contributing to the development of trichokeratin intermediate filament bundles (macrofibrils) (Fig. 10.7).

Gap junctions allow intercellular communication via electronic signal transmission and the interchange of solutes and small molecules [119]. The transmembrane proteins of gap junctions are connexins, six of which form a channel termed a connexon. When the connexons of two adjacent cells are aligned, they form open channels that allow the passage of ions and small particles (up to approximately 1200 Da in size) between the cells. The permeability of gap junctions can be rapidly controlled by changes in pH, and Ca^{2+} concentration and by extracellular signals. Gap junction numbers increase significantly in wool cortex cells after Zone A [52], and are also a feature of human follicles [120]. Orwin, Thomson and Flower estimated the percentage area of the cortex cell surface covered by gap junctions, and found the maximum in Zone B (6.89 ± 1.4%, c.f. 4.45 ± 1.4% in Zone A and 3.99 ± in Zone C). Transfer of low molecular weight cell signalling substances via gap junctions in the Zone B cortex appears to be critically important for coordinating and synchronizing the activities of these cells.

The continuity of this network has been investigated using fluorescent dyes [121] and autoradiographically using S-35 labelled cysteine. Radiolabelled cysteine was incorporated first into Zone B cells and then spread rapidly upwards (~30 min in mouse and rat pelage, and ~60 min in sheep wool) into Zones C and D at a rate far in excess of cell movement [122].

Tight junction proteins in the cortex of human hair are only found at low levels [123] and are notably absent in transmission electron micrographs from wool [124].

10.3.6 Cell Junctions, Extracellular Matrix and Plasma Membranes Are Remodelled into the Cell Membrane Complex

During their movement through Zone B, cortex cells are likely to be constantly moving against one another as their individual shapes and relative positions change. In Zone C, with most of these changes complete, the inter-cellular space begins to change into what will become the permanently fixed cell membrane complex (CMC). The process by which plasma membranes, extracellular matrix, desmosomes, gap junctions and other junctions become the multi-layer CMC is not clearly understood. However, studies have revealed some of the ultrastructural and molecular changes (from cytochemical and immunohistochemical studies) associated with CMC development.

In human scalp hair follicles, the membranes widen as does the inter-cellular distance from 12–15 nm in Zone B to 20–30 nm in Zone C [3, 108]. Associated with this is a change in the nature of junctions between cells in Zone C and above. In wool follicles an observed reduction in desmosome numbers in upper Zone B [84] appears to correlate with a drop off in Desmoglein 2 and 3 [117]. The remaining Desmoglein 4 protein is, however, still strongly detected in Zone C and, in human follicles, is spatially associated with the intracellular desmosome plaque protein desmoplakin [117]. Curiously, in wool follicles, freeze-etch and lanthanum stain methods have failed to show desmosomes in Zone C and above. The same study showed that gap junctions are still present in Zone C of wool although at about half the density seen in Zone B [52].

Studies of Romney wool follicles indicate that changes in extracellular matrix composition also occur during Zone C, indicated by gradually reducing staining intensities for transmission electron microscopy (TEM) methods sensitive to polysaccharides and glycoproteins [56, 58, 109]. Acid phosphatase activity is observed in the plasma membrane surface material of the intercellular space of cortex cells close to the transition between Zones B and C, and continues distally [58]. This is suggestive of some kind of molecular breakdown or reorganisation process.

Starting in Zone C there is an apparent interaction between the intracellular surface of the developing cell membrane complex and the IFs of adjacent macrofibrils. In wool follicles this is observed by TEM as an increasing number of short fibrils that extend from macrofibrils to the membrane surface, with this effect becoming more pronounced in Zone D (Fig. 10.12) [125]. By the distal end of the zone, the keratin has entirely filled the intracellular gap between macrofibrils and membrane. In TEM micrographs the effect is such that the trilaminar appearance of the membranes is lost and only the intercellular material is visible (typically as an electron-lucent line) [50].

Acid phosphatase activity continues to be high in the intercellular space through Zone D. Conventional electron staining remains patchy, but finely granular in the intercellular space indicating that if protein is present, it is not present in large structures [5]. Despite the lack of ultrastructurally-identifiable desmosomes in wool follicles, and the now uniform nature of the membrane complex ultrastructure, fluorescently immunolabeled transmembrane protein desmoglein 4 [117] and intracellular plaque proteins (desmoplakin and plakoglobin) are still clearly present along the membranes of the cortex (Joy L. Woods, Kevin D. Hurren, Warren G. Bryson, unpublished). What role these retained, but reconfigured cell junction proteins may play, and where precisely they locate within the fully formed CMC is unclear.

The CMC finalizes its transformation in Zone E to become what is the only continuous structure that runs the entire length and breadth of each hair. By the beginning of Zone E the cortex CMC of wool is devoid of ultrastructurally-identifiable cell junctions [5, 52, 84] and adjacent macrofibrillar material has become fused with the intracellular surface of the CMC. A measurement-based study (n = 50–250 cells per cell type per zone) by Orwin and Thomson found that, unlike in human scalp follicles (see above), the cortex CMC in wool follicles maintains a similar average width throughout Zones B, C and D (~26.5 nm), but in Zone E the width drops significantly (~14 nm) and much of this is due to a reduction in the intercellular gap from 11.7 down to 6.6 nm [125]. Associated with this change is a transformation of the appearance of the membranes of apposing cells and the condensation of an intercellular band that stains darkly using conventional TEM stains [3, 125].

The observations of development of CMC correspond well to studies carried out in fully keratinized fibres to try and establish CMC composition and structure (see Chap. 1, Fig. 1.4) [126–128].

10.3.7 Nuclear and Other Cytoplasmic Changes

The transformation of the cortex from biologically active transient amplifying cells in Zone A, to elongated husks filled with chemically cross-linked keratin and connected to each other by a condensed cell membrane complex is accompanied by significant changes to the remaining cellular machinery. These changes are complex, can easily be mistaken for apoptosis due to a large overlap in the molecular mechanisms involved, and have a variety of functional roles in fibre development that are not well understood.

It has been long known that the nuclei of cortical and other fibre and IRS cell lines are degraded during the process of fibre development (see Chap. 8, Fig. 8.1), and that nuclear DNA becomes degraded with respect to various microscopy stains distal of the bulb region [5, 82, 104, 129–131]. In preserved samples, this degradation in DNA staining is typically observed only from Zone C and is complete before full

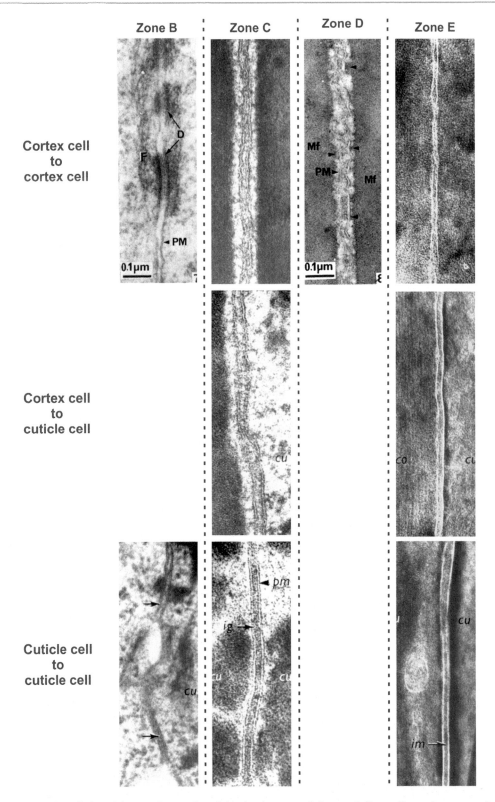

Fig. 10.12 Transmission electron micrographs of the development of the wool fibre cell membrane complexes. Modified after original plates by D. F. G. Orwin.

keratinization in Zone E [97]. However, a recent study in which live tissue imaging was used [96], suggests that changes to nuclear function occur before the top of the bulb (within Zone B) and this matches up with changes in nuclear shape which occur in the distal bulb as part of the major cell reshaping in the cortex [97], and with the observation that single-strand DNA damage is clearly detected in embryonic rat hair shafts during Zone B just distal of the dermal papilla [51]. During the course of development the cortex cell nuclei change from circular to highly elongated, sometimes developing nuclear pockets [82]. Nuclear pockets are common in the cortex and the outer-root sheath of wool follicles in the keratogenous region (Zone C and distal), but can occur from proximal Zone B, adjacent to the germ cells. The situation is not shared by the other cell lines.

Irrespective of the precise location (Zone B or C) where the nuclear DNA becomes incapable of gene expression, we do know that this point occurs proximal of the first appearance of some keratin and KAP species within Zone C (see section on keratin structures above). Presumably a form of post-transcriptional gene regulation is involved, but this is an area that has hitherto received little attention. It is also worth noting that although the same pattern of nuclear degradation occurs in the cuticle as the cortex [51], the process onset is later and in Zone B cuticle nuclei are surrounded by a peri-nuclear basket of unknown filaments [94, 132].

The process of nuclear degradation itself appears to differ in nature between the cortex and other cell lines of the follicle, in particular the IRS, which has a pattern of nuclear degradation which resembles that observed in the developing *stratum corneum* [97]. Many other organelle structures disappear between Zones B and E, and there is an increase in acid phosphatase activity in wool follicles. Acid phosphatase activity is associated with degradation of cell components and correlates with increases in observed lysosome numbers [58]. Within Zone D and into Zone E, at the distal parts of the keratogenous region of the follicle, the emergent picture in the cortex follicles from all species is one of a general degradation of non-essential cytoplasmic and nuclear components, effectively the living machinery of the cell and its protein production mechanisms. It has been hypothesized [5] that this may be the basis of a diameter decrease of around 25% seen in the wool cortex during Zone D [15]. Such a diameter change would presumably come about by compacting and de-watering of bulky organelles, and therefore the proportion of the fibre cross-section that is composed of keratin macrofibrils would increase. The degraded remains of nucleus and cytoplasmic constituents collect together into remnants that sit between the macrofibrils within every cortex cell. These cytoplasmic remnants tend to fill the centre of cells (see Chap. 1, Fig. 1.5).

The functional importance of this controlled degradation of structures in the cortex cells is indicated by a study of a knockout mouse lacking the non-apoptotic keratinocyte-specific endonuclease DNase1-like 2 [133]. Fischer and colleagues established that when absent, unusually large and atypical cytoplasmic remnants occur within hairs, which they suggest (based on immunohistochemistry) interferes with the ability of some keratins (e.g., K31) to integrate into cortex cells. The resulting hair was more easily broken up in a bead mill, thus suggesting a link between the degradation process and hair structural integrity.

Importantly this degradation is not apoptosis. While the apoptosis of a cell has a clear result in the complete destruction of a cell, what happens in follicle cells just prior to hardening is more restrained because some cellular constituents are degraded but other essential materials (e.g., variously assembled KAPs, keratins and cell membrane components) are entirely spared. The process uses many of the same biochemical tools involved in apoptosis.

Degradation of organelles is also by no means complete. The spaces between macrofibrils in the cortex cells of fully hardened mature fibres of all species contain mitochondrial DNA which is intact enough to form the basis of genetic analyses used in forensics and archaeology [134]. Parts of, or entire organelles are encountered within cytoplasmic remnants [109] (Fig. 10.13).

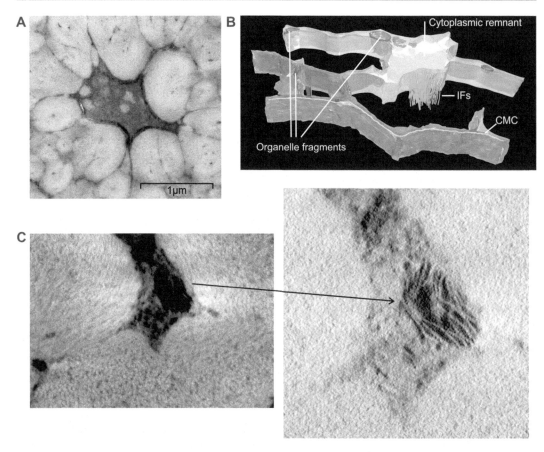

Fig. 10.13 Structures within cytoplasmic remnants of mature hairs. (**a**) Transmission electron micrograph of wool cytoplasmic remnant with unidentified cell components in centre and surrounded by what may be a degraded membrane structure. (**b**) Rendering of three-dimensional model reconstructed from electron tomography data of a cytoplasmic remnant in wool with a boundary that links it to nearby CMC and which contains membrane bound fragments of unknown origin. In this figure, except one example, macrofibrils are not shown. (**c**) Electron tomography image from human scalp hair, averaged across many sections illustrating heterogeneity of cytoplasmic remnant. Detail of one slice of the data (right) shows internal structure of a large inclusion, which resembles a partially degraded Golgi complex

10.3.8 Mitochondrial Degradation Is Peculiar in the Cortex

Mitochondria, in particular, appear to degrade in an unexpected way if the aim of degradation is simply to crush them up to save space. Changes start very early with a sudden drop in mitochondrial membrane potential ($\Delta\Psi$), as measured by live-cell dyes for mitochondrial function, in living bovine and human scalp follicles occurring somewhere in the upper bulb [135]. Understanding these early changes is also at an early stage. Most observations are from the later stages of the process, distal of the bulb (Zones C–E) in human, rat or wool follicles which clearly identify by TEM, changes to the mitochondrial membrane [5, 51], confirming that it is unlikely that they have normal function.

One consideration, especially given the early changes in mitochondria during Zone B, is cytoplasmic metabolism. A potential proxy for the biochemical activity of a cell might be the increase or decrease of coated vesicles as an indication of protein production versus use as differentiation progresses. In the cortex of wool follicles, TEM counting surveys suggest that there is an increase in free floating and membrane associated vesicles between Zones A and B which then reduces in C and D. This is mirrored by the prevalence of Golgi complexes [136].

Deactivating mitochondria almost certainly places limitation on what kind of processes can happen, so why do it early? If there is a functional explanation, it is one we don't yet fully understand, however, two hypotheses have been put forward. Early deactivation of mitochondria may play a role in preventing full-on apoptosis by removing mitochondrial proteins (Bcl-2 family proteins, cytochrome c and others) that normally play an important role in the apoptotic cascades [137]. Another is based on mitochondria being a potential source of reactive oxygen species as a waste product of their normal metabolic function, and that the controlled degradation of mitochondria in the cortex and cuticle are to control the release of these toxic waste products, potentially with implications for disulfide bond formation [97].

10.4 Development of the Medulla

10.4.1 Early Differentiating Medulla Cells Migrate Up the Dermal Papilla

Like the cortex, cuticle, IRS and Companion Layer, the medulla originates from stem cells aligned along the basement membrane of the proximal half of the dermal papilla. At approximately the widest point of the dermal papilla there is a change in the nature of the basement membrane [6] and it is likely that this point represents the end of the line of sequential niches within the dermal papilla that set the fate of stem cell offspring [30]. It appears that cells that divide from the medulla-specific stem cells migrate up dermal papilla membrane while starting to differentiate and then merge seamlessly into a medulla column (Fig. 10.14). This seems to be a common feature observed in wool [15], mouse and rat pelage [51, 138], and human scalp and beard hair [25, 26], and are sometime referred to as the pre-medulla [26].

Pre-medulla cells differ from the stem cells that are attached to the more proximal basement membrane because they lack both tight junctions between adjacent cells [123] and hemidesmosomes connecting cells to the basement membrane [6]. Keratin expression also begins in the pre-medulla. In human beard follicles this begins with a diverse group of epithelial keratins (K5, K7, K14, K25, K27, K28 and K75) [26]. This pattern of early keratin expression demonstrates two key differences between the medulla and other cell compartments: no other follicle cell line produces such a large diversity of keratins from the onset, and no other contains the range of different keratin types (normally associated with follicle and non-follicle epidermal tissues).

10.4.2 The Keratin Lottery Approach to Medulla Development

Above the apex of the dermal papilla, medulla cells are easily differentiated from the adjacent cortex cells because they are larger and cuboid, in small medullae, or more laterally flattened than the cortex cells in larger medullae [15]. The pre-medulla keratins listed above continue their expression and are joined in the distal half of the bulb by another wave of epithelial keratins, K16, K17 and K6, and six hair keratins, K31, K36, K37, K38, K81 and K86. It is a proteomically, unique pattern found only in the medulla; we now have a mixture of 16 different keratins within the medulla in the distal bulb and above the bulb, in Zone C, there is the addition of more hair keratin species and one more epithelial keratin [26].

The key to understanding the strange proteome of the medulla is that any individual medulla cell will express only a small subset of these many keratins, but which keratins are expressed appears to be random. In all other cell lines of the growing follicle keratins are expressed in compatible Type I-Type II pairs (e.g., K31 + K85 in the cortex), which form heterodimers. The random expression within individual medulla cells seems to force normally incompatible keratin species together. What significance this apparent randomness might have is unclear,

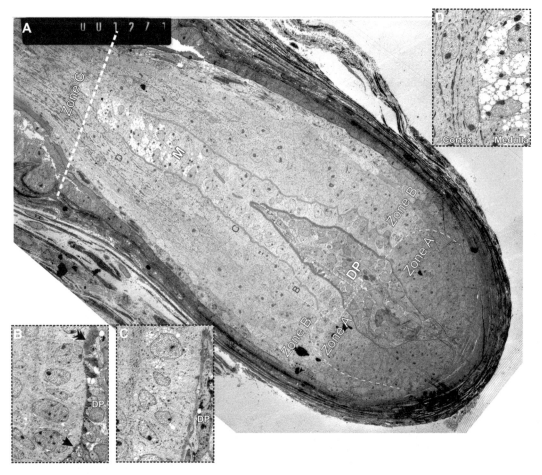

Fig. 10.14 Longitudinal transmission electron micrograph of medullated wool follicle. (**a**) Overview showing Zones and extent of medulla cells (orange lines). Dermal papilla (DP) basement membrane (BM) is shown as a thin red line in Zone A and a thicker line in Zone B. Higher magnification micrographs indicated with boxes. (**b**) Change in BM between DP and stem cells across Zone A-Zone B boundary. Zone A BM is single layer (arrow) while in Zone B it is considerably thicker and convoluted (double arrow-head). (**c**) Differentiation of medulla cells starts while cells are still attached to the basement membrane. (**d**) Towards distal region of Zone B, medulla cells are typically oriented horizontally and contain many cytoplasmic vacuoles which are either electron lucent or are electron dense trichohyalin granules. (Original micrographs courtesy of Joy L. Woods)

but one possibility is that it plays a role in disrupting the formation of forms of keratin such as intermediate filaments (as found in cortex and IRS) or layers of material such as is found in the fibre cuticle [26], and it has been observed that IF structures within these developing medulla cells are minimal.

In addition to keratins, medulla cells of all species so far investigated also produce dense amorphous granules [15, 26, 79, 139, 140]. Vesicles and vacuoles begin to form a little in advance of formation of the medullary granules [79], and these become larger and more numerous as the cells differentiate throughout Zone B. The granules are composed of trichohyalin, also found in the IRS cells [141, 142], and are high in the amino acids glutamate, arginine, glutamine and leucine, and low in cysteine [143]. *In situ* hybridization demonstrated the synthesis of the enzyme peptidylargenine deiminase in medullary granules located immediately above the dermal papilla [142]. This enzyme converts arginine to citrulline as the granules differentiate further up the follicle.

Studies which indicate the precise timing of events in medulla development in comparison with those of other cell lines is patchy in human and rodent studies, but seem to agree broadly with studies of wool follicles which conclude that by the distal end of Zone B (correlated with hardening of Henle's cells), medulla cells are typically horizontally-arranged, have an irregular cushion-like morphology and contain considerable numbers (~50% of cytoplasm) of granules and vacuoles.

In the medullae of developing scalp hair [144] and wool [5] events that began in Zone B continue above the bulb, throughout Zone C. That is, medullary vesicles continue to enlarge and sometimes merge, and trichohyalin granules continue to appear dispersed among them.

10.4.3 Some Interactions Occur Between Medulla and Cortex

Limited exocytosis of trichohyalin granules into neighbouring cortex cells has been observed from TEM observation of rats [51] and human scalp follicles [145]. While the amount of trichohyalin transferred into the cortex is assumed to be small, it is worth noting that the enzyme involved in cross-linking trichohyalin and keratin within IRS cells and trichohyalin to itself in the cornifying medulla [146–148] has also been shown to occur in the cortex [110, 149].

Around the transition from Zone B to C appears to be where "medullary-cortex" cells are first observed. These are cortex cells which invade and persist within the developing medulla. They begin in both human beard follicles [26] and rat pelage follicles [51] as a partial intrusion of cortex cells into the medullary column. The importance of noting the similarity between human and rat medullary-cortex cells is that rats (along with many mammal species) have hairs which have highly structured medullas, while human hair is typically less organised, and even possibly vestigial [144].

There are many types of medullae that appear to form essential components of hairs with specific functions [42, 150]. The simple repeating "uni-serial ladder" medulla of the pelage underfur fibres of rats and mice are the best understood in terms of their development. The ingress of medullary cortex cells in rat pelage follicles may also be associated with limited transfer of cytoplasm from medulla cells to apposing cortex cells [51]. The ingrowths of cortex cells into rat medulla cells continue to develop in Zone D to produce a distinct "groove" or "waist" around cells (Fig. 10.15a, b). Notably, both in-growth and exocytosis events occur about halfway along medulla cells, never close to where medulla cells appose. Although this may seem like an entirely academic aside with respect to human hair, similar observations have recently been made in the development of human beard hair medulla [26].

10.4.4 Terminal Differentiation of the Medulla

Hardening of medulla structures and the preparative cytoplasmic degradation has not been clearly described with respect to how it coordinates with events in other cell lines, but it seems likely that these events occur during Zones D and E. In addition to the medullary vesicles and trichohyalin granules which fill the cells during this late stage, there are increasing numbers of vacuoles (Fig. 10.15c) that are apparently derived from degenerating mitochondria, which, in mouse pelage follicles, increase in size sometimes by coalescence [79]. It is notable that the disposal of mitochondria within lysosomes in the medulla cells differs from the more gradual disintegration observed in the cortex [97]. In the medulla cells of rat pelage the cytoplasm becomes a tightly packed menagerie of vesicles and granules that include lysosomes and some vesicles of unknown composition and function [51].

In Zone D there is a change to the cell membranes of apposed medulla cells in rodents [51, 79], which involves an increase in intercellular space.

Like terminal hardening of the IRS layers, the event in the medulla is rapid and a fairly sudden transition. The cell constituents fuse into what is often described as an amorphous mass, usually

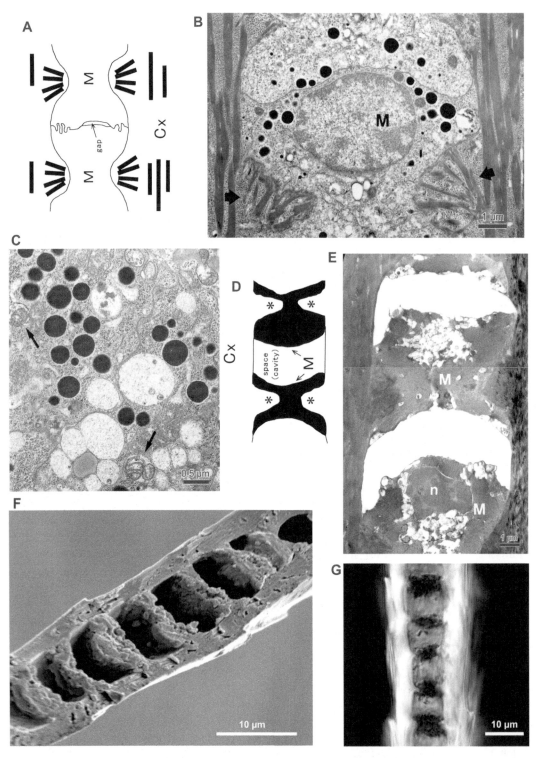

Fig. 10.15 Formation of a "waist" in rodent ladder medulla cells. (**a**) After initial formation in Zone C, the cortex ingrowths fill with macrofibrils. (**b**) Transmission electron micrograph of same region. (**c**) Transmission electron micrograph of vesicles and granules in the Zone D rat medulla. Black trichohyalin granules are common, as are lysosomes. Arrows indicate mitochondria being degraded in lysosomes. (**d**) Schematic and (**e**) transmission electron micrograph of Zone E in which cortex and medulla contents harden, the medulla splits along the cell membranes and the cell contents collapse into the septa forming a series of regular chambers. (**f**) and (**g**). uniserial ladder medullae in keratinized under fur fibres of Australian brush-tailed possum, scanning electron and phase light micrograph (showing non-polarising melanin granules). ((**a–e**) reprinted from Hair Follicle, Differentiation Under The Electron Microscope, 2005, Morioka [51], with permission from Springer)

around the periphery of the cells [15, 79, 140]. However, TEM studies suggest that the mass is not entirely amorphous and can include an apparently randomly organized filamentous texture [108].

Isopeptide bonds, as established by the presence of ε-(γ-glutamyl) lysine bonded fragments following digestion, are catalysed by transglutaminase [146–148, 151]. In rat follicles, the nucleus undergoes modification, with chromatin first aggregating and then diffusing, with the result an amorphous mass [51]. Finally, the nucleus and all remaining cytoplasm remnants condense leaving an air space. The process has not been well studied at a cellular level in human hair or wool, being generally described as a "collapse" or similar wording that describes a relatively disorderly event. However, in rat pelage hairs [51], and those of many other mammal species [150], the medulla is organized into a ladder-like series of regular chambers that arises from events that occurred in Zones C and D. The increase in intracellular space between apposed medulla cells in Zone D may now function to create a splitting point in Zone E, where in rodent ladder medullae the split occurs at the widest points in the medulla (Fig. 10.15d, e). The remains of the cells (degraded nucleus and cytoplasm reinforced with isopeptide bonds), often filled with melanin granules, then condense around the groove, forming a periodically reinforced air space structure reminiscent of bamboo (Fig. 10.15f, g).

10.5 Development of the Cuticle and IRS Cuticle

10.5.1 Cuticle Layers and Scale Edges Are Defined by Cell Shaping in the Bulb

The first clear ultrastructural sign of differentiation of both IRS cuticle and fibre cuticle cells occur as a change in shape, in proximal Zone B, before cells are level with the top of the dermal papilla. For fibre cuticle cells, Zone B is a region of significant cell shape change and cell reorganization [15, 94, 144, 152]. In wool follicles [94], fibre cuticle cells become cuboid early in Zone B but then become gradually laterally elongated (i.e., around the periphery of the fibre) as they migrate up the bulb (Fig. 1.6). By the distal end of Zone B the cells are elongated axially as well as peripherally, and the apposed cuticle/cuticle membranes are tilted towards the outside of the follicle. This effect of tilting and overlapping is more pronounced in human scalp hair follicles (Fig. 10.16a) where there are multiple overlapping cuticle cell layers [144, 152]. Irrespective of the number of eventual cuticle layers that a fibre may have (~9 in human scalp hair), during Zone B the cuticle develops as a single column of cells, and it is changes in cell shape and inclination along the vertical axis of the follicle that result in either a single or multi-layered cuticle.

The cells of the IRS cuticle layer are intimately associated with the fibre cuticle cells by mid-Zone B [94]. In sheep follicles, proximal Zone B IRS cuticle cells are irregularly shaped, similar to early fibre cuticle cells, but by mid Zone B are cuboid when viewed in longitudinal section, lined up one-to-one with the adjacent fibre cuticle cells, and are very elongated peripherally. In the distal Zone B of wool, IRS cuticle cells bulge into the fibre cuticle cells and the part of the cuticle cell membrane adjacent to the apposed IRS cuticle cell membranes is shaped into a point when viewed in sections cut longitudinally along the fibre axis (Fig. 10.16b). Less attention to the fine details of IRS cuticle cell shape development has been given to human follicles, but general observations from micrographs and early studies show that the one-to-one matching of cuticle cells from fibre and IRS is not as strict as in wool [144, 152].

10.5.2 Complex Patterns of Cell Junctions and Cytoplasmic Fibril Networks Develop

Amongst the first signs of differentiation in human scalp hair cuticle cells is a "smoothing out" of the membranes between adjacent cuticle cells associated with the introduction of what

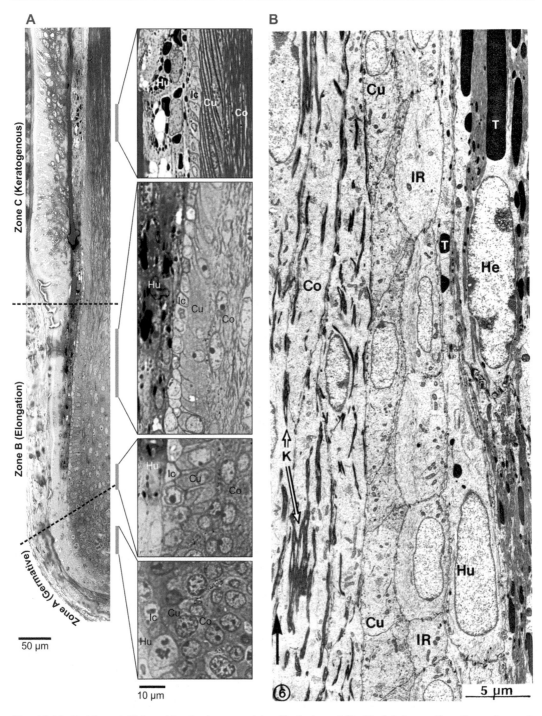

Fig. 10.16 Cuticle and IRS cuticle development. (**a**) Toludine blue stained light micrograph of thin section of human scalp hair follicle, illustrating changes in cuticle cell morphology. Hu, Huxley's layer; Ic, IRS cuticle; Cu, fibre cuticle; Co, Cortex. (**b**) Transmission electron micrograph of longitudinal section of distal Zone B of a wool follicle. Co, cortex; Cu, fibre cuticle; IR, IRS cuticle; Hu, Huxley's; and He, Henle's layers. Cu cells are flattened, connected by flattened interfaces and are beginning to tilt outwards. IR cells match cuticle cells in a one-to-one relationship. Other features: T, trichohyalin granules; K, keratin macrofibrils in cortex. ((**b**) reprinted from J. Ultrastruct. Res., 1982, Woods and Orwin [94], with permission from Elsevier)

appear in transmission electron micrographs to be adherens junctions [152]. These smooth and lengthy junctions are not found between cuticle and IRS cuticle or between cuticle and cortex.

Tight junction proteins have been weakly detected immunohistochemically in hair cuticle cells in the lower part of human scalp hairs [123], but tight junctions have not been demonstrated ultrastructurally in the fibre cuticle cells of wool follicles [124]. Taken together, these observations suggest that the proteins may not be organized into typical tight junction structures in the fibre cuticle in this zone. In contrast, β-catenin was found to be strongly expressed along the cell membranes of human scalp fibre cuticle cells at this stage [123], indicating the presence of adherens junctions, and implying actin-mediated cell shaping processes. Gap junctions were shown ultrastructurally on fibre cuticle cells in wool follicles [52], but were not common, and in particular there is a lack of gap junctions between cuticle and cortex cells.

Although desmosomes are present along the fibre cuticle membranes, as demonstrated ultrastructurally in guinea pig and sheep follicles [79, 84, 94] and immunohistochemically in human scalp follicles [123], they are much less numerous than in the adjacent cortex cells, and consequently cytoskeletal IFs are not abundant.

However, desmosomes are abundant in the IRS cuticle, and associated with these an apparent IF network develops, with a semi-random arrangement reminiscent of the network that forms in other IRS cell lines and in epidermal keratinocytes during the early stages of stratum corneum development. Work from human beard-hair follicles indicates that a mixture of Type I (K25, 27, 28) and Type II (K71 and 73) keratins are initially expressed. By distal Zone B, two additional keratins (Type I K26 and Type II K72) appear and the IF network becomes concentrated on the side of the cells that are adjacent to the fibre cuticle cells (Fig. 10.17). The nuclei, mitochondria and other cell organelles are displaced to the outer side of the cells. Small droplets of the protein trichohyalin begin to appear in the IRS cuticle cells of wool [94] and scalp hair [144], always associated with filaments [141].

A different kind of filament appears in fibre cuticle cells as they move through the bulb. These are bundles of approximately 7 nm diameter filaments of unknown type, but are not conventionally structured IFs [79, 153]. The most in-depth description of these filaments comes from wool follicles in which they are first observed as bundles free within the cytoplasm almost as soon as cuticle cells pass out of Zone A [94, 132]. Beginning with no obvious preferred orientation, these filaments quickly become lined up with the cell nucleus and may be linked together to form something like a cage, or perinuclear basket. By mid-Zone B most filament bundles are oriented with the long axis of the cell, which is perpendicular to the fibre axis. By distal Zone B the bundles have become oriented more parallel with the fibre axis, and they continue this way into Zone C where keratin begins to accumulate in close association with them.

These filament bundles are evidently intimately associated with cuticle development but their composition and function are somewhat disputed. In human hair they have been described as desmosome associated intermediate filaments [144]. In wool, some filaments are associated with rare cuticle-cell desmosomes, but it is not clear that these are the same as the free-floating filament bundles [94], and in rat cuticle it has been suggested that there are no intermediate filaments at all [51]. It has been suggested, but not confirmed, that these filaments contain actin [132]. Morioka et al. [81] provide some immunohistochemical evidence for myosin and actin in cells near the DP of rat hair follicles, but only myosin in cuticle cells higher in the follicle. Given the considerable changes in cuticle cell shape and size, and the well-documented role of the actin/myosin cytoskeleton with respect to cell shape and movement in other cell types [154], if the 7 nm filaments observed in fibre cuticle cells are actin/myosin, then they are highly likely to be contractile.

Although the moulding of fibre cuticle cell shape by IRS cuticle cells had been suggested in earlier studies [15, 155, 156], the study of Woods and Orwin [94] demonstrated the precise interaction of the IRS cuticle and fibre cuticle

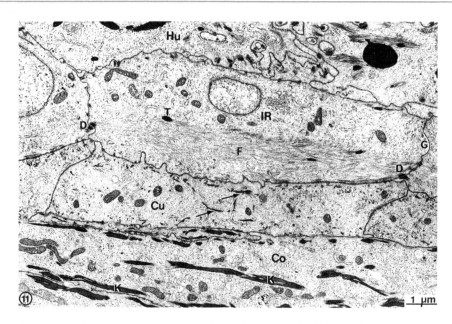

Fig. 10.17 Transmission electron micrograph of longitudinal section of cuticle and IRS cuticle cell (IR) in distal Zone B of a wool follicle. Bundles of filaments (arrows) in a fibre cuticle cell (Cu) are often oriented parallel to the fibre axis. F, Filaments abundant on the fibre cuticle side of an IRS cuticle cell, with associated trichohyalin (T). D, desmosomes appear the point of origin of the fibre network. G, gap junction link adjacent IRS cuticle cells. K, developing macrofibrils in cortex (Co). (Reprinted from J. Ultrastruct. Res., 1982, Woods and Orwin [94], with permission from Elsevier)

cells, and suggested that the presence of cytoskeletal proteins and their association with cell shape changes indicated that they were involved in directing and controlling cell shape, and hence the final appearance of the fibre cuticle scale pattern.

10.5.3 Fibre Cuticle Has a Specific Pattern of Early Keratin Expression

Despite the lack of IFs, the cytoplasm of fibre cuticle cells of Zones A and B are (similarly to cortex cells) packed with ribosomes. Keratin expression appears to differ with species, or perhaps hair type; the number of hair types and species in which keratin expression order is known are few. In human scalp hair follicles keratin expression begins early in Zone B (Fig. 10.10a), with genes *KRT32, KRT35* and *KRT85*. Expression starts at around Auber's critical level, while *KRT82* expression starts in the middle of the bulb. All continue into the middle of Zone B above the bulb before ceasing [157, 158]. In the wool follicle (Fig. 10.10b), the expression pattern is slightly more complex, with *KRT35* and *KRT85* appearing around at Auber's critical level, followed shortly after by *KRT32, KRT82* and *KRT84* [99]. *KRT84* expression ceases very early, distal bulb, while proteins K35 and K85 continue to appear past this stage and well into the middle of Zone C or higher.

Whether cuticle keratins are expressed at these more distal points in the process or are produced by post-transcriptional expression is not clear. While cuticle nuclei degrade in a similar pattern to that seen in cortex cells, the cellular biology and mode of keratin structure assembly in the fibre cuticle is both delayed and also differs markedly from that in the cortex. Nevertheless, some kind of post-nuclear protein expression must occur in the developing fibre cuticle because late appearing keratin K40 continues appearing well above the top of the bulb.

KAP24.1 is the first cuticle KAP to express in human scalp and beard hairs [101]. Expression begins at the apex of the dermal papilla (mid Zone B) and thus is the first KAP in the cuticle that is available to interact with K35, K85 and K32 and may be critical for setting up the initial structures on which later appearing keratins and KAPs will deposit. This hypothesised special role is also implicated by KAP24.1's unusual (for a KAP) composition – relatively low cysteine content and a number of tyrosine-rich repeat sequences at the C-terminus. HSPs and UHSPs are thought to interact with the keratins by way of disulfide interactions, whereas HGTPs do so by way of hydrophobic interactions. Thus, expressed as it is prior to the appearance of the other classes, KAP24.1 may have a dual functionality.

Importantly, and perplexingly, despite this coordinated pattern of keratin expression in the fibre cuticle cells throughout Zone B, there is no obvious ultrastructural signs of keratinous or KAP24.1-based layers or other forms of assembly throughout the bulb. This is in stark contrast with neighbouring cortex and IRS cuticle cells, where intermediate filament structures are clearly evident. It could be that, in the fibre cuticle cells, keratins and KAP24.1 are simply building up in concentration in the cytoplasm until Zone C.

Through Zone C cells of the IRS-cuticle continue to differentiate with the addition of more keratin fibrils to the network that is roughly aligned along the follicle axis within the cytoplasm on the side closest to the fibre cuticle. Cell size and shape change little, and the nucleus and organelles are increasingly squeezed into a smaller space on cells' outer periphery. Increasing numbers of small trichohyalin droplets appear and become associated closely with the fibrils, such that fibrils appear to be attached, pass through, or emerge from the droplets [141, 142, 159]. This mode of deposition differs markedly from the random spherical droplets of trichohyalin observed in the medulla.

10.5.4 Cuticle Shape Finalisation and Keratin Structures Formation

The reshaping of cuticle cells into a series of thin overlapping plates is in an advanced stage by the beginning of Zone C, and by the beginning of Zone D the cuticle cell shapes in wool follicles have stabilised and the approximate final shape of the cuticle cells is achieved due to interactions with the IRS cuticle cells, which, in wool, mould the fibre cuticle cell surface [5, 94]. Fine details of cuticle moulding differ slightly in human hair because the cuticle cells are considerably more overlapped by this stage and each IRS cuticle cell is associated with approximately two or three fibre cuticle cell edges (Fig. 10.16a), compared to an approximately one-to-one relationship in wool (Fig. 10.16b).

At the beginning of Zone C, the cytoplasm of cuticle cells probably contain significant amounts of keratin and KAP protein, but it is not until this point that the first signs of keratin-related ultrastructure become visible. Droplets presumed to be keratin, which are about 30 nm in diameter, appear in the cytoplasm [132, 152]. Although the droplets appear across the extent of the cytoplasm, they appear to migrate toward the side of the cell closest to the IRS. On arrival they transform, or are integrated, into what appears to be a random network of highly folded globular sheets close to, but not attached to, the plasma membrane.

Initially this developing keratin layer has a patchy appearance, but spreads laterally and thickens over the course of the cuticle cells' movement through Zone C. At the distal end of Zone C the entire surface of the cell membrane closest to the IRS-cuticle is coated in the structure, although there is still a narrow gap between keratin and membrane. In wool, the point at which the layer of keratin is continuous along the cuticle-IRS boundary marks the beginning of Zone D (see Chap. 8). Cuticle development has not been studied in detail, but it is assumed that this first complete layer of keratin becomes the highly resistant exocuticle A-layer [160, 161], while later thickening of the structure (primarily during Zone D) forms the remainder of the exocuticle.

The keratins that form this first layer of exocuticle must be a combination of one or more of Type I keratins K32 and K35, Type II keratins K82, K84 (in sheep only) and K85, and KAP 5.5 (Fig. 10.10). In other layers, heterodimers of

Type I and Type II keratins form the basis of intermediate filaments. In the cuticle, where there are no apparent filaments, the nature of dimerization is unclear. Two more keratins, K39 and K40, appear late in the development of the initial keratin layer and probably contribute mostly to the subsequent exocuticle layer development [162].

The number of KAPs increases in early Zone C, although the exact onset of new KAPs has not been well correlated with landmarks such as the hardening of Henle's layer. KAP24.1, is joined by KAP5, KAP10 and KAP12, which are said to begin being expressed about 20–25 cells above the apex of the dermal papilla, placing them probably in early Zone C [100]. While it has been established using immunolabelling transmission electron microscopy (TEM) and proteomics that the exocuticle contains KAPs [160, 163], little is known about the relationship between all these proteins at a nanostructure level. The only study to investigate the exocuticle layer development in detail used TEM and classed the material incorporated into the structure as various types of droplets based on size [132], but the composition of the droplets was unknown.

10.5.5 Hardening of the Cuticle

Cuticle keratin coalesces at the beginning of Consolidation/Hardening zone, at the top of Zone D or start of Zone E, to form the amorphous exocuticle layer. It is at this point that the A-layer can be discriminated from the rest of the exocuticle because in TEM it stains slightly darker using conventional heavy metal stains, especially silver-based stains [131]. The narrow gap between the exocuticle and the cuticle cell membrane disappears [152]. The remaining cytoplasm condenses during Zone E to form the endocuticle which contains the recognizable, but oft degraded, remnants of the nuclei and other organelles [109, 152].

Transglutaminase 3 (TGA3) is now known to occur throughout Zone C and D of the cuticle and cortex. In Zone D, data from immunolabelling of longitudinal sections of human scalp hair suggest that TGA3 occurs in higher concentrations in the cuticle than in the cortex [148, 149]. However, there also appears to be no established epidermal substrate of TGA3 present in the cuticle or cortex in Zone D [163].

Degradation of the living machinery of the cell occurs in cuticle cells in Zone D, but not to the same extent as observed in the cortex [5]. In wool this is evidenced by lower levels of acid phosphatase activity in Golgi complexes, lysosomes and vesicles than in the cortex at the same level [58], and the autophagy that does occur before keratinization is not as complete as that seen in the cortex. This lesser degradation is reflected in observations of considerably more cytoplasmic fragments and heterogeneous texture to the contents of the endocuticle compared with cytoplasmic remnants in the cortex, and the particular susceptibility of the endocuticle to attack by enzymes such as trypsin and pronase [152, 164], but not to keratolytic agents such as peracetic acid/ammonia [165].

Terminal keratinization of the Huxley's layer and IRS cuticle begins in early Zone E, and has broad similarities to the keratinization of Henle's layer in human scalp hair [108, 166], guinea pig [140], mouse [4] and wool follicles [5].

10.5.6 The Cuticle Cell Membrane Complex and Development of Fibre Surface

During Zone B, the plasma membrane between adjacent fibre cuticle cells develops extensive cell junctions similar to adherens junctions. Like the cortex cells, cell junctions probably play an important role in cell reshaping, and then, later in the process they transform into a cell membrane complex (CMC), which probably contains repurposed molecular constituents.

Cuticle cell reshaping extends into Zone C, above the top of the bulb, and is still occurring at a point where similar changes have long ceased in the adjacent cortex cells. Our understanding of the transformation from active plasma membrane to solid CMC is less clear in the cuticle than it is in the cortex. In human scalp hair, with its multiple overlapping cuticle cells, early investigators reported an increase in intercellular gap from 12–15 nm to 20–30 nm, in line with cortex cells [3]. In wool, with overlapping that results in only 1–2 cuticle layers in the final fibre, this increase is not apparent [52]. During its passage through Zones B, C and D, and similar to the cortex-cortex and cortex-cuticle CMC, the width of the cuticle-cuticle CMC in wool remains at around 26 nm before reducing in Zone E to around 14 nm [125]. However, it is difficult to match up where each study was examining, especially as human hair studies were often less precise about identifying their observations with respect to developmental landmarks such as the hardening of Henle's layer.

In mature fibres, the structure of the cuticle-cuticle CMC is thought to differ from the lipid double-bilayer arrangement of the cortex-cortex or cortex-cuticle CMC; the cuticle-cuticle CMC is thought to have a simpler structure in which the cell membranes have converted to monolayers and the protein δ-layer has a different composition from that of the cortex CMC δ-layer [127, 128] (see Chap. 1, Fig. 4). The differences between the cuticle-cuticle CMC and others likely arises because of the specific cell junctions that develop between cuticle cells in Zones B and C. While cortex-cortex CMC contains a wide range of cell junctions [5], the interface between adjacent cuticle cells becomes closely apposed and dominated by what is probably a single type of specialised adherens cell junction [125, 152].

How adherens-like junctions and associated plasma membrane transforms into CMC is not clear but it is also likely to be slightly different than what happens in CMC between cortex cells. In wool, unlike the membranes between cortex cells and between cortex and cuticle cells, the intercellular space between adjacent fibre cuticle cells are free of acid phosphatase [58]. However, changes in electron staining suggest some kind of differentiation occurs in Zone C (Fig. 10.12) because, in wool, the presumptive nascent δ-layer becomes darker [125]. By the beginning of Zone E, the cuticle-cuticle CMC already has a dense central layer (that began to appear in Zone C) that stains darkly using conventional TEM stain methods.

In addition to cuticle-to-cuticle CMC, each fibre cuticle cell also has boundaries with the cortex and with the IRS cuticle, with each interface developing differently. In mature fibres, ultrastructural differences emerge between CMC of adjacent cuticle cells, the CMC between cuticle and cortex cells, and the CMC found between two cortex cells. These differences emerge during Zone C and D [125, 127].

The cortex-cuticle CMC develops more similarly to the cortex-cortex CMC than it does to the cuticle-cuticle CMC, but still differs slightly in appearance from that of the cortex-cortex CMC in both human scalp follicles and in wool follicles (Fig. 10.12).

The remaining boundary is between the fibre cuticle and the IRS cuticle. This interface has two functions which are very different from those of the other membrane complexes, because it is the development of this boundary that defines the surface of the mature fibre. It is along this boundary that the fibre breaks away from the follicle within the pilary canal (Zone F). Therefore, the boundary between the fibre cuticle and IRS cuticle must develop as a weak point in order to provide a clean break. Transmission electron microscope studies indicate that, at a point where the exocuticle is in a relatively advanced state of formation, the plasma membrane on the outer surface of the fibre cuticle cells appears to be disrupted and coated with a double laminar membrane a few nanometres thick with a third thinner layer between the two [168]. Then, at a point which is probably in Zone E, the laminar structure cleaves along the central layer, leaving one of the dark membranes on the IRS cuticle surface and the other on the fibre cuticle. This process occurs along all the fibre external surfaces including the scale edges and also between adjacent fibre cuticle cells to about 1–2 μm from the fibre edge (Fig. 10.18).

This newly formed surface has a second function of presenting a dense surface composed of a proteolipid layer in which the covalently bound fatty acid (+)-18-methyleicosanoic acid, typically called 18-MEA, is oriented to provide a hydrophobic surface [167] which is in contrast to the CMC between fibre cuticle cells, which allows

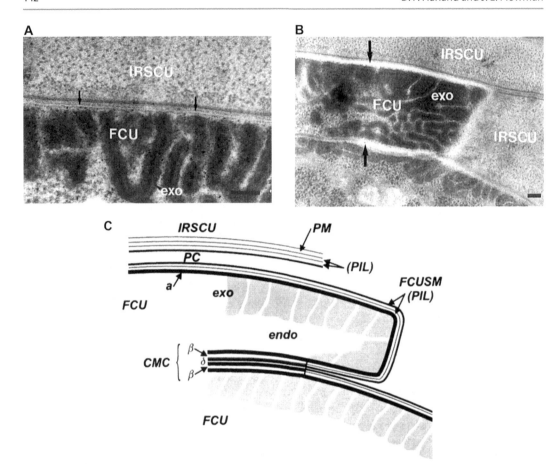

Fig. 10.18 Development of the fibre surface. (**a**) Transmission electron micrograph from a transversely sectioned wool follicle of the cuticle (FCU) and IRS cuticle (IRSCU) interface (distal of the bulb in Zone C). The developing exocuticle lamellae are darkly stained (exo). The CMC between FCU and IRSCU is composed of four distinct layers (arrows). (**b**) A cuticle scale edge (late Zone C or Zone D), illustrating the cleavage process which extends slightly under the scale (big arrows). (**c**) Schematic diagram of the key features of the fibre cuticle surface membrane (FCUSM) development, in which the plasma membrane (PM) of the IRSCU is replaced by intercellular laminae (PIL) which define a splitting point at which the fibre separates from the IRS to enable its release in the pilary canal (PC). The FCUSM containing 18-MEA is in direct contact with the A-Layer (**a**) of the exocuticle. (Reprinted from, Micron, 1997, Jones and Rivett [167], with permission from Elsevier)

ingress of water vapour into the hair. Research on patients of maple syrup urine disease, in which there is a defect in the process of producing 18-MEA, has revealed that 18-MEA is also present on the outer surface of internal cuticle cells (part of the CMC between adjacent fibre cuticle cells), indicating that it is the formation process between fibre cuticle and IRS cuticle cells that forms the weak point for fibre release from the cuticle and not the 18-MEA in itself [169].

10.6 Development of the Inner Root Sheath

10.6.1 Huxley's and Henle's Layers

Huxley's layer is the Inner Root Sheath (IRS) layer adjoining the IRS cuticle. Although generally starting in the lower bulb (low Zone B) as a single layer, in the follicles of hairs of some species, e.g., mice and rabbits, it is sometimes multi-layered depending on the hair type being produced [43, 170]. During their transition through Zone B, Huxley's cells tend to become progressively more vertically elongated. In contrast, Henle's layer, the outermost IRS layer is always a monolayer and the cells tend to be larger than adjacent Huxley's cells.

Droplets of trichohyalin are one of the earliest ultrastructural indicators of development in both Huxley's and Henle's cell lines, beginning at the start of Zone B in the proximal half of the bulb [141]. Droplets and intermediate filaments (IFs) develop in Henle's cells more rapidly than in Huxley's, such that the amount of trichohyalin and of associated IFs is much greater in Henle's cells at any given level in Zone B. Despite this difference in developmental rate, the cytoplasmic and membrane development of both Huxley's and Henle's layers are highly similar and it is, therefore, not surprising that their keratin protein complement is also similar (Fig. 10.10a).

Similar to all other cell lines, cell junctions are an important part of the developmental process in Zone B. During transition through the bulb there is increasing apposition of cell membranes between cells of the same cell line. Desmosomes, tight junctions and gap junctions have been observed in wool follicles [5]. Desmosomes appear to be particularly important because they form anchor points for the developing cytoplasmic keratin IF network. Intermediate filaments in Henle's and Huxley's cells form a loosely aggregated network throughout the cell, which is aligned more-or-less in the direction of the follicle axis. There are also differences in the organisation of desmosomes that join cells of the same type compared with those between cell lines. In particular desmosomes are concentrated on the Huxley's side of the interface with the IRS cuticle, and on the Henle's side of developing Companion layer cells. The desmosomes along the cell membranes of apposed Henle's cells in this Zone are particularly numerous and large [84] and in the run up to Henle's cornification at the distal end of the bulb region, there is either an increase or conformational change in the nature of the transmembrane protein desmocollin 1 (as observed by TEM immunolabelling) [171].

The hardening of Henle's layer is a good marker for the end of the bulb and beginning of the keratogenous region (See Chap. 8), and marks the transition from Zone B to Zone C. It is notable that this point, or threshold, in Orwin's zoning scheme, represents a functional point at which the fibre diameter is set and many biologically driven processes of fibre development are reduced and chemically driven processes (such as non-enzymatic disulfide formation) become more prevalent.

The hardening of Henle's layer occurs very rapidly – often within one cell length. During this process the extensive droplets of trichohyalin which are intersected and attached to the IF network collapse and form a solid amorphous mass.

Huxley's layer continues to build up its own IF network and large trichohyalin granules. Although the exact point is unclear, the enzyme transglutaminase 1 (TGA1) begins to appear around or just before the Zone B-C boundary and is strongly present in IRS-cuticle, and Huxley's cells throughout Zone C [172]. However, the transglutaminases found in the IRS (TGA1 and 5) are different than the one (transglutaminase 3) found in the hair cell lines [148, 149].

Windows in the sheath of hardened Henle's cells (Fig. 10.19) become apparent in Zone C through which parts of some Huxley's cells protrude and make contact with the companion layer to which they are attached by gap junctions and large desmosomes [173]. Originally called *Flügelzellen* (lit. winged-cell) when first discovered using light microscopy in 1927 [173], these gaps are known from electron microscopy studies of both scalp hair [159] and wool [5, 153]. Recent work using immunologically labelled Huxley's specific keratins has established that the *Flügelzellen* form early in Zone B [173, 174], and are attached to Henle's cells by considerable numbers of desmosomes, but being relatively small and uncommon they are not easily visible until Henle's layer has hardened.

Similar to the developing fibre cortex and cuticle, there is also a programme of cytoplasmic and nuclear degradation that occurs within IRS cells. While nuclei of cortical cells are also degraded and the appearance of this degradation appears

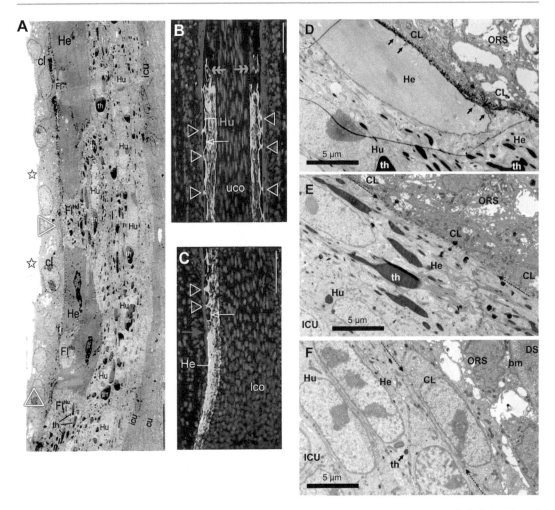

Fig. 10.19 IRS and companion layer interactions. (**a–c**) *Flügelzellen*, or windows in Henle's layer through which Huxley's cells make direct contact with companion layer cells. (**a**) Transmission electron micrograph (probably distal Zone C, or Zone D) showing follicle cell layers from companion layer (cl) through to hair cuticle (cu). *Flügelzellen* (fl) indicated by open arrowheads. Note that plucked follicle has broken from skin along the junction between companion layer and outer root sheath (stars). (**b**) Confocal micrograph showing Zone E with double-labeling of IRS-specific keratin (green) and companion layer keratin (red). Henle's layer is keratinized and not fluorescently labeled. *Flügelzellen* indicated by open head arrows. (**c**) Boundary of Zones B and C (indicated by red arrow). Keratinized Henle's cells indicated by white arrow. Same staining as (**b**). Scale bars: (**a**), 10 μm; (**b** and **c**), 150 μm. (**d–f**) development of CL to He interface. (**d**) When He hardens at the top of the bulb, the interface contains densely stained cytoskeletal material along the CL side and notches invaginating the hardened He (arrows). (**e**) In mid-Zone B, the CL-He interface (dashed line arrows) is highly convoluted and contains many desmosomes. (**f**) At the beginning of Zone B (lower bulb), the CL-He interface is smooth, but CL cells take on their elongated shape. ((**a–c**) reprinted from, J. Invest. Dermatol., 2002, Langbein et al. [173], with permission from Elsevier)

similar to that of IRS cells under the microscope, in the IRS there is evidence of a strong nucleophagy process which is absent in the cortex [97]. This reinforces that the processes underpinning cornification in the IRS are more like those in the epidermis rather than in the hair fibre.

The actual terminal hardening of the Huxley's layer and IRS cuticle begins in early Zone E (at Adamson's Fringe in human follicles) and has broad similarities with keratinization of the Henle's layer in a range of species [4, 5, 108, 140, 166]. By the distal end of Zone D, much of the cytoplasm of Huxley's and IRS cuticle cells have filled with IFs arranged in masses that are aligned roughly in the direction of the follicle/fibre axis. Similar to the IF network that develops in other IRS lines, the organisation is much looser than in the cortex.

Spherical or oblong trichohyalin droplets become closely associated with the fibrillar material which is often seen to emerge from the trichohyalin. Also, similar to Henle's cells, the remaining cytoplasmic constituents (e.g., ribosomes and nucleus) are apparently squeezed into spaces between the fibril bundles and show signs of degradation just prior to keratinization, which is also correlated to lysosomal activity in membranous organelles as indicated by acid phosphatase activity [175, 176]. Keratinization is relatively rapid in Huxley's and IRS cuticle cells, and is characterized by the disappearance of all trichohyalin droplets and a filling of the cytoplasm with a continuous material within which tubular filaments can be observed in a matrix of low electron density. The filament texture is aligned parallel to the follicle axis. Degradation of cytoplasmic constituents is generally less complete in the Huxley's and IRS-cuticle cells than in the cortex and there are organelles and membranous bodies that can be identified in the hardened material, and often active acid phosphatase is still initially associated with these structures [109]. Likewise, the IRS CMCs that develop (all similar) appear to be less specialized than those in the fibre cell lines and often contain recognizable remnants of desmosomes.

10.6.2 IRS, Especially Huxley's Layer, Have a Role in Along-Fibre Shape Changes

The importance of the IRS is easy to underestimate; it's more than just cheap packing material surrounding the growing hair. Huxley's layer, in particular, might be seen as not playing an important functional role, but the effect of genetic mutations on proteins localized in IRS cells highlight the role they play, and also how little we understand. For example, a woolly hair condition in which developing hairs in adult anagen scalp follicles are distorted and poorly attached to the IRS is associated with mutations of the *P2RY5* gene. The P2RY5 trans-membrane protein of unknown function is localized only in the Huxley's and Henle's layers from Zone A onwards [177].

Point mutations to IRS-unique keratin genes also affect fibre structure. For example, a single amino acid substitution, in the *Krt71* gene gives rise to wavy coat hairs and curly vibrissae in mouse mutant strains Rco12 and 13. Although the K71 keratin is present in the hair Henle's and Huxley's layers of both wildtype and mutant mice, in the mutants the IFs in the Henle's and Huxley's layers were in not linear arrays but in large whorled aggregates, and the mature hair has the unusual alterations in shape [178].

These observations support earlier studies [15, 166] that suggested the IRS forms part of a moulding process for not just supporting, but also forming hair fibre shape, and these mutations also emphasize the critical role that IFs have in directing and maintaining cell shape. Loose anagen hair syndrome in humans is associated with a similar disruption of the arrangement of IFs in Huxley's and Henle's layers, as well as many vacuoles in Huxley's and IRS cuticle cells [179]. The Maf family of basic leucine zipper transcription factors, which are important in cell type specific gene expression in a variety of tissues, have also been linked to the IRS and vibrissae morphology in mutant mice [180]. In wildtype mice with normal whiskers, cMaf and MafB are expressed in Huxley's and Henle's cells, respectively, but not in the fibre-related cell lines. Mutant mice that lacked the genes for these proteins were found to have abnormal cuticle scales on their whiskers.

These mutations clearly connect Henle's and Huxley's layers to fibre shape. Many mammals appear to use this connection in a functional way to create complex patterns of along fibre morphology change, but not humans. Human scalp hair is, for the most part, a cylinder of uniform width, with the same arrangement of medulla, cuticle and cortex cells being churned out over most of anagen. Wool is similar. The hair of many mammals is more complex, with significant changes in hair morphology and coloration occurring during a single anagen VI phase [42] (Fig. 10.3a). The underpinning mechanism of this along-fibre developmental programme has been studied in rat underfur (zig-zags) [43] and rabbit guard hairs [16, 44].

The zig-zag hairs of mice are a simple example of such a programme, because the hairs are a repeating structure of circular cross-sections linked by narrower elliptical waists which result in a series of zigs and zags. Each zig or zag has a circular cross-section and contains a medulla. In a cross-section of the follicle, the cell lines form a nice set of concentric rings. When producing the waist, the follicle retains its circular cross section, but the cortex, cuticle and IRS cuticle are displaced to one side and take up only half the area of the zig or zag region [45]. The rest of the follicle is packed out with additional Huxley's cells (Fig. 10.3b). The production of the full fibre then can be generated by a cyclical process of stem cell reprograming by changes that might occur within the stem-cell controlling microniches along the dermal papilla [30].

The pattern of IRS and Huxley's cells is also found when fibres adopt a complex cross-sectional shape (irrespective of whether that shape changes over the length of the fibre); for example, rabbit guard hairs [16], or the trilobite shaped human moustache hair [181] (Fig. 10.3c–d).

10.7 Development of the Companion Layer

Older texts consider the companion layer to be part of the outer root sheath (ORS). However, the companion layer is a distinctly functional layer that is closely associated with Henle's Layer in the inner root sheath [182, 183]. Unlike the ORS cells, the companion layer moves alongside the Henle's cells distally with the growing hair [83], and, beginning in Zone B (lower bulb), has a cell-line specific four keratin complement that differs markedly from that of the ORS or IRS [184]. However, like the ORS, cornification of the companion layer occurs above Zone E.

In Zone B, the companion layer cells quickly differentiate to become flattened and associate closely with the Henle's layer cells. In wool they are attached to both ORS and Henle's cells by desmosomes [5] and appear to strongly express Desmoglein 1, 2 and 3, but not 4 (which is associated with cornification and is found in cortex and cuticle) [117]. Tight junction protein also appears to be strongly expressed, from what can be made out at histological resolution, between adjacent companion cells starting in proximal Zone B [123]. Although it forms a distinct monolayer that can be clearly tracked from Zone A in wool follicles, transmission electron microscopy evidence from rat follicles indicate that, especially around the proximal half of the bulb, that ORS cells sometimes make direct contact with IRS cells and the companion layer is apparently patchy or absent [51].

The main likely function of the companion layer only becomes apparent at the top of the bulb, and in particular with the hardening of Henle's layer. This event, marking the start of Zone C, is associated with changes to both the shape of companion layers cells, as they widen and become more box like, and changes to their relationship with Henle's layer [182, 183]. Through the bulb, up until the beginning of Zone C, the main developmental features apparent at the ultrastructural level in the companion layer is a loose IF network tending to be oriented close to the Henle's layer membrane. From the beginning of Zone C this IF network becomes more developed and in human scalp hair and wool follicles it becomes associated with companion-Henle's layer desmosomes, which are caught in the now hardened Henle's layer wall [125, 183]. However, in rat pelage follicles this association has not been found [51], possibly because there are fewer desmosomes on the companion-Henle's junction in these hairs.

As the cells progress through Zone C and D (the keratogenous region), the IF network becomes more closely associated with the companion-Henle's membrane and other cytoplasmic constituents become increasingly excluded. The junction between companion and Henle's cells is smoother in Zone C than in Zone B, although the smooth interface is punctuated by ingrowths of companion cells that, finger-like, reach inside the Henle's cells in wool, rat pelage and human scalp follicles (Fig. 10.19d–f).

On the ORS side of companion cells, intercellular gaps, often in stretches of plasma membrane between two desmosomes become increasingly apparent in wool follicles [182], and although this effect is less evident in rat companion cells, the companion-ORS interface takes on a pleated appearance with loops of membrane bulging outwards between adjacent desmosomes [51]. The tighter relationship between companion and Henle's than between companion and ORS cells is evident in studies using plucked hair where this junction is often broken, prompting some researchers to suggest that the companion layer should be counted as part of the IRS [173].

The purpose of this set of features and specialisations around cytoskeleton and cell junctions, along with cell shape and delayed cornification appears to be to allow the companion layer to act as a sliding interface between the distally moving Henle's layer (and all cell lines within) and the immobile ORS.

Keratinization of the companion layer never appears to be as consistent or complete as that of the IRS or of the hair, in follicles [5, 183, 185]. By the beginning of Zone E where the IRS and fibre cornifies, the companion cells are almost entirely filled with their IF network. Whether the nucleus and mitochondria of the companion layer continue to function or not has not been established. Final keratinization and cornification occurs much later (following that of the IRS layers), some distance up Zone E or F and is preceded by the appearance of trichohyalin granules [183]. The final keratinization of the companion layer may well be a part of the poorly understood process by which Henle's and Huxley's cells are partially extracted of material prior to their collapse and disintegration in Zone F.

10.8 Emerging Hair

In wool [5], Zone F contains the bulge of the ORS, the *arrector pili* muscle attachment point and the narrowing referred to as "the isthmus". Being approximately adjacent to the bulge, it is Zone F to which the follicle (minus DP) retreats during the telogen phase of the hair cycle (see Chap. 2). In anagen VI, by Zone F, the hair (medulla, cortex and cuticle) is fully keratinized and it is here that the IRS and companion layers break down to release the hair into the pilary canal, or infundibulum [159].

This IRS breakdown occurs in a species-specific manner [185]. In human scalp [183] and wool follicles [159], Huxley's layer is the first to break down and disappear; which Ito and colleagues describe as occurring in the middle of the isthmus while Henle's cells are retained (attached to the recently keratinized companion layer) until the top of Zone F, at which point both the Henle's and companion cells appear to slough.

In wool follicles from merino sheep, following cornification of the wool fibre and Huxley's cells, small protrusions develop from Henle's cells (and presumably Huxley's cells via *Flügelzellen*) into the cytoplasm of the companion cells. Soon after, slightly less common multilaminate membranous vacuoles appear, also protruding into companion cell cytoplasm [159]. In some cases, these multilaminar vesicles are adjacent to what may be the small finger-like protrusions from companion cells into Henle's cells described earlier. Both protrusions and multilaminar vesicles become increasingly frequent and more closely apposed through Zone F. The Henle's and Huxley's layers begin to take on an extracted appearance, and as this becomes more pronounced, the cells appear to compact, distort and finally break apart. Remnants of the process are shed into the pilary canal along with the IRS-cuticle cells, which appear only slightly affected [159]. The breakdown of the IRS, as described by Gemmel and Chapman for merino wool, is very similar to the less detailed descriptions of the process in human scalp hair [185, 186].

In contrast to humans and merino sheep, the IRS breakdown in the wool follicles in some other breeds [15] and also the pelage follicles of mice [186] appear to differ because often in these follicles the IRS can become highly folded during Zone F. However, Ito and colleagues describe the opposite situation for mice in which the mouse vibrissae and pelage IRS tapers cleanly down to a state of complete extraction, bringing the ORS almost in contact with the emerging hair in the distal portion of the isthmus [185].

As the mature hair, now free of the constraints of the IRS, emerges from the follicle, it is coated with sebum, primarily lipids from the sebaceous glands, the ducts of which join the pilary canal just above Zone F and also sweat or suint from the apocrine glands, if these are present [5, 129]. Differences between the lipid composition of sebum and hair internal lipid (e.g., occurring in the CMC) is reviewed in the following references [127, 187, 188].

In addition to sweat and sebum, cellular debris become integrated into sebum and adhere to emerging hairs [129, 159, 176]. In some species, hormones and other specialised volatile proteins can be included in the material deposited on hair, taking advantage of the high surface area. An example is the small secretoglobin protein dimer Fel d, which is the major source of human allergies to domestic cats [189].

Because sheep don't shampoo, wool sebum (called wool grease) is a significant component of what is shorn from sheep, and is separated and purified to produce lanolin. The sebum contains a wide range of debris that has been studied using transmission electron microscopy [190] (Table 10.1). Although the repeated use of shampoos is known to remove all the non-bound surface lipids [188] along with the various types of debris, the findings of wool are nevertheless relevant to hair from a range of species because they give a picture of the range of likely material that is deposited on hair as it leaves the follicle.

Table 10.1 Summary of proteinaceous debris associated with wool fibres as they emerge from the skin (summarized from [190])

Debris size	Types of debris
Whole cells or large fragments (occasionally on hair surface)	Stratum corneum of the epidermis
	Outer root sheath cells from a ring of cells 5–10 deep that surround the orifice of the follicle and had relatively high levels of presumed keratin
	Outer root sheath cells from the pilary canal wall, which were similar to epidermal cells
	IRS cuticle cells
Small cornified fragments showing little cellular detail (occasionally on hair surface)	Bits of Huxley's, Henle's and companion layer cells relatively rich in keratin
	Fragments of keratinized ORS cells from pilary canal or follicle orifice
	Lysed sebaceous gland cells
Small non-keratin cell fragments with cellular or sub cellular detail (common on hair surface)	Fragments of sebaceous gland cells lysed during release of contents
	Inter-cellular material from sloughed keratinized (probably IRS) cells
	Membranous nuclear and other unidentified cellular fragments
	Melanin granules (either in cell fragments or loose)
Micro-organisms	Bacteria and possible fungi, typically inside fibers close to damage

References

1. Alonso, L., & Fuchs, E. (2003). Stem cells of the skin epithelium. *Proceedings of the National Academy of Sciences of the United States of America, 100*(Suppl. 1), 11830–11835.
2. Hardy, M. H. (1992). The secret life of the hair follicle. *Trends in Genetics:TIG, 8*(2), 55–61.
3. Birbeck, M. S. C., & Mercer, E. H. (1957). The electron microscopy of the human hair follicle. Part1. Introduction and the hair cortex. *Journal of Biophysical and Biochemical Cytology, 3*, 203–214.
4. Roth, S. I., & Helwig, E. B. (1964). The cytology of the dermal papilla, the bulb, and the root sheaths of the mouse hair. *Journal of Ultrastructure Research, 11*, 33–51.
5. Orwin, D. F. G. (1979). The cytology and cytochemistry of the wool follicle. *International Review of Cytology, 60*, 331–374.
6. Orwin, D. F., & Woods, J. L. (1982). Number changes and development potential of wool follicle cells in the early stages of fiber differentiation. *Journal of Ultrastructure Research, 80*(3), 312–322.
7. Hashimoto, K., & Shibazaki, S. (1976). Ultrastructural study on differentiation and function of hair. In T. M. Kobori & W. Montagna (Eds.), *Biology and disease of hair* (pp. 23–57). Baltimore: University Park Press.
8. Nutbrown, M., & Randall, V. A. (1995). Differences between connective tissue-epithelial junctions in human skin and the anagen hair follicle. *Journal of Investigative Dermatology, 104*(1), 90–94.
9. Lee, K. (2001). Characterization of silk fibroin/S-carboxymethyl kerateine surfaces: Evaluation of biocompatibility by contact angle measurements. *Fibers and Polymers, 2*(2), 71–74.
10. Joubeh, S., et al. (2003). Immunofluorescence analysis of the basement membrane zone components in human anagen hair follicles. *Experimental Dermatology, 12*(4), 365–370.
11. Millard, M. M. (1972). Analysis of surface oxidized wool fiber by x-ray electron spectrometry. *Analytical Chemistry, 44*(4), 828–829.
12. Cotsarelis, G., Sun, T., & Lavker, R. M. (1990). Label-retaining cells reside in the bulge area of pilosebaceous unit: Implications for follicular stem cells, hair cycle, and skin carcinogenesis. *Cell, 61*, 1329–1337.
13. Hynd, P. I., et al. (1986). Mitotic activity in the cells of the wool follicle bulb. *Australian Journal of Biological Science, 39*, 329–339.
14. Scobie, D. R. (1992). The short term effects of stress hormones on cell division rate in wool follicles. In *Department of Animal Sciences* (p. 207). Adelaide: The University of Adelaide.
15. Auber, L. (1951). The anatomy of follicles producing wool-fibres, with special reference to keratinization. *Transactions of the Royal Society of Edinburgh, 62*, 191–254.
16. Straile, W. E. (1965). Root sheath-dermal papilla relationships and the control of hair growth. In A. G. Lyne & B. F. Short (Eds.), *Biology of the skin and hair growth* (pp. 35–57). Sydney: Angus and Robertson.
17. Hynd, P. I. (1994). Follicular determinants of the length and diameter of wool fibres I. Comparison of sheep differing in fibre length/diameter ratio at two levels of nutrition. *Australian Journal of Agricultural Research, 45*, 1137–1147.
18. Van Scott, E. J., & Ekel, T. M. (1958). Geometric relationships between the matrix of the hair bulb and its dermal papilla in normal and alopecic scalp. *Journal of Investigative Dermatology, 31*, 281–287.
19. Fontaine, S., et al. (2005). Characterization of roughness–friction: Example with nonwovens. *Textile Research Journal, 75*(12), 826–832.
20. Silva, T., et al. (2005). Effect of deamidation on stability for the collagen to gelatin transition. *Journal of Agricultural and Food Chemistry, 53*, 7802–7806.
21. Tobin, D. J., et al. (2003). Plasticity and cytokinetic dynamics of the hair follicle mesenchyme: Implications for hair growth control. *Journal of Investigative Dermatology, 120*(6), 895–904.
22. Adelson, D. L., Kelley, B. A., & Nagorcka, B. N. (1992). Increase in dermal papilla cells by proliferation during development of the primary wool follicle. *Australian Journal of Agricultural Research, 43*, 843–856.
23. Pierard, G. E., & de la Brassinne, M. (1975). Modulation of dermal cell activity during hair growth in the rat. *Journal of Cutaneous Pathology, 2*, 35–41.
24. Tobin, D. J., et al. (2003). Plasticity and cytokinetic dynamics of the hair follicle mesenchyme during the hair growth cycle: Implications for growth control and hair follicle transformations. *Journal of Investigative Dermatology Symposium Proceedings, 8*(1), 80–86.
25. Langbein, L., & Schweizer, J. (2005). Keratins of the human hair follicle. *International Review of Cytology, 243*, 1–78.
26. Langbein, L., et al. (2010). The keratins of the human beard hair medulla: The riddle in the middle. *Journal of Investigative Dermatology, 130*(1), 55–73.
27. Legué, E., & Nicolas, J.-F. (2005). Hair follicle renewal: Organization of stem cells in the matrix and the role of stereotyped lineages and behaviors. *Development, 132*(18), 4143–4154.
28. Rogers, G. E. (2006). Biology of the wool follicle: An excursion into a unique tissue interaction system waiting to be re-discovered. *Experimental Dermatology, 15*(12), 931–949.
29. Epstein, W. L., et al. (1967). Cell proliferation and movement in human hair bulbs. In W. Montagna, & R. L. Dobson (Eds.), *Advances in biology of skin* (Hair growth, Vol. IX) pp. 83–97). New York: Pergamon Press.
30. Yang, H., et al. (2017). Epithelial-mesenchymal micro-niches govern stem cell lineage choices. *Cell, 169*(3), 483–496. e13.

31. Pripis-Nicolau, L., et al. (2001). Automated HPLC method for the measurement of free amino acids including cysteine in musts and wines; first applications. *Journal of the Science of Food and Agriculture, 81*(8), 731–738.
32. Oro, A. E., & Higgins, K. (2003). Hair cycle regulation of Hedgehog signal reception. *Developmental Biology, 255*(2), 238–248.
33. Gambardella, L., et al. (2000). Pattern of expression of the transcription factor Krox-20 in mouse hair follicle. *Mechanisms of Development, 96*(2), 215–218.
34. Jamora, C., et al. (2003). Links between signal transduction, transcription and adhesion in epithelial bud development. *Nature, 422*(6929), 317–322.
35. Kaufman, C. K., et al. (2003). GATA-3: An unexpected regulator of cell lineage determination in skin. *Genes and Development, 17*(17), 2108–2122.
36. Rezza, A., Wang, Z., Sennett, R., Heitman, N., Mok, K. W., Clavel, C., Ma'ayan, A., & Rendl, M. (2016). Signaling networks among stem cell precursors, transit-amplifying progenitors, and their niche in developing hair follicles. *Cell Reports, 14*(12), 3001–3018. https://doi.org/10.1016/j.celrep.2016.02.078
37. Oliver, R. F. (1966). Whisker growth after removal of the dermal papilla and lengths of follicle in the hooded rat. *Journal of Embryology and Experimental Morphology, 15*, 331–347.
38. Reynolds, A. J., & Jahoda, C. A. (1996). Hair matrix germinative epidermal cells confer follicle-inducing capabilities on dermal sheath and high passage papilla cells. *Development, 122*(10), 3085–3094.
39. Schmidt-Ullrich, R., & Paus, R. (2005). Molecular principles of hair follicle induction and morphogenesis. *BioEssays, 27*(3), 247–261.
40. Alonso, L., & Fuchs, E. (2006). The hair cycle. *Journal of Cell Science, 119*(3), 391–393.
41. Driskell, R. R., et al. (2011). Hair follicle dermal papilla cells at a glance. *Journal of Cell Science, 124*(8), 1179–1182.
42. Wildman, A. B. (1954). *The microscopy of animal textiles fibres*. Leeds: Wool Industries Research Association.
43. Priestley, G. C. (1967). Histological studies of the skin follicle types of the rat with special reference to the structure of the Huxley layer. *Journal of Anatomy, 101*, 491–504.
44. Durward, A., & Rudall, K. M. (1955). The axial symmetry of animal hairs. In *Proceedings of the 1st International wool textile conference*.
45. Priestley, G. C., & Rudall, K. M. (1965). Modifications in the huxley layer associated with changes in fibre diameter and output. In A. G. Lyne & B. F. Short (Eds.), *Biology of the skin and hair growth* (pp. 165–182). Sydney: Angus and Robertson.
46. Kassenbeck, P. (1981). Morphology and fine structure of hair. In E. E. Orfanos, W. Montagna, & G. Stuttgen (Eds.), *Hair research, status and future aspects* (pp. 52–64). Berlin: Springer.
47. Hess, W. M., et al. (1990). Human hair morphology: A scanning electron microscopy study on a male caucasoid and a computerized classification of regional differences. *Scanning Microscopy, 4*(2), 375–386.
48. Jahoda, C. A. B. (2003). Cell movement in the hair follicle dermis – More than a two-way street? *Journal of Investigative Dermatology, 121*(6), ix–xi.
49. Ohyama, M., et al. (2010). The mesenchymal component of hair follicle neogenesis: Background, methods and molecular characterization. *Experimental Dermatology, 19*(2), 89–99.
50. Orwin, D. F. G., & Thomson, R. W. (1973). Plasma membrane differentiations of keratinizing cells of the wool follicle (4. Further membrane differentiations). *Journal of Ultrastructure Research, 45*(1), 41–49.
51. Morioka, K. (2005). *Hair follicle, differentiation under the electron microscope – An atlas* (p. 152). Tokyo: Springer.
52. Orwin, D. F. G., Thomson, R. W., & Flower, N. E. (1973). Plasma membrane differentiations of keratinizing cells of the wool follicle. (1. Gap Junctions). *Journal of Ultrastructure Research, 45*(1), 1–14.
53. Carlsen, R. A. (1974). Human fetal hair follicles: The mesenchymal component. *Journal of Investigative Dermatology, 63*(2), 206–211.
54. Iguchi, M., et al. (2003). Communication network in the follicular papilla and connective tissue sheath through gap junctions in human hair follicles. *Experimental Dermatology, 12*(3), 283–288.
55. Nanba, D., Nakanishi, Y., & Hieda, Y. (2003). Establishment of cadherin-based intercellular junctions in the dermal papilla of the developing hair follicle. *The Anatomical Record Part A: Discoveries in Molecular, Cellular, and Evolutionary Biology, 270A*(2), 97–102.
56. Orwin, D. F. G. (1970). A polysaccharide-containing cell coat on keratinizing cells of the Romney wool follicle. *Australian Journal of Biological Science, 23*, 623–635.
57. Elliott, K., Stephenson, T. J., & Messenger, A. G. (1999). Differences in hair follicle dermal papilla volume are due to extracellular matrix volume and cell number: Implications for the control of hair folicle siize and androgen responses. *Journal of Investigative Dermatology, 113*, 873–877.
58. Orwin, D. F. G. (1976). Acid phosphatase distribution in the wool follicle. I. Cortex and fiber cuticle. *Journal of Ultrastructure Research, 55*, 312–324.
59. Chase, H. B. (1958). The behavior of pigment cells and epithelial cells in the hair follicle. In W. E. Montagna & R. A. Ellis (Eds.), *The biology of hair growth* (pp. 229–237). New York: Academic Press.
60. Slominski, A., et al. (2005). Hair follicle pigmentation. *Journal of Investigative Dermatology, 124*(1), 13–21.
61. Forrest, J. W., Fleet, M. R., & Rogers, G. E. (1985). Characterization of melanocytes in wool-bearing skin of Merino sheep. *Australian Journal of Biological Science, 38*, 245–257.
62. Snell, R. S. (1972). An electron microscopic study of melanin in the hair and hair follicles. *Journal of Investigative Dermatology, 59*, 144–154.

63. Jimbow, K., et al. (1971). Ultrastructural and cytochemical studies of melanogenesis in melanocytes of normal human hair matrix. *Journal of Electron Microscopy, 20*, 87–92.
64. Raposo, G., & Marks, M. S. (2007). Melanosomes – Dark organelles enlighten endosomal membrane transport. *Nature Reviews of Molecular Cell Biology, 8*(10), 786–797.
65. Jimbow, K., et al. (2000). Assembly, target-signaling and intracellular transport of tyrosinase gene family proteins in the initial stage of melanosome biogenesis. *Pigment Cell Research, 13*(4), 222–229.
66. Slominski, A., & Paus, R. (1993). Melanogenesis is coupled to murine anagen: Toward new concepts for the role of melanocytes and the regulation of melanogenesis in hair growth. *Journal of Investigative Dermatology, 101*(S1), 90S–97S.
67. Dubucq, M., et al. (1981). Enzymoimmunoassay of the main core protein (p28) of mouse mammary tumour virus (MMTV). *European Journal of Cancer (1965), 17*(1), 81–87.
68. Enshell-Seijffers, D., et al. (2010). b-Catenin activity in the dermal papilla regulates morphogenesis and regeneration of hair. *Developmental Cell, 18*(4), 633–642.
69. Vancoillie, G., et al. (2000). Kinesin and kinectin can associate with the melanosomal surface and form a link with microtubules in normal human melanocytes. *Journal of Investigative Dermatology, 114*(3), 421–429.
70. Van Den Bossche, K., Naeyaert, J.-M., & Lambert, J. (2006). The quest for the mechanism of melanin transfer. *Traffic, 7*(7), 769–778.
71. Nordlund, J. J. (2007). The melanocyte and the epidermal melanin unit: An expanded concept. *Dermatologic Clinics, 25*(3), 271–281.
72. Plonka, P. M. (2009). Electron paramagnetic resonance as a unique tool for skin and hair research. *Experimental Dermatology, 18*, 472–484.
73. Straile, W. E. (1964). A comparison of x-irradiated melanocytes in the hair follicles and epidermis of black and dilute-black Dutch rabbits. *Journal of Experimental Zoology, 155*, 325–342.
74. Straile, W. E. (1964). A study of the hair follicle and its melanocytes. *Developmental Biology, 10*, 45–70.
75. Downes, A. M., & Lyne, A. G. (1959). Measurement of the rate of growth of wool using cystine labelled with sulphur-35. *Nature, 184*, 1834–1885.
76. Daly, R. A., & Carter, H. B. (1956). Fleece growth of young Lincoln, Corriedale, Polwarth and fine Merino maiden ewes grazed on an unimproved paspalum pasture. *Australian Journal of Agricultural Research, 7*(1), 76–83.
77. Daly, R. A., & Carter, H. B. (1955). The fleece growth of young Lincoln, Corriedale, Polwarth, and fine Merino maiden ewes under housed conditions and unrestricted and progressively restricted feeding on a standard diet. *Australian Journal of Agricultural Research, 6*, 476–513.
78. Chapman, R. E., Downes, A. M., & Wilson, P. A. (1980). Migration and keratinization of cells in wool follicles. *Australian Journal of Biological Science, 33*, 587–603.
79. Roth, S. I., & Helwig, E. B. (1964). The cytology of the cuticle of the cortex, the cortex, and the medulla of the mouse hair. *Journal of Ultrastructure Research, 11*, 52–67.
80. Harland, D. P., et al. Intrinsic curvature in merino wool. In Prep. Target, in preparation (Sumitted 9 Feb 2017).
81. Morioka, K., Matsuzaki, T., & Takata, K. (2006). Localization of myosin and actin in the pelage and whisker hair follicles of rat. *Acta Histochemica et Cytochemica, 39*(4), 113–123.
82. Orwin, D. F. G. (1969). New ultrastructural features in the wool follicle. *Nature, 223*, 401–403.
83. Chapman, R. E. (1971). Cell migration in wool follicles of sheep. *Journal of Cell Science, 9*(3), 791–803.
84. Orwin, D. F. G., Thomson, R. W., & Flower, N. E. (1973). Plasma membrane differentiations of keratinizing cells of the wool follicle (2. desmosomes). *Ultrastructure Research, 45*, 15–29.
85. Candi, E., Schmidt, R., & Melino, G. (2005). The cornified envelope: A model of cell death in the skin. *Nature Reviews of Molecular Cell Biology, 6*(4), 328–340.
86. Fuchs, E., & Raghavan, S. (2002). Getting under the skin of epidermal morphogenesis. *Nature Reviews Genetics, 3*(3), 199–209.
87. McKinnon, A. J., Harland, D. P., & Woods, J. L. (2016). Relating self-assembly to protein expression in wool cortical cells. *Journal of Textile Engineering, 62*(6), 123–128.
88. Matsunaga, R., et al. (2013). Bidirectional binding property of high glycine-tyrosine keratin-associated protein contributes to the mechanical strength and shape of hair. *Journal of Structural Biology, 183*(3), 484–494.
89. Fraser, R. D. B., Rogers, G. E., & Parry, D. A. D. (2003). Nucleation and growth of macrofibrils in trichocyte (hard-α) keratins. *Journal of Structural Biology, 143*, 85–93.
90. Chapman, R. E., & Gemmell, R. T. (1971). Stages in the formation and keratinization of the cortex of the wool fiber. *Journal of Ultrastructure Research, 36*(3–4), 342–354.
91. Plowman, J. E., et al. (2015). The proteomics of wool fibre morphogenesis. *Journal of Structural Biology, 191*(3), 341–351.
92. Yu, Z., et al. (2009). Expression patterns of keratin intermediate filament and keratin associated protein genes in wool follicles. *Differentiation, 77*(3), 307–316.
93. McKinnon, J., & Harland, D. P. (2011). A concerted polymerization-mesophase separation model for formation of trichocyte intermediate filaments and macrofibril templates 1: Relating phase separation to structural development. *Journal of Structural Biology, 173*(2), 229–240.
94. Woods, J. L., & Orwin, D. F. (1982). The cytology of cuticle scale pattern formation in the wool follicle. *Journal of Ultrastructure Research, 80*(2), 230–242.

95. Zhao, B. S., Roundtree, I. A., & He, C. (2017). Post-transcriptional gene regulation by mRNA modifications. *Nature Reviews. Molecular Cell Biology, 18*(1), 31–42.
96. Lemasters, J. J., et al. (2017). Compartmentation of mitochondrial metabolism in hair follicles: A ring of fire. *Journal of Investigative Dermatology, 137*, 1434–1444.
97. Jones, L., Harland, D., Jarrold, B., Connolly, J., & Davis, M. (2018). The walking dead: Sequential nuclear and organelle destruction during hair development. *The Brititsh Journal of Dermatology*. Accepted Author Manuscript. https://doi.org/10.1111/bjd.16148
98. Forslind, B., & Swanbeck, G. (1966). Keratin formation in the hair follicle I. An ultrastructural investigation. *Experimental Cell Research, 43*, 191–209.
99. Yu, Z., et al. (2011). Annotations of sheep keratin intermediate filament genes and their patterns of expression. *Experimental Dermatology, 20*(7), 582–588.
100. Rogers, M. A., et al. (2006). Human hair keratin-associated proteins (KAPs). *International Review of Cytology, 251*, 209–263.
101. Rogers, M. A., et al. (2007). Characterization of human KAP24.1, a cuticular hair keratin-associated protein with unusual amino-acid composition and repeat structure. *Journal of Investigative Dermatology, 127*(5), 1197–1204.
102. Powell, B. C., Nesci, A., & Rogers, G. E. (1991). Regulation of keratin gene expression in hair follicle differentiation. *Annals of the New York Academy of Sciences, 642*, 1–20.
103. Mercer, E. H. (1949). Some experiments on the orientation and hardening of keratin in the hair follicle. *Biochimica et Biophysica Acta, 3*, 161–169.
104. Mercer, E. H. (1961). Keratin and keratinization (1st ed., Vol. 12, International series of monographs on pure and applied biology, p. 316) Oxford: Pergamon Press.
105. Rafik, M. E., et al. (2006). In vivo formation steps of the hard α-keratin intermediate filament along a hair follicle: Evidence for structural polymorphism. *Journal of Structural Biology, 154*(1), 79–88.
106. Bornschlögl, T., et al. (2016). Keratin network modifications lead to the mechanical stiffening of the hair follicle fiber. *Proceedings of the National Academy of Sciences, 113*(21), 5940–5945.
107. Rogers, G. E. (1959). Electron microscopy of wool. *Journal of Ultrastructure Research, 2*(3), 309–330.
108. Rogers, G. E. (1964). Structural and biochemical features of the hair follicle. In W. Montapna & W. C. Lobitz (Eds.), *The epidermis* (pp. 179–236). New York: Academic Press.
109. Orwin, D. F. G. (1976). Acid phosphatase distribution in the wool follicle. III. Fate of organelles in keratinized cells. *Journal of Ultrastructure Research, 55*, 335–342.
110. Thibaut, S., et al. (2008). Transglutaminase-3 enzyme: A putative actor in human hair shaft scaffolding? *Journal of Investigative Dermatology, 129*(2), 449–459.
111. Caldwell, J. P., et al. (2005). The three-dimensional arrangement of intermediate filaments in Romney wool cortical cells. *Journal of Structural Biology, 151*(3), 298–305.
112. Burdett, I. A. A. (1998). Aspects of the structure and assembly of desmosomes. *Micron, 29*(4), 309–328.
113. Garrod, D. R., et al. (1999). Desmosomal adhesion. *Advances in Molecular and Cell Biology, 28*, 165–202.
114. Thomason, H. A., et al. (2010). Desmosomes: Adhesive strength and signalling in health and disease. *Biochemical Journal, 429*(3), 419–433.
115. Mils, V., et al. (1992). The expression of desmosomal and corneodesmosomal antigens shows specific variations during the terminal differentiation of epidermis and hair follicle epithelia. *Journal of Histochemistry and Cytochemistry, 40*(9), 1329–1337.
116. Kurzen, H., et al. (1998). Compositionally different desmosomes in the various compartments of the human hair follicle. *Differentiation, 63*, 295–304.
117. Bazzi, H., et al. (2006). Desmoglein 4 is expressed in highly differentiated keratinocytes and trichocytes in human epidermis and hair follicle. *Differentiation, 74*(2–3), 129–140.
118. Runswick, S. K., et al. (2001). Desmosomal adhesion regulates epithelial morphogenesis and cell positioning. *Nature Cell Biology, 3*, 823–830.
119. Tsvetkov, E. A. (2001). Gap junctions: Structure, functions, and regulation. *Journal of Evolutionary Biochemistry and Physiology, 37*(5), 457–468.
120. Arita, K., et al. (2004). Gap junction development in the human fetal hair follicle and bulge region. *British Journal of Dermatology, 150*(3), 429–434.
121. Kam, E., & Hodgins, M. B. (1992). Communication compartments in hair follicles and their implication in differentiative control. *Development, 114*, 389–393.
122. Downes, A. M., Lyne, A. G., & Clarke, W. H. (1962). Radioautographic studies of the incorporation of [35S]cystine into wool. *Australian Journal of Biological Science, 15*, 713–719.
123. Brandner, J. M., et al. (2003). Expression and localization of tight junction-associated proteins in human hair follicles. *Archives of Dermatological Research, 295*(5), 211–221.
124. Orwin, D. F. G., Thomson, R. W., & Flower, N. E. (1973). Plasma membrane differentiations of keratinizing cells of the wool follicle (3. Tight junctions). *Journal of Ultrastructure Research, 45*, 30–40.
125. Orwin, D. F. G., & Thomson, R. W. (1972). An ultrastructural study of the membranes of keratinizing wool follicle cells. *Journal of Cell Science, 11*(1), 205–219.
126. Nogues, B., et al. (1988). New advances in the internal lipid composition of wool. *Textile Research Journal, 58*, 338–342.
127. Robbins, C. R. (2009). The cell membrane complex: Three related but different cellular cohesion components of mammalian hair fibers. *Journal of the Society of Cosmetic Chemistry, 60*(4), 437–465.
128. Bryson, W. G., et al. (1995). Characterisation of proteins obtained from papain/dithiothreitol digestion

of merino and romney wools. In *Proceedings of the 9th International Wool textile research conference.* Biella, Italy.
129. Montagna, W., & Parakkal, P. F. (1974). *The structure and function of skin* (3rd ed.). New York: Academic Press.
130. Downes, A. M., et al. (1966). Proliferative cycle and fate of cell nuclei in wool follicles. *Nature, 212*, 477–479.
131. Swift, J. A. (1977). The histology of keratin fibers. In R. S. Asquith (Ed.), *Chemistry of natural protein fibers* (pp. 81–146). London: Wiley.
132. Woods, J. L., & Orwin, D. F. G. (1980). Studies on the surface layers of the wool fibre cuticle. In D. A. D. Parry & L. K. Creamer (Eds.), *Fibrous proteins: Scientific, industrial and medical aspects* (pp. 141–149). London: Academic Press.
133. Fischer, H., et al. (2011). Essential role of the keratinocyte-specific endonuclease DNase1L2 in the removal of nuclear DNA from hair and nails. *Journal of Investigative Dermatology, 131*(6), 1208–1215.
134. Gilbert, M. T. P., et al. (2007). Whole-genome shotgun sequencing of mitochondria from ancient hair shafts. *Science, 317*(5846), 1927–1930.
135. Lemasters, J. J., et al. (2017). Compartmentation of mitochondrial and oxidative metabolism in growing hair follicles: A ring of fire. *Journal of Investigative Dermatology, 137*(7), 1434–1444.
136. Orwin, D. F., & Thomson, R. W. (1972). The distribution of coated vesicles in keratinizing cells of the wool follicle. *Australian Journal of Biological Science, 25*(3), 573–583.
137. Yang, J., et al. (1997). Prevention of apoptosis by Bcl-2: Release of cytochrome c from mitochondria blocked. *Science, 275*(5303), 1129–1132.
138. McGowan, K. M., & Coulombe, P. A. (2000). Keratin 17 expression in the hard epithelial context of the hair and nail, and its relevance for the Pachyonychia Congenita phenotype. *Journal of Investigative Dermatology, 114*(6), 1101–1107.
139. Hojiro, O. (1972). Fine structure of the mouse hair follicle. *Journal of Electron Microscopy, 21*, 127–138.
140. Parakkal, P. F., & Matoltsy, A. G. (1964). A study of the differentiation products of the hair follicle cells with the electron microscope. *Journal of Investigative Dermatology, 43*, 23–34.
141. Rothnagel, J. A., & Rogers, G. E. (1986). Trichohyalin, and intermediate filament-associated protein of the hair follicle. *Journal of Cell Biology, 102*, 1419–1429.
142. Rogers, G., et al. (1997). Peptidylarginine deiminase of the hair follicle: Characterization, localization, and function in keratinizing tissues. *Journal of Investigative Dermatology, 108*(5), 700–707.
143. Fietz, M. J., et al. (1993). Analysis of the sheep trichohyalin gene: Potential structural and calcium-binding roles of trichohyalin in the hair follicle. *Journal of Cell Biology, 121*(4), 855–865.
144. Hashimoto, K. (1988). The structure of human hair. *Clinics in Dermatology, 6*(4), 7–21.
145. Ito, M., & Hashimoto, K. (1982). Trichohyaline granules in hair cortex. *Journal of Investigative Dermatology, 79*, 392–398.
146. Harding, H. W., & Rogers, G. E. (1971). (γ-glutamyl) lysine cross-linkage in citrulline-containing protein fractions from hair. *Biochemistry, 10*, 624–630.
147. Harding, H. W. J., & Rogers, G. E. (1972). The occurrence of the (γ-glutamyl)lysine cross-link in the medulla of hair and quill. *Biochimica et Biophysica Acta, 257*(1), 37–39.
148. Rogers, G. E. (1989). Special biochemical features of the hair follicle. In G. E. Rogers, P. J. Reis, K. A. Ward, & R. C. Marshall (Eds.), *The biology of wool and hair* (pp. 69–85). London/New York: Chapman and Hall.
149. Cheng, T., van Vlijmen-Willems, I. M. J. J., Hitomi, K., Pasch, M. C., van Erp, P. E. J., Schalkwijk, J., & Zeeuwen, P. L. J. M. (2008). Colocalization of cystatin M/E and its target proteases suggests a role in terminal differentiation of human hair follicle and nail. *Journal of Investigative Dermatology, 129*(5), 1232–1242.
150. Brunner, H., & Coman, B. J. (1974). *The identification of mammalian hair.* (1st ed.p. 176). Melbourne: Inkata Press.
151. Harding, H. W. J., & Rogers, G. E. (1976). Isolation of peptides containing citrulline and the cross-link (γ-glutamyl)lysine, from hair medulla protein. *Biochimica et Biophysica Acta, 427*, 315–324.
152. Birbeck, M. S. C., & Mercer, E. H. (1957). The electron microscopy of the human hair follicle. Part 2. The hair cuticle. *Journal of Biophysical and Biochemical Cytology, 3*, 215–221.
153. Happey, F., & Johnson, A. G. (1962). Some electron microscope observations on hardening in the human hair follicle. *Journal of Ultrastructure Research, 7*, 316–327.
154. Gardel, M. L., et al. (2010). Mechanical integration of actin and adhesion dynamics in cell migration. *Annual Review of Cell and Developmental Biology, 26*(1), 315–333.
155. Kassenbeck, P. (1959). The kinetics of the keratinization process and the formation of the keratinized fibre. *Bulletin de l'institut Textile de france, 83*, 26–40.
156. Bradbury, J. H., & Rogers, G. E. (1963). The theory of shrinkproofing of wool. Part IV. Electron and light microscopy of polyglycine on the fibers. *Textile Research Journal, 33*, 452–458.
157. Langbein, L., et al. (1999). The catalog of human hair keratins. I. Expression of the nine type I members in the hair follicle. *Journal of Biological Chemistry, 274*(28), 19874–19884.
158. Langbein, L., et al. (2001). The catalog of human hair keratins. II. Expression of the six type II members in the hair follicle and the combined catalog of human type I and II keratins. *Journal of Biological Chemistry, 276*(37), 35123–35132.
159. Gemmell, R. T., & Chapman, R. E. (1971). Formation and breakdown of the inner root sheath and features of the pilary canal epithelium in the wool follicle. *Journal of Ultrastructure Research, 36*, 355–366.

160. Bringans, S. D., et al. (2007). Characterization of the exocuticle a-layer proteins of wool. *Experimental Dermatology, 16*(11), 951–960.
161. Swift, J. A., & Bews, B. (1974). The chemistry of human hair cuticle-II: The isolation and amino acid analysis of the cell membranes and A-layer. *Journal of the Society of Cosmetic Chemistry, 25*, 355–366.
162. Langbein, L., et al. (2007). Novel type I hair keratins K39 and K40 are the last to be expressed in differentiation of the hair: Completion of the human hair keratin catalogue. *Journal of Investigative Dermatology, 127*, 1532–1535.
163. Jones, L. N., et al. (2010). Location of keratin-associated proteins in developing fiber cuticle cells using immunoelectron microscopy. *International Journal of Trichology, 2*(2), 89–95.
164. Swift, J. A., & Holmes, A. W. (1965). Degradation of human hair by papain. Part III: Some electron microscope observations. *Textile Research Journal, 35*, 1014–1019.
165. Rogers, G. E. (1959). Electron microscope studies of hair and wool. *Annals of the New York Academy of Sciences, 83*, 378–399.
166. Birbeck, M. S. C., & Mercer, E. H. (1957). The electron microscopy of the human hair follicle Part 3. The inner root sheath and trichohyaline. *Journal of Biophysical and Biochemical Cytology, 3*, 223–230.
167. Jones, L. N., & Rivett, D. E. (1997). The role of 18-methyleicosanoic acid in the structure and formation of mammalian hair fibres. *Micron, 28*(6), 469–485.
168. Jones, L. N., Horr, T. J., & Kaplin, I. J. (1994). Formation of surface membranes in developing mammalian hair fibres. *Micron, 25*(6), 589–595.
169. Jones, L. N., et al. (1996). Hairs from patitents with maple syrup urine disease show a structural defect in the fiber cuticle. *Journal of Investigative Dermatology, 106*(3), 461–464.
170. Priestly, G. C. (1967). Seasonal changes in the inner root sheath of the primary follicles in Herdwick sheep. *Journal of Agricultural Science, 69*, 9–12.
171. Donetti, E., et al. (2004). Desmocollin 1 expression and desmosomal remodeling during terminal differentiation of human anagen hair follicle: An electron microscopic study. *Experimental Dermatology, 13*(5), 289–297.
172. Tamada, Y., et al. (1995). Expression of transglutaminase 1 in human anagen hair follicles. *Acta Dermatovenerologica, 75*, 190–192.
173. Langbein, L., et al. (2002). A novel epithelial keratin, hK6irs1, is expressed differentially in all layers of the inner root sheath, including specialized Huxley cells (flugelzellen) of the human hair follicle. *Journal of Investigative Dermatology, 118*(5), 789–799.
174. Langbein, L., et al. (2003). K6irs1, K6irs2, K6irs3, and K6irs4 represent the inner-root-sheath-specific type II epithelial keratins of the human hair follicle. *Journal of Investigative Dermatology, 120*(4), 512–522.
175. Orwin, D. F. G. (1974). Acid phosphatase in keratinising wool follicle cells. In *The 8th International Congress of* electron microscopy.
176. Orwin, D. F. G. (1976). Acid phosphatase distribution in the wool follicle. II. Henle's layer and outer root sheath. *Journal of Ultrastructure Research, 55*, 325–334.
177. Shimomura, Y., et al. (2008). Disruption of P2RY5, an orphan G protein-coupled receptor, underlies autosomal recessive woolly hair. *Nature Genetics, 40*(3), 335–339.
178. Runkel, F., et al. (2006). Morphologic and molecular characterization of two novel Krt71 (Krt2-6g) mutations: Krt71 rco12 and Krt71 rco13. *Mammalian Genome, 17*, 1172–1182.
179. Mirmirani, P., Uno, H., & Price, V. H. (2011). Abnormal inner root sheath of the hair follicle in the loose anagen hair syndrome: An ultrastructural study. *Journal of the American Academy of Dermatology, 64*(1), 129–134.
180. Miyai, M., et al. (2010). c-Maf and MafB transcription factors are differentially expressed in Huxley's and Henle's layers of the inner root sheath of the hair follicle and regulate cuticle formation. *Journal of Dermatological Science, 57*(3), 178–182.
181. Hess, W. M., et al. (1990). A scanning electron microscopy study of laser-cut hair. *Proceedings of the International Congress for Electron Microscopy, 12*, 730–731.
182. Orwin, D. F. (1971). Cell differentiation in the lower outer sheath of the Romney wool follicle: A companion cell layer. *Australian Journal of Biological Science, 24*(5), 989–999.
183. Ito, M. (1986). The innermost cell layer of the outer root sheath in anagen hair follicle: Light and electron microscopic study. *Archives of Dermatological Research, 279*, 112–119.
184. Winter, H., et al. (1998). A novel human type 2 cytokeratin, K6hf, specifically expressed in the companion layer of the hair follicle. *Journal of Investigative Dermatology, 111*(6), 955–962.
185. Ito, M. (1988). Electron microscopic study on cell differentiation in anagen hair follicles in mice. *Journal of Investigative Dermatology, 90*, 65–72.
186. Montagna, W. (1962). *The structure and function of skin* (2nd ed.). New York: Academic Press.
187. Ro, I. B., & Dawson, T. L. (2005). The role of sebaceous gland activity and scalp microfloral metabolism in the etiology of seborrheic dermatitis and dandruff. *Journal of Investigative Dermatology Symposium Proceedings, 10*(3), 194–197.
188. Robbins, C. R. (1994). *Chemical and physical behavior of human hair* (3rd ed.p. 330). New York: Springer.
189. Kaiser, L., et al. (2003). The crystal structure of the major cat allergen Fel d 1, a member of the secretoglobin family. *Journal of Biological Chemistry, 278*(39), 37730–37735.
190. Orwin, D. F. G., & Woods, J. L. (1985). Cellular debris in the grease of wool fibres. *Textile Research Journal, 55*, 84–92.

Macrofibril Formation

11

Duane P. Harland and A. John McKinnon

Contents

11.1	**Macrofibrils Vary in Architecture**	156
11.2	**Macrofibril Development**	158
11.2.1	Cortical Keratin Structures Form as Dense Bundles and Coalesce	158
11.2.2	Unit Length Filaments Are the Monomers of IFs	158
11.3	**Different Models of Macrofibril Assembly**	159
11.3.1	Fraser-Parry Model	159
11.3.2	Mesophase Model	160
11.3.2.1	Rod-Based Mesophase Formation	160
11.3.2.2	Mesophase Theory Requirements	161
11.3.2.3	Key Features of Mesophase Formation	163
11.3.2.4	The Model for ULF Polymerization into Tactoids	164
11.3.2.5	Aligning the Mesophase Model with Keratin Expression in the Follicle	165
11.3.2.6	Mesophase Model and Different Macrofibril Architecture	166
References		167

Abstract

Macrofibrils are the main structural component of the hair cortex, and are a composite material in which trichokeratin intermediate filaments (IFs) are arranged as organised arrays embedded in a matrix composed of keratin-associated proteins (KAPs) and keratin head groups. Various architecture of macrofibrils is possible, with many having a central core around which IFs are helically arranged, an organisation most accurately described as a double-twist arrangement. In this chapter we describe the architecture of macrofibrils and then cover their formation, with most of the material focusing on the theory that the initial stages of macrofibril formation are as liquid crystals.

D. P. Harland (✉)
AgResearch Ltd., Lincoln, New Zealand
e-mail: Duane.Harland@agresearch.co.nz

A. J. McKinnon
Institute of Fundamental Sciences, Massey University, Palmerston North, New Zealand

Keywords

Hair follicle · Trichokeratin · Keratin intermediate filament · Intermediate filament polymerization · Macrofibril · Supramolecular assembly · Mesophase

11.1 Macrofibrils Vary in Architecture

Hair cortex is a multiscalar structure. Chap. 6 covered the structure of the keratin intermediate filaments (IFs) which are the primary regular structural component of the cortex at a nanometre scale, and Chap. 7 covered the keratin associated proteins (KAPs) which surround IFs within fibres and have a highly variable molecular structure. In Chap. 1 we saw that, at a scale of hundreds of nanometres, IFs and KAPs take on the form of elongated bundles called macrofibrils. In Chap. 10, keratin bundles (proto-macrofibrils) were described as forming in the lower region of follicles, and coalescing to create long macrofibrils. Precisely how proto-macrofibrils form is not fully understood. The key problem is not just the formation of the biocomposite material (7.6 nm diameter IFs spaced about 10 nm apart and embedded in a matrix of KAPs and protruding keratin head groups), but that the internal architecture and the volume fraction of matrix also varies.

Cortical cell types in wool were initially defined based on their propensity to differentially take up light microscopy stains [1, 2]. Fine diameter (15–25 μm) wool also contains three predominant macrofibril architectures [3, 4, 5] (Fig. 11.1), which correlate well to cell type. Orthocortical cells tend to contain macrofibrils which are cylindrical columns composed of IFs arranged along a central linear axis around which the filaments are helically pitched with a linear increase in helical angle for filaments in the radial dimension. It is as if there is one central straight IF surrounded by a shell of six filaments which are gently wound around the core, the next concentric ring of filaments is wound gently around the inner shell, and so on to the macrofibril periphery. A term borrowed from liquid crystal studies describes the architecture perfectly as a double-twist arrangement. Paracortical cells have more widely spaced filaments, but contain filaments that are all at least roughly parallelly aligned along the fibre axis; the liquid crystal term for this architecture is nematic. Mesocortical cells are often on the border between clusters of orthocortex and paracortex, and while they contain macrofibrils that are predominantly ortho-like or para-like, they contain regions in which IFs are highly aligned and packed laterally into a tight hexagonal array with a spacing intermediate between orthocortical and paracortical cells.

The close relationship between cell type and macrofibril structure in fine diameter wools has become a *de facto* definition applied to hairs from many species. However, in human scalp hair [6, 7], non-sheep wool fibres such as goats, alpaca and rabbits [8], deer hairs [9], higher-diameter sheep fibres (25–50 μm) and lustre

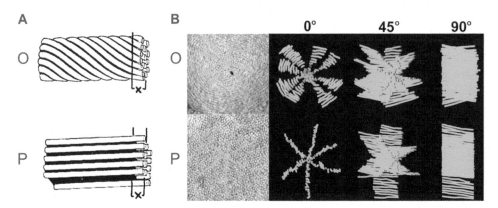

Fig. 11.1 Macrofibril architecture types. (**a**) Architecture of macrofibrils from orthocortical (O) and paracortical (P) cells of wool as hypothesised by George Rogers in the 1950s from transmission electron micrographs through sections (indicated as x). (**b**) Macrofibril architecture as experimentally measured by tracing filaments through electron tomograms of from wool in the early 2000s. ((**a**) reprinted from J. Ultrastruct. Res., 1959, Rogers [3]; (**b**) reprinted from J. Struct. Biol., 2005, Caldwell, Mastronarde [10], both with permission from Elsevier)

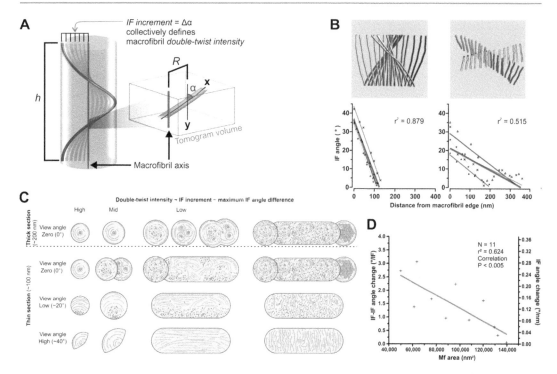

Fig. 11.2 Helical pitch in macrofibrils. (**a**) Helical pitch (α) increases linearly from centre to periphery. (**b**) The slope of this change (intensity) varies between different macrofibrils. IFs traced from a high-intensity double-twist macrofibril are shown on the left and from a low-intensity macrofibril on the right. (**c**) In projection (e.g., macrofibrils viewed in cross sections of fibres using transmission electron microscopy) double-twist structure of different intensities take on particular patterns, which change with section thickness. (**d**) There is a rough inverse relationship between macrofibril diameter and intensity in human scalp hair. (Reprinted from J. Struct. Biol., 2014, Harland, Walls [7] with permission from Elsevier)

mutant wools (Harland and Plowman, unpublished results) single cortex cells can contain macrofibrils of multiple architectures. Also across all fibres, including those of fine diameter wools, double-twist architecture varies in the extent of helical twist.

Extent of helical twist, or the intensity of a specific macrofibril, is measured as a change in helical angle between adjacent IFs along an axis from centre to edge (Fig. 11.2a). Some macrofibrils have an intense helical structure (e.g., 3.0°/shell) resulting in IFs at the periphery of the macrofibril being about 40° different compared to the central IF. Intensity can also be very low (e.g., 0.5°/shell), and although there appears to be an inverse relationship between macrofibril diameter and intensity, the outermost IFs in a large low intensity macrofibril might only be about 15° different from the centre (Fig. 11.2b). This situation has historically led to difficulty in determining the extent of low intensity macrofibrils using transmission electron microscopy (TEM) because of geometrical artefacts (Fig. 11.2c). It is a problem in interpreting repeating structures of complex geometry from projections through sections, especially in extracellular biology [11]. In macrofibrils, low intensity double-twist macrofibrils can be difficult to distinguish from nematic (paracortex) macrofibrils in very thin section (e.g., typical TEM sections are between 50 and 100 nm), and often the same macrofibril could be classified as orthocortical in a thick section and paracortical in a thinner section [7].

Variable architecture and variable double-twist intensity may influence fibre properties such as swelling, fibre bending stiffness, and cell length—with implications for natural curvature [12] and how that curvature changes with different external conditions (e.g., humidity). Consequently, an understanding of how macrofibrils form must also account for their variable structure.

11.2 Macrofibril Development

11.2.1 Cortical Keratin Structures Form as Dense Bundles and Coalesce

Rogers, who first described the ultrastructure of macrofibrils of wool fibres [3], found it striking that even in the earliest stages, no precursors of IFs, or macrofibrils (such as isolated IFs) were to be observed in the cytoplasm [13]. Other authors have found the same in species from human to rat, and the first signs of IFs in the proximal regions of Zone B are invariably described as aggregations or bundles that are either associated with desmosomes, and typically oriented along cell membranes, or appear as isolated proto-macrofibrils in the cytoplasm [14–18]. The cytoplasm of cortex cells in early Zone B is characterized by a large concentration of free-floating ribosomes, numerous mitochondria, but relatively small amounts of Golgi complexes, endoplasmic reticulum and vacuoles [17]. Keratins and KAPs are likely to be produced directly in the cytoplasm where they transfer directly to developing macrofibrils. Keratin protein expression begins with Type I K35 and Type II K85, with the more abundant Type I K31 expressing within a few cell lengths of onset of K35 [19]. It is likely that the first filaments are composed predominantly of K31/K85 heterodimers [20].

There has been relatively little work carried out on these early stages of IF and macrofibril assembly in hairs. Partly this is due to the difficulty of extracting intact trichocyte IFs from mature fibres and reforming them under biologically realistic conditions in vitro [21, 22]. The approach of Jones and Pope involved disrupting the dissected cortex elongation region from the bulbs of human hair plucks and then observing the resulting structures [23]. Apparent macrofibrils in the process of being disrupted were observed, and contained a complex mixture of loose filaments and short tight bundles, of 100–200 nm length, and 50 nm wide (Fig. 11.3). What level of macrofibril assembly these structures represent is not clear. Other approaches have investigated interactions in vitro using bacterially expressed keratins and KAPs and desmosome proteins [24]. Interactions between KAPs and IFs as well as with desmosomes have been observed, but fairly harsh conditions (DTT and high pH) were required in order to keep KAPs in a soluble form. While these studies show the challenges of experimental determination of macrofibril formation, they also indicate the importance of dense aggregates, as opposed to a network of IFs, in the development process.

Irrespective of how the initial proto-macrofibrils form, their growth into macrofibrils is primarily by fusion with one another (see Chap. 10 Fig. 10.9). How macrofibrils fuse may depend heavily on their underlying architecture. Macrofibrils which contain IFs that are hexagonally packed (e.g., mesocortex and paracortex) can merge both end-to-end and laterally with adjacent macrofibrils. However, those with double-twist architecture are limited to end-to-end fusion, because of the incompatible directions of IFs at their periphery [10, 25, 26].

11.2.2 Unit Length Filaments Are the Monomers of IFs

The details of IF structure were covered in Chap. 6, and it was noted that there is a lack of certainty over the polymerization of IFs from expressed keratin heterodimers in the hair keratins. The assumption usually made in models of hair assembly is that the process is similar to that which has been established for other types of IF [27].

Research on vimentin (a cytoskeletal IF relative of keratin) has revealed that homodimer-based IFs of that protein form from monomeric unit length filaments (ULFs) [28, 29] and this also seems to be the case with some epithelial keratins [30]. Key parameters of ULFs are

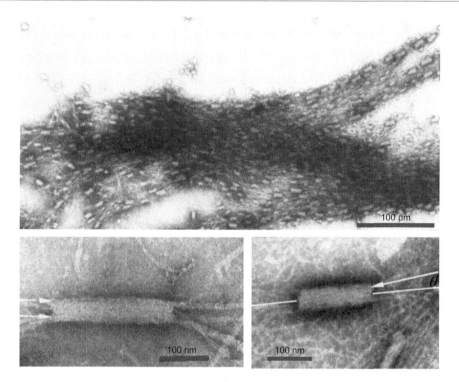

Fig. 11.3 Transmission electron micrographs of dense aggregates observed within disrupted developing macrofibrils from the upper bulb cortex cells of human scalp hair. (Reprinted from J. Cell Biol., 1985, Jones and Pope [23], with permission from The Rockefeller University Press)

that they are side-by-side agglomerates which are (in the case of keratins) almost universally visualised as a 32-chain structure, consisting of eight tetramers, or 16 dimers. One ULF has a somewhat disordered lateral structure and is about 50–60 nm long and approximately 10–15 nm diameter, making it anisometric with a length:diameter ratio of about 5:1 [31, 32].

11.3 Different Models of Macrofibril Assembly

All this forms a backdrop to initial macrofibril formation. Much of the focus in trying to understand macrofibril assembly during Zones B-D, has been on explaining how the different macrofibril architectures found in the cortex of fine diameter wool fibers may arise. Two papers published in 2003 [25, 33] suggested the first detailed approach to establishing how macrofibrils might form, and hypothesised a mechanism based on layering of sheets of parallel interacting IFs. More recently, another approach, the principal focus of this chapter, suggests that the ULFs of cortical keratins polymerise into IFs within dense liquid crystal aggregates (tactoids) that grow and coalesce into mature macrofibrils [26, 34, 35].

11.3.1 Fraser-Parry Model

In the Fraser-Parry model of macrofibril assembly (Fig. 11.4), fully formed IFs interact laterally to form sheets. These lateral interactions are mediated by an early-expressing keratin associated KAP or possibly inter-filament head-group interactions [33]. Intermediate filaments in these sheets are oriented with respect to neighbors. Interaction between sheets is then based on regular patterns of IF surface features across the sheets. Links between IFs in a sheet are considered to be reasonably flexible and depending on the pattern of surface features, or

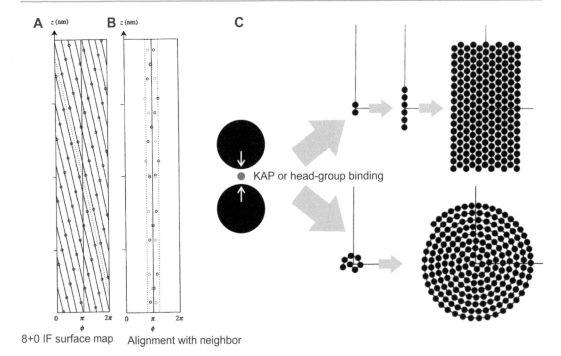

Fig. 11.4 Summary of the Fraser-Parry model of macrofibril nucleation. (**a**) Surface groups (black circles) on a Zone B (unoxidised) 8 + 0 IF (**b**) have some level of potential alignment with those of a neighbouring IF (grey dashed-circles) which allows (**c**) binding of oriented IFs via KAP or head-group interactions into a sheet. Sheets then interact with neighbouring sheets or themselves (possibly mediated by more KAP-IF interactions) which result in linear (top) or tilted (bottom) longitudinal alignment, and meso/paracortical or orthocortical macrofibrils respectively. ((**a**, **b**) reprinted from, J. Struct. Biol., 2003, Fraser and Parry [33]; (**c**) reprinted from, J. Struct. Biol., 2003, Fraser, Rogers [25], both with permission from Elsevier)

interactions between IFs and KAPs, the sheets will align in different ways that result in different macrofibril architectures [25]. In the simplest case, sheets align longitudinally parallel, and the resulting macrofibril is mesocortical/paracortical like. If alignment is such that there is a slight rotation between adjacent sheets longitudinally, then they also wrap around one another to form a cylinder with increasing angular tilt. Addition of more slightly tilted sheets and wrapping results in a double-twist structure as associated with the orthocortex. The model is underpinned by significant theoretical consideration of IF supramolecular structure (Chap. 6), especially regarding the implications of alignment of headgroups between adjacent IFs, and is supported by limited transmission-electron microscopy observations of cross-sections of fully keratinized fibres.

11.3.2 Mesophase Model

11.3.2.1 Rod-Based Mesophase Formation

Liquid crystals (mesophases) come in different types: thermotropic mesophases form when certain crystalline materials are elevated in temperature to just above their melting point, and lyotropic mesophases form from anisometric solutes or dispersions in a liquid medium (in this case cytoplasm). The model here considers that the precursors of macrofibrils form as lyotropic mesophase droplets (tactoids) composed of relatively large mesogens (compared to most studied), namely keratin unit length filaments, and short polymers derived from their longitudinal aggregation. Keratin proteins have a molecular weight in the order of 50,000 Da, thus making the mesogen mass some low multiple of 1.6 M Da.

Early work on liquid crystals was carried out almost entirely in the context of physics and physical chemistry, starting in the late nineteenth century with the work of Reinitzer, Lehmann [36] and Friedel [37], summarised by Mitov [38]. A key contribution to the thermodynamics of mesophase formation relevant to macrofibril formation came in 1949 with the prediction that phase separation can occur in dispersions of anisometric particles, based on their entropy (translational, rotational, orientational) as a function of particle concentration [39]. By this time liquid crystal tactoids formed from large biological rods in the form of tobacco mosaic virus had been observed [40]. To some extent, the subject then merged with the developing topic of the physical chemistry of polymer solutions, as developed using lattice models, initially by Flory and Huggins [41]. Subsequently, Flory's work contributed to an alternative way of developing the theory of lyotropic mesophase separation [42, 43]. Flory's theory of phase separation has been supplanted, at least among physicists, by more modern theories which are capable of producing more precise results especially for well-studied mesogens.

Within a series of papers from Flory's group in 1978, Flory and Frost recognised that the formation of a mesophase from a polymer solution provided an appealing mechanism for self-assembly of biological macromolecules. We can do no better than quote the essential passage, with a few inessential elisions, from Flory and Frost [44].

Assume for simplicity that the species of size x is a polymer M_x comprising x units joined in rigid, rodlike arrangement. Let the chemical potential of structural units be fixed (as in an open chemical system). Random interchange of units between species under the aegis of a catalyst mobilizes the array of processes typified by

$$M_{x_1} + M_{x2} \leftrightarrow M_{x_1+x_2}$$

For purposes of exposition, we assume the catalyst to be effective in the isotropic phase. That this arbitrary restriction can have no effect on the final state of the equilibrium is assured by the second law of thermodynamics...... But the coexisting anisotropic phase will sequester the larger species. Replenishment of these species through chemical exchange perpetuates transfer to the ordered phase. It follows that the polymer must ultimately be transferred in its entirety to the anisotropic phase for any sustained concentration sufficiently above the hypothetical limit ....

The combination of a chemically mobile aggregation process potentially capable of yielding highly anisometric particles and phase separation provides a singularly simple mechanism for self-organisation with possible relevancy to biological systems....

Thus, the tendency for concatenation need not be large for complete conversion, in the presence of a catalyst, to a highly ordered phase.

The mesophase model for macrofibrils started independently but converged with the Flory-Frost approach [26]. Notably, in the keratin model, polymerization continues within the anisotropic phase (the macrofibril template) even in the absence of a catalyst in that phase, in the approach to a most probable distribution.

While the mesophase model as developed here is specific for intracellular keratin structures, it has close affinities with other such models that describe self-organising repeating structures in a range of fibrous tissues, including collagen and chitin [11, 45]. Bouligand, the long-time leader of the important French school exploring the liquid crystal origins of many biological tissues, summed it up thus [46]:

These geometrical similarities between fibrous tissues and liquid crystals were at the origin of numerous works defining a new type of self-assembly involving two successive steps: a phase transition from an isotropic liquid phase to a liquid-crystalline one, followed by a sol-gel stabilization of the liquid crystal ...

11.3.2.2 Mesophase Theory Requirements

Flory's writings contain other statements and predictions on aspects of polymer solution thermodynamics which are also central to the stability in solution of the precursors of macrofibrils. One such issue relates to whether isolated helical macromolecules can exist in solution, which was under serious debate in the 1960s. Flory and Leonard [47] showed quite clearly that the stability of isolated helices in solution could be achieved by the release of the configurational and mixing entropy of the regular side chains on the polymer, upon solution. In the macrofibril scenario, helical parts of the IF structure are quite highly hydrophobic, and it is a principle of the model that their stability in solution is only achieved by the entropy gain in solution of the pendant head and tail groups. This principle is displayed also by many of the well-known polymeric mesogens, such as classic examples from polysaccharide science, schizophyllan and xanthan [48], which have regular pendant mono- or tri-saccharide side chains, in contrast to cellulose which has limited solubility in water and is not in itself a mesogen.

Such a role (among others) for the head groups is a vital aspect of the mesophase model. Most of the primary sequence diversity occurs within the head groups of macrofibril forming keratins. It is also important to consider how ULFs are internally structured, especially how the dimers overlap to generate different end effects. Filament formation has been demonstrated in vimentin IFs to be a result of end-to-end annealing of fully formed ULFs, by addition of ULFs to existing filaments, and by end-to-end association of filaments [31, 32]. Other options such as growth of filaments by tetramer addition to filament ends, were ruled out. Therefore, the arrangement of the ends and changes to that arrangement will likely influence IF formation.

Detailed analysis of the periodic charge distributions on Type I and Type II keratin chains [49, 50] led to understanding of three preferred modes of chain overlap in which electrostatic charge interactions are maximised—A_{11}, A_{22}, and A_{12} (see Chap. 6, Fig. 6.4). In forming a ULF, interaction mode is important because in A_{11} or A_{22} mode the ends will have protruding regions of keratin (in the form of dimers) at each end giving them a "shaggy ended" appearance (Fig. 11.5). Work on assembly of vimentin IFs indicates that these "shaggy ended" ULFs interdigitate together via presumed enthalpic electrostatic interactions to form filaments, with inter-ULF bonds involving the two regions of individual dimers [51].

While cortical hair keratin IF polymerization and vimentin IF polymerization have much in common they also differ significantly. Vimentin

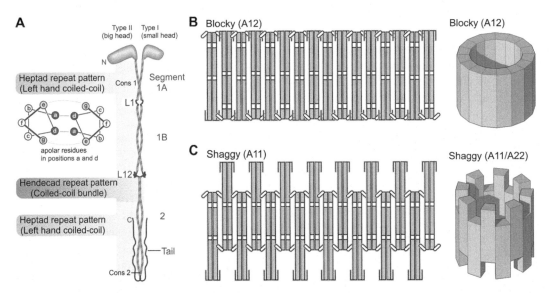

Fig. 11.5 Keratins and ULF activation state. (**a**) Schematic of a hair cortex keratin dimer showing elements of structure, including non-helical head and tail groups. (**b**) Stable or blocky form ULF with antiparallel tetramers in A12 arrangement, and potential steric repulsion between head and tail regions, and between associating ULFs. (**c**) Shaggy-end form of ULF, as activated for polymerization, possibly A11 (shown) or A22 arrangement with steric interactions alleviated by dimer translation. In both (**b**) and (**c**) an "unrolled" version of the cylindrical ULF is shown for clarity. ((**b and c**) reprinted from J. Text. Eng., 2016, McKinnon, Harland [20], with permission from the Textile Machinery Society of Japan)

IFs, skin epidermal keratin IFs, and IFs of the inner root sheath of the follicle form as isolated single filaments which form networks, while the keratin of the cortex forms within a densely packed elongated tactoid droplet.

Another requirement of the mesophase theory is that concentration of the mesogen (in this case a ULF or small ULF polymer) can increase within the cytoplasm (or isotropic phase) to a critical point at which tactoids (or anisotropic phase) forms. Temperature can also be a variable, but in follicles, the temperature remains within a tight mammalian range (35–37°C), meaning that the concentration of ULF monomers, along with that of the initiator species which creates an axial shift in the ULF, in the isotropic phase, are the primary variables.

The final requirement is that the keratins must have features (in this case head and tail groups) that stabilise the solubility of their central helix regions, maintain a ULF organisation (A_{12}) that will not polymerise into IFs at physiological temperatures, and that can be modified to transition, without destroying the ULF, into a form that can be polymerised (A_{11} or A_{22}).

Type I trichokeratins such as, for example, K31 and K33 isoforms have relatively small head groups (56 residues), which, we hypothesise, by their configurational and mixing entropy, along with that of all the other head and tail appendages, assist to stabilise the rod structure in an aqueous medium but do not so detract from the rod-like properties that neighbouring molecules cannot sense their mutual rod-like character. The head-group selection in the early stage of keratin fibre biosynthesis is thus quite critical to the creation and maintenance of a mesophase. So that a concentration of stable non-polymerised ULFs can be built up, and then a mesophase triggered which involves reconfigured ULFs, can be controlled by mixing and matching specific keratins within dimers.

By contrast, this also explains why cytokeratins do not readily form mesophases. Essentially the head and tail groups of the cytokeratins (homodimers) are such that ULFs are formed from tetramers in an A_{11} or A_{22} configuration, which are ready to proceed directly to filaments by a process of dimer interdigitation. There is thus no opportunity for a build-up of ULF concentration leading to phase segregation, of the sort we describe for trichokeratins, once fibre formation is triggered. None of the cytokeratins have Type I head groups as short as the 56 residues of the K31 and K33 isoforms, together responsible for most of the Type I complement of keratin fibres. The shortest cytokeratin heads are on Type I K20 (70 residues) and K23 (72) [52].

11.3.2.3 Key Features of Mesophase Formation

Thermodynamics and structural features combine to ensure that hair keratins interact in a pathway to consolidated fibre formation. Here we summarise in compact form those key features which are peculiar to fibre formation.

- Keratin dimers with head and tail groups that enable them (and their aggregates) to remain soluble in an aqueous environment (including that in a mesophase).
- ULFs which are assembled initially from short-head Type I proteins in the A_{12} (blocky) configuration to prevent premature IF formation.
- Controlled expression or availability of Type I long-head) proteins (K35, K34 and K38) which can be incorporated into ULFs to displace the ULF into a shaggy form, acting as a polymerization initiator.
- Critical concentrations of ULFs in both initiator (long-head Type I) and monomer (short-head) forms being reached, to bring about a sudden filament formation event leading to phase transfer and macrofibril template formation.
- After the initial tactoid formation stage, the composition of the co-existing anisotropic and isotropic phases will remain approximately constant, with the volume fraction of the former rapidly increasing, until almost all mesogens are transferred to the anisotropic phase.
- The anisotropic phase is much enriched in the longer filaments, which are almost completely absent from the isotropic phase.

- This diversity of filament length assists the initial formation of tactoids by minimising their interfacial energy.
- Tactoid coalescence removes surface energy constraints and leads to filament extension requiring no catalyst other than filament ends.

11.3.2.4 The Model for ULF Polymerization into Tactoids

The mesophase model of macrofibril formation [26, 35] takes the concept introduced above from Flory and Frost [44] that an equilibrium polymerization process can be coupled to phase transfer as a possible route to self-assembly in biological systems. We further introduce into the model features of a "sudden event" polymerization, as typified by floor and ceiling temperature polymerizations, as classically developed by Tobolsky and Eisenberg [53, 54], and which in their technical terms is a Type Ia equilibrium polymerization, with two equilibrium constants, the first relating to an activation event. In more modern soft matter physics parlance, the polymerization model is a co-operative nucleated supramolecular polymerization [55]. This modern field still includes application of the classic Tobolsky-Eisenberg small-monomer theory to macromolecular phenomena [56].

In this case a "blocky" ULF has to be activated to shift into a "shaggy" ULF through the incorporation into the ULF of a long Type I head-containing tetramer (which might contain two long-head Type I chains). The mechanism is as follows, with U representing an abundant blocky (A_{12}) ULF species, U* representing an activated shaggy (A_{11} or A_{22}) ULF containing some long-head Type I chains, and U_n representing a polymeric species formed from n ULFs (Fig. 11.6).

Activation $U \rightarrow U^*$ equilibrium constant K
(incorporation of long-head Type I keratin into the ULF)
Propagation $U^* + U \rightarrow {}^*UU^*$ equilibrium constant K_3
 ${}^*U_n^* + U \rightarrow {}^*U_{n+1}^*$ equilibrium constant K_3
Equilibrium ${}^*U_n^* + {}^*U_m^* \leftrightarrow {}^*U_{n+m}^*$ equilibrium constant K_5
$K_3 = KK_5$ as can easily be shown.

The K, K_3 and K_5 nomenclature is carried over from Tobolsky and Eisenberg. For a sudden event ceiling temperature phenomenon to occur, the ratio K_3/K must be very high, which is to say that the probability of the species U* must be low. Polymer formation under these conditions will not occur until, at biological temperature, critical values of U* and U are reached, at which point sudden polymer formation will occur. In the absence of such conditions the ULF polymers (i.e. IFs) are effectively thermally depolymerised – biological temperature is above the ceiling temperature for the polymerization. We must of course always keep in mind that this process is coupled to a phase transfer, that converts the early stages of the equilibrium polymerization to a non-equilibrium process. The equilibrium conditions of such a ceiling temperature are thus in fact never quite reached – phase transfer

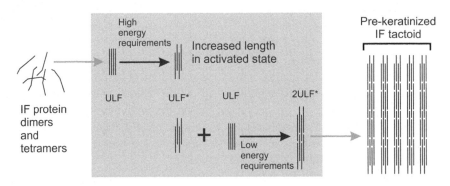

Fig. 11.6 Stylised illustration IF formation, showing our proposed mechanism of generation of multi-ULF precursors in central box. Activated ULFs are indicated by an asterisk

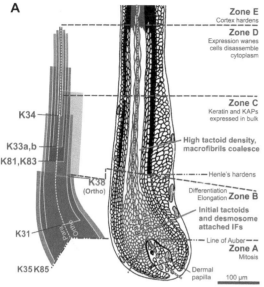
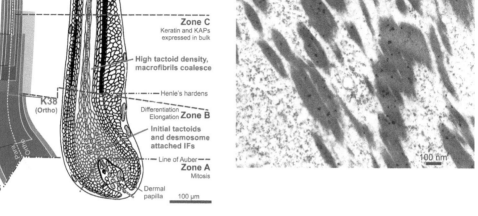

Fig. 11.7 Keratin expression pattern and tactoids. (**a**) Expression pattern of major wool cortex keratins [59] with estimated matching to ultrastructural events. Keratins: Type I, red; Type II, blue. Protein labels indicate the approximate point in the follicle at which expression of that protein begins. Horizontal levels in the expression diagram match those in the follicle diagram. Expression of any protein continues up the corresponding vertical bar, and ceases at the top. (**b**) An oblique longitudinal section through an array of IF tactoids in a wool follicle, showing their tactoidal shape. This micrograph depicts an ultra-thin section (ca. 50–80 nm) through the tactoid array, and electron-dense particles on tactoids are from immunolabeling for keratin protein. ((**a**) reprinted from, J. Text. Eng., 2016, McKinnon, Harland [20], with permssion from The Textile Machinery Society of Japan, (**b**) reprinted from, J. Cell Biol., 1986, French and Hewish [60], with permission from The Rockefeller University Press)

intervenes, as postulated by Flory, and consistent with our original conception.

One parameter of the above formulation is critical – a ceiling temperature can only be observed if the critical chain extension step is exothermic, meaning that both the free energy and the enthalpy of the polymerization reaction are negative [57]. In a biological context, in which many reactions are driven predominantly by entropy changes, this is a reasonably demanding test for the model. The work of Ishii, Abe [58] studying the stepwise formation of keratin IFs by isothermal titration calorimetry, confirmed that the length-wise aggregation of dimers to form displaced tetramers is indeed exothermic. As this is essentially the key reaction in creating filaments from shaggy ULFs with exposed dimers protruding from the ends, we may enjoy some relief that the model passes this test. A similar conclusion has emerged from work on vimentin [51].

11.3.2.5 Aligning the Mesophase Model with Keratin Expression in the Follicle

The stepwise expression and timing of keratins during cortex development was described in Chap. 10 (Sect. 10.3) and for the cortical keratins is summarised in Fig. 11.7a. The first expressed Type I keratin is K35 (with a long head), which soon gives way to abundant expression of K31 (a short-head keratin). The matching Type II keratin over this time is K85. K35/K85 dimers are thought to fulfil some early stage cytoskeletal functions partly associated with desmosomes. It is also relevant that IFs undergo scission and reformation and it is possible for free K35-based IF fragments to form in this way. Our conception is that "blocky" K31/K85 ULFs build up in the cytoplasm as monomer, along with ULF or IF fragments incorporating K35 as potential initiator, until the levels of each are high enough to trigger filament formation. This will first occur just below the true ceiling temperature [53]. We also note the presence of long-head forms of K34 and of K38 being expressed in the filament formation zone. The latter may well be supplementary sources of initiator species, expressed when the K35 supply is in itself too low to reach the critical level for filament formation.

Because of the high concentration of anisometric ULFs in the cytoplasm (the isotropic phase), growing filaments will very quickly separate as small tactoids of anisotropic phase, the mesophase. Well-formed spindle-shaped tactoids are clearly visible dispersed in the cytoplasm with lengths as short as 700 nm (about 12 ULFs long) (Fig. 11.7b). The interfacial energy of the tactoid is an important factor at this stage and the distribution of filament lengths is partly constrained by the interfacial energy. As tactoids coalesce, surface energy becomes a less limiting factor and filaments grow in accord with the trend to attain an equilibrium distribution of polymer (filament) lengths. This process was foreseen by Flory. We have pointed out elsewhere (McKinnon and Harland [35], McKinnon et al. [20]) that this process can be usefully envisaged as an example of a one-dimensional Ising problem in statistical mechanics, as first described in a polymer context by Tobolsky [57]. This places our model for IF formation on the same mathematical basis as that of other protein transitions, such as helix-coil and $\alpha \rightarrow \beta$ transitions.

11.3.2.6 Mesophase Model and Different Macrofibril Architecture

One particular appeal of the liquid crystal approach to self-assembly is that a single mesogen can form different shaped tactoids with differing internal architecture depending on differing formation conditions. Architectures can include both double twist and nematic, corresponding to the classical description of orthocortex and paracortex macrofibrils respectively from fine-diameter wools.

In the model, there are two factors that combine to determine tactoid architecture and external morphology, the concentrations of initiator and of primary monomer (shaggy and blocky ULFs respectively). If in a given cell the level of initiator is low, phase separation will occur at a higher ULF concentration: and *vice versa*. Variations in these levels give rise to different lengths of IF entering the tactoids. If the ULF concentration is high, phase separation will involve shorter IFs. A high initiator concentration will result in separation at lower ULF concentrations and result in longer IFs being transferred to the mesophase.

Short IFs can approach one another more closely in the mesophase, providing an opportunity for chiral twisting forces to be exerted, and thus resulting in a double-twist structure in the mesophase. Very long IFs are only able to line up in a hexagonal fashion but maintain a greater spacing. This difference in populations of long and short filaments is controlled by the relative production rate of, in the first instance K35 and K31, and therefore is under the direct control of conventional biological gene expression mechanisms. By varying the rate, various lengths, and populations of lengths can be generated. The tendency of similar-length IFs to accumulate within the same tactoids to minimise internal strain enables different tactoids (even within a single cell) to have graduations of double-twist intensity (Fig. 11.8).

Finally, the idealised model suggests tactoids (whatever the twist intensity) which are perfect structures. Hypothetically such structures would form in an open system in which they remain liquid, but within the developing cortex cells they are rapidly constrained and distorted (easily done as they are liquid) by cell boundaries, organelles and other macrofibrils, as well as by dehydration and oxidative processes (keratinisation). What we see in mature fibres is akin to fossils of the original liquid crystals.

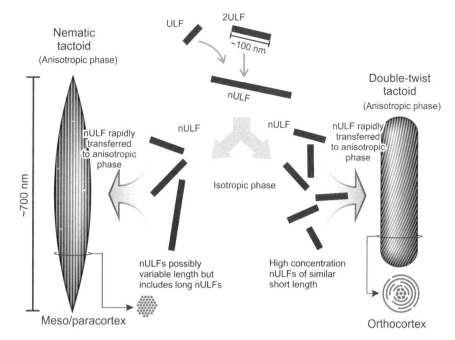

Fig. 11.8 Schematic summary of relationship between population of different short ULF polymers in the cytoplasm and the type of tactoids generated. (Reprinted from, J. Struct. Biol., 2011, McKinnon and Harland [26], with permission from Elsevier)

References

1. Horio, M., & Kondo, T. (1953). Crimping of wool fibers. *Textile Research Journal, 23*(6), 373–387.
2. Swift, J. A. (1977). The histology of keratin fibers. In R. S. Asquith (Ed.), *Chemistry of natural protein fibers* (pp. 81–146). London: Wiley.
3. Rogers, G. E. (1959). Electron microscopy of wool. *Journal of Ultrastructure Research, 2*(3), 309–330.
4. Harland, D. P., et al. (2011). Arrangement of trichokeratin intermediate filaments and matrix in the cortex of Merino wool. *Journal of Structural Biology, 173*, 29–37.
5. Whiteley, K. J., & Kaplin, I. J. (1977). The comparative arrangement of microfibrils in ortho, meso, and paracortical cells of Merino wool fibres. *Journal of the Textile Institute, 68*(11), 384–386.
6. Bryson, W. G., et al. (2009). Cortical cell types and intermediate filament arrangements correlate with fiber curvature in Japanese human hair. *Journal of Structural Biology, 166*(1), 46–58.
7. Harland, D. P., et al. (2014). Three-dimensional architecture of macrofibrils in the human scalp hair cortex. *Journal of Structural Biology, 185*(3), 397–404.
8. Thomas, A., et al. (2012). Interspecies comparison of morphology, ultrastructure and proteome of mammalian keratin fibres of similar diameter. *Journal of Agricultural & Food Chemistry, 60*(10), 2434–2446.
9. Woods, J. L., et al. (2011). Morphology and ultrastructure of antler velvet hair and body hair from red deer (Cervus elaphus). *Journal of Morphology, 272*(1), 34–49.
10. Caldwell, J. P., et al. (2005). The three-dimensional arrangement of intermediate filaments in Romney wool cortical cells. *Journal of Structural Biology, 151*(3), 298–305.
11. Neville, A. C. (1993). *Biology of fibrous composites: Development beyond the cell membrane* (1st ed.). New York: Cambridge University Press. 214.
12. Harland, D. P., Vernon, J. A., Woods, J. L., Nagase, S., Itou, T., Koike, K., Scobie, D. A., Grosvenor, A. J., Dyer, J. M., & Clerens, S. (2018). Intrinsic curvature in wool fibres is determined by the relative length of orthocortical and paracortical cells. *The Journal of Experimental Biology, 221*(6), jeb172312.
13. Rogers, G. E. (1964). Structural and biochemical features of the hair follicle. In W. Montagna & W. C. Lobitz (Eds.), *The epidermis* (pp. 179–236). New York: Academic.

14. Orwin, D. F. G., Thomson, R. W., & Flower, N. E. (1973). Plasma membrane differentiations of keratinizing cells of the wool follicle (2. desmosomes). *Journal of Ultrastructure Research, 45*, 15–29.
15. Roth, S. I., & Helwig, E. B. (1964). The cytology of the cuticle of the cortex, the cortex, and the medulla of the mouse hair. *Journal of Ultrastructure Research, 11*, 52–67.
16. Birbeck, M. S. C., & Mercer, E. H. (1957). The electron microscopy of the human hair follicle. Part1. Introduction and the hair cortex. *Journal of Biophysical and Biochemical Cytology, 3*, 203–213.
17. Orwin, D. F. G. (1979). The cytology and cytochemistry of the wool follicle. *International Review of Cytology, 60*, 331–374.
18. Morioka, K. (2005). *Hair follicle, differentiation under the electron microscope – an atlas*. Tokyo: Springer. 152.
19. Langbein, L., & Schweizer, J. (2005). Keratins of the human hair follicle. *International Review of Cytology, 243*, 1–78.
20. McKinnon, A. J., Harland, D. P., & Woods, J. L. (2016). Relating self-assembly to protein expression in wool cortical cells. *Journal of Textile Engineering, 62*(6), 123–128.
21. Wang, H., et al. (2000). In vitro assembly and structure of trichocyte keratin intermediate filaments: A novel role for stabilization by disulfide bonding. *Journal of Cell Biology, 151*(7), 1459–1468.
22. Paton, L. N. (2005). *Proteomic analysis of wool intermediate filament proteins*. Christchurch: School of Biological Sciences, University of Canterbury.
23. Jones, L. N., & Pope, F. M. (1985). Isolation of intermediate filament assemblies from human hair follicles. *Journal of Cell Biology, 101*(4), 1569–1577.
24. Matsunaga, R., et al. (2013). Bidirectional binding property of high glycine–tyrosine keratin-associated protein contributes to the mechanical strength and shape of hair. *Journal of Structural Biology, 183*(3), 484–494.
25. Fraser, R. D. B., Rogers, G. E., & Parry, D. A. D. (2003). Nucleation and growth of macrofibrils in trichocyte (hard-α) keratins. *Journal of Structural Biology, 143*, 85–93.
26. McKinnon, A. J., & Harland, D. P. (2011). A concerted polymerization-mesophase separation model for formation of trichocyte intermediate filaments and macrofibril templates 1: Relating phase separation to structural development. *Journal of Structural Biology, 173*(2), 229–240.
27. Parry, D. A., et al. (2007). Towards a molecular description of intermediate filament structure and assembly. *Experimental Cell Research, 313*(10), 2204–2216.
28. Sokolova, A. V., et al. (2006). Monitoring intermediate filament assembly by small-angle X-ray scattering reveals the molecular architecture of assembly intermediates. *Proceedings of the National Academy of Science of USA, 103*, 16206–16211.
29. Franke, W. W., Schiller, D. L., & Grund, C. (1982). Protofilamentous and annular structures as intermediates during reconstitution of cytokeratin filaments in vitro. *Biology of the Cell, 46*, 257–268.
30. Herrmann, H., et al. (2002). Characterisation of early assembly intermediates of recombinant human keratins. *Journal of Structural Biology, 137*, 82–96.
31. Kirmse, R., et al. (2010). Plasticity of intermediate filament subunits. *PLoS One, 5*(8), e12115.
32. Portet, S., et al. (2009). Vimentin intermediate filament formation: In vitro measurement and mathematical modeling of the filament length distribution during assembly. *Langmuir, 25*(15), 8817–8823.
33. Fraser, B. R. D., & Parry, D. A. D. (2003). Macrofibril assembly in trichocyte (hard α-) keratins. *Journal of Structural Biology, 142*(2), 319–325.
34. McKinnon, A. J. (2006). The self-assembly of keratin intermediate filaments into macrofibrils: Is this process mediated by a mesophase? *Current Applied Physics, 6*, 375–378.
35. McKinnon, A. J., & Harland, D. P. (2010). The role of liquid-crystalline structures in the morphogenesis of animal fibers. *International Journal of Trichology, 2*(2), 101–103.
36. Lehmann, O. (1889). Über fliessende Krystalle. In *Zeitschrift für Physikalische Chemie* (p. 462).
37. Friedel, G. (1922). Les états mésomorphes de la matière. *Annals of Physics, 9*(18), 273–474.
38. Mitov, M. (2014). Liquid-crystal science from 1888 to 1922: Building a revolution. *Chemphyschem, 15*(7), 1245–1250.
39. Onsager, L. (1949). The effects of shape on the interaction of colloidal particles. *Annals of the New York Academy of Sciences, 51*(4), 627–659.
40. Bernal, J. D., & Fankuchen, I. (1941). X-ray and crystallographic studies of plant virus preparations. *The Journal of General Physiology, 25*(1), 111.
41. Flory, P. J. (1953). *Principles of polymer chemistry* (1st ed., p. 688). The George Fisher Baker Non-Resident Lectureship in Chemistry at Cornell University, Ithaca, United States Cornell University Press.
42. Flory, P. J. (1984). Molecular theory of liquid crystals. *Advances in Polymer Science, 59*, 1–36.
43. Flory, P. J. (1956). Phase equilibria in solutions of rod-like particles. *Proceedings of the Royal Society of London. Series A. Mathematical and Physical Sciences, 234*(1196), 73.
44. Flory, P. J., & Frost, R. S. (1978). Statistical thermodynamics of mixtures of rodlike particles. 3. The most probable distribution. *Macromolecules, 11*(6), 1126–1133.
45. Brown, A. I., Kreplak, L., & Rutenberg, A. D. (2014). An equilibrium double-twist model for the radial structure of collagen fibrils. *Soft Matter, 10*(42), 8500–8511.
46. Bouligand, Y. (2008). Liquid crystals and biological morphogenesis: Ancient and new questions. *Comptes Rendus Chimie, 11*(3), 281–296.

47. Flory, P. J., & Leonard, W. J. (1965). Thermodynamic properties of solutions of helical polypeptides. *Journal of the American Chemical Society, 87*(10), 2102–2108.
48. Sato, T., & Teramoto, A. (1996). Concentrated solutions of liquid-crystalline polymers. *Advances in Polymer Science, 126*, 85–161.
49. Parry, D. A. D. (1990). Primary and secondary structure of IF protein chains and modes of molecular aggregation. In R. D. Goldman & P. M. Steinert (Eds.), *Cellular and molecular biology of intermediate filaments* (pp. 175–204). New York: Plenum.
50. Parry, D. A. D., & Steinert, P. M. (1999). Intermediate filaments: Molecular architecture, assembly, dynamics and polymorphism. *Quarterly Reviews of Biophysics, 32*(2), 99–187.
51. Köster, S., et al. (2015). Intermediate filament mechanics in vitro and in the cell: From coiled coils to filaments, fibers and networks. *Current Opinion in Cell Biology, 32*, 82–91.
52. Szeverenyi, I., et al. (2008). The human intermediate filament database: Comprehensive information on a gene family involved in many human diseases. *Human Mutation, 29*(3), 351–360.
53. Tobolsky, A. V., & Eisenberg, A. (1962). Transition phenomena in equilibrium polymerization. *Journal of Colloid Science, 17*(1), 49–65.
54. Tobolsky, A. V., & Eisenberg, A. (1960). A general treatment of equilibrium polymerization. *Journal of the American Chemical Society, 82*(2), 289–293.
55. De Greef, T. F. A., et al. (2009). Supramolecular polymerization. *Chemical Reviews, 109*(11), 5687–5754.
56. Douglas, J. F., Dudowicz, J., & Freed, K. F. (2008). Lattice model of equilibrium polymerization. VII. Understanding the role of "cooperativity" in self-assembly. *The Journal of Chemical Physics, 128*(22), 224901.
57. Tobolsky, A. V., & MacKnight, W. J. (1965). *Polymeric sulfur and related polymers*. New York/London: Interscience.
58. Ishii, D., et al. (2011). Stepwise characterization of the thermodynamics of trichocyte intermediate filament protein supramolecular assembly. *Journal of Molecular Biology, 408*(5), 832–838.
59. Yu, Z., et al. (2011). Annotations of sheep keratin intermediate filament genes and their patterns of expression. *Experimental Dermatology, 20*(7), 582–588.
60. French, P. W., & Hewish, D. R. (1986). Localization of low-sulfur keratin proteins in the wool follicle using monoclonal antibodies. *Journal of Cell Biology, 102*(4), 1412–1418.

Part IV

Hair Chemistry and Thermodynamics

Crosslinking Between Trichocyte Keratins and Keratin Associated Proteins

12

Santanu Deb-Choudhury

Contents

12.1 Introduction 173
12.2 Assembly of Intermediate Filaments 175
12.3 Keratin-KAP Interactions 175
12.4 Accessibility of Cysteines 176
References 182

Abstract

Trichocyte keratins differ considerably from their epithelial cousins in having a higher number of cysteine residues, of which the greater proportion are located in the head and tail regions of these proteins. Coupled with this is the presence of a large number of keratin associated proteins in these fibres that are high in their cysteine content, the high sulfur proteins and ultra-high sulfur proteins. Thus it is the crosslinking that occurs between the cysteines in the keratins and KAPs that is an important determinant in the functionality of wool and hair fibres. Studies have shown the majority of the cysteine residues are involved in internal crosslinking in the KAPs leaving only a few specific cysteines to interact with the keratins, with most evidence pointing to interactions between these KAP cysteines and the keratin head groups.

Keywords

Trichocyte keratins · Keratin associated proteins · Protein crosslinks · Disulfide bridges · Cysteine accessibility

12.1 Introduction

The most significant structural component of hair and wool fibres is protein [1–3]. Fibres such as wool and hair can be described mainly as an assembly of longitudinally arranged trichocyte keratin proteins that form the intermediate filaments (IF) which are embedded in an amorphous matrix composed mainly of keratin associated proteins (KAPs). The IFs and the KAPs form a

S. Deb-Choudhury (✉)
AgResearch Ltd., Lincoln, New Zealand
e-mail: Santanu.Deb-choudhury@agresearch.co.nz

dense cortex that is further covered by a relatively thin layer of overlapping cuticle cells having a peripheral curvature. Medulla cells found at the centre of wool and hair fibre consist of cell remnants that are bound together by proteins found in desmosomes, gap junctions and tight junctions [4]. Keratins and KAPs have an extensive network of inter and intramolecular protein-protein crosslinks that act as molecular bridges providing them with natural strength [5]. The physico-mechanical properties of these fibres are therefore intrinsically related to the frequency, pattern and position of these crosslinks within the network [6–8]. Wool and hair also contain a number of non-keratinous proteins including cell adhesion proteins such as desmoplakin, desmoglein and proteins found in intermacrofibrillar material and the cellular remnants of the cortex, cuticle and medulla. These proteins also contribute to forming the network, for example, the C-terminus of desmoplakin attaches to intracellular trichocyte keratin proteins and the N-terminal domain binds a transmembrane protein, desmoglein, forming intracellular bridging [9–11]. An important role of these proteins within the network is to help disperse the mechanical stress that the fibre may be exposed to [12, 13]. Isopeptide bonds have also been reported to be present in hair. The presence of enzymes that catalyses the formation of these bonds such as transglutaminase supports other evidence of the existence of these in the network of proteins specifically contributing to the association of keratins and desmosome proteins [3, 14–17]. The cuticle of hair has been long known to contain isopeptide bonds [18, 19] and it is highly likely that the presence of these bonds is important for defining the high mechanical strength and chemical resistance of the exocuticle layers. Likewise the medulla is also known to contain isopeptide bonds, albeit between different proteins than in the cuticle [20, 21].

An important interaction in wool and hair fibres is between the trichocyte keratins and the considerably diverse KAPs (Chap. 3, Table 3.5). The type and extent of trichocyte keratins and KAP interactions or crosslinks underpin the stability of the fibre structure. Keratin-KAP and KAP-KAP interactions, through the formation of inter and intramolecular disulfide bonds, form the main constituent of the fibre crosslink network. The consistency of these crosslinks so far remains inconclusive and it is unclear whether the network is dynamic in nature and subject to change through disulfide interchange reactions. Structure prediction algorithms predict that KAPs are unusual in that they lack any secondary structural features [22]. This is in sharp contrast to the highly organized keratin proteins. Keratin heterodimers form tetramers and then protofilaments with a molecular sub-unit packing length of 47 nm [23]. These are surrounded by an amorphous matrix composed of KAP aggregates that stabilize the structure via a covalent network of disulfide bridges giving structure to the cytoplasm of the cortical cells [24]. The multiprotein network of keratin and KAP proteins provide hair and wool their mechanical strength. The building blocks of intermediate filaments (IFs) are composed of eight tetramers and each tetramer contains a pair of heterodimers arranged in an antiparallel manner. Heterodimers are laterally arranged with a 7 + 1 configuration, where seven of these are located around a central heterodimer [25] (Chap. 6, Sect. 6.5). Adjacent heterodimers are arranged with a 19.8 nm stagger and the ones on the periphery of the IFs are thought to have their head groups protruding from the filament into the matrix in a spiral manner with a repeat of 47 nm [26, 27]. The helical manner of the arrangement of the head groups has been hypothesized to have a unique molecular mechanism [28]. It has been postulated that if the head and tail regions of trichocyte keratins recognize each other, then they will be forced to align in a helical manner rather in a parallel fashion. It is therefore highly likely that the interactions in the IF-matrix composite are between the less structured head and tail regions of the keratin dimers with KAPs through the formation of disulfide bridges with the highly structured rod domains of the keratin dimers possibly playing a less vital role in these crosslinks [29].

12.2 Assembly of Intermediate Filaments

Earlier studies have reported that the assembly of trichocyte keratins into intermediate filaments are initiated in a reducing environment, for instance in the cortical cells above the dermal papilla of the hair follicle [2]. Molecular alignments of reduced trichocyte keratins indicate the possibility of formation of two or three disulfide bonds due to the appropriate alignment of cysteine residues. At a later stage upon oxidation, which occurs further up the hair follicle during terminal differentiation and cell death, disulfide bond formation occurs. Instead of a possible three, a much larger number of disulfide bridges forms upon oxidation, that include six between the cysteines present in the rod regions [30, 31] and four between the rod domain and the head domain cysteines. The formation of these bonds reflect a molecular shift in the antiparallel alignment of the keratin proteins during the formation of the protofilament [32]. This molecular shift is thought to maximise the number of intermolecular disulfide bridges thus having a net effect of strengthening and stabilizing the trichocyte intermediate filament structure and the tissue containing them [32].

Studies conducted on the effect of sulfur on the swelling and supercontraction of wool fibres have shown that decrease in sulfur content, through poor nutrition, reduces the stress relaxation of wool but also has a greater effect on the swelling and supercontraction of wool. Sulfur content has been shown to be indirectly proportional to swelling and supercontraction of wool. This could be indicative of an increase in crosslinking occurring in the fibre matrix but not in the crystalline areas or fibrils in the presence of sulfur [33].

12.3 Keratin-KAP Interactions

Intermolecular disulfide crosslinks between KAP-KAP and also intramolecular crosslinks have been reported earlier [29]. The intramolecular crosslinks are thought to confer some limited tertiary structure to these KAP proteins [34]. High glycine tyrosine proteins (HGTPs) have been postulated to be good candidates to regulate the association between the head and tail domains of the keratins and the formation of the orthocortex and paracortex. The unique amino acid sequence of these KAPs which are high in tyrosine and the presence of high amounts of arginine in the head domains of keratins favors this association which are presumed to be via cation-π interactions. These KAPs are also hypothesized to bind to the C-terminus of cell adhesion proteins such as desmoplakin contributing to the overall protein network and the strength of the fibre.

HGTPs such as KAP8.1 has been found to bind to the head domain of Type II keratin K85 in a two-step exothermic interaction [28]. The aromatic residues in KAP8.1 and the basic residues in the head domain of K85 are thought to contribute towards the strength of the association of these proteins [28]. Among the high sulfur family of KAP proteins, KAP11.1 is considered a unique member due to the lack of the highly conserved primary structure and also possessing six repeats of cysteine-glutamine-proline motif at the C-terminus [35]. KAP11.1 also contains a high percentage of cysteine residues (12%) that form covalent crosslinks with the trichocyte keratin proteins. Studies conducted on expressed KAP11.1 revealed that the association with Type I keratins and other KAPs is not dependent on intermolecular disulfide crosslinks but are instead first initiated by sulfide-independent association which is then followed by stabilization of these complexes by disulfide crosslinks [36]. The association of KAP11.1 is also highly dependent on the amino acid sequence of the head domains of Type I keratins. KAP11.1 has been shown to bind to the head domains of recombinant forms of K31, 33a, 33b, 34 but not 39. All these hair Type I keratins have a highly conserved N-terminal head, the central rod domain and a C-terminus that is unstructured. K39 differs from the rest of the Type I keratins in its amino acid composition of the head domain and in the fact that it is a lot longer than the other Type I keratins. *In vitro* experiments have shown the ability of KAP11.1 to self-assemble without any disulfide bridges [36]. Self-assembly has also been reported in case of proteins belonging to the KAP2 family [29]. KAP2 proteins are expressed largely in the keratinizing zone of the human hair cortex along with several trichocyte hair keratins such as K85, K86 and K34 [2, 37]. *In vitro* studies clearly indicated that KAP2.1 protein is capable of interactions with hair trichocyte keratins but not with epithelial keratins. These results were confirmed with co-immunoprecipitation assays and glutathione S-transferase pull-down assays using truncated K86 proteins to define the binding domains of the K86 protein with KAP2.1. The head domain of K86 was determined to be crucial for this interaction [29]. The sulfur content of the head domains of hair keratins is higher compared to epithelial keratins and therefore is predicted to contribute to disulfide bond formation towards the stability of these interactions [29].

12.4 Accessibility of Cysteines

The location, pattern and the accessibility of cysteines either in their free form or in crosslinks have so far been difficult to map and understand due to the traditional techniques used during protein extraction that involve disruption of these disulfide bonds. In order to circumvent this problem, sequential extraction and labelling techniques have been employed to expose cysteine residues in the wool fiber based on their ease of accessibility. It is imperative to understand how the distribution of these crosslinks, contributed both by inter and intra molecular reactivity of the cysteine residues in the keratin proteins and KAPs influence the macroscopic nature of keratin materials.

The physical and mechanical properties of these fibres are highly influenced by the disulfide bond network present within the cortex. Intramolecular disulfide bridges and intra and intermolecular polar and non-polar interactions between amino acids in the fibre are critical in maintaining the three-dimensional structure of IFs and KAPs. Previous study indicated that only a limited number of cysteine residues are involved in crosslinks with keratin proteins or other KAPs. Peroxide bleaching of wool proteins was used to arrive at this conclusion ([38]). An initial linear increase of oxidation of cysteines to cysteic acid with an eventual levelling off into a plateau was noticed with an increasing level of peroxide treatment of merino wool proteins. Effects on proteins from the KAP1 family was particularly noticeable in 2D electrophoresis, wherein a reduction of the intensity of the spots belonging to this family of proteins resulted with increasing peroxide concentration. In conjunction with this, a discrete train of spots at a lower pH to the original KAP1 spots appeared. This effect was possible if relatively few cysteines in the KAP1 proteins were oxidized to cysteic acid; an extensive oxidation of cysteine residues would have resulted in a broad smear instead of discrete spots. This was surprising as the KAP family contains a relatively large number of cysteines that can vary from 34 to 41 residues. Low levels of cysteine conversion to cysteic acid implied that most of the cysteines in the KAP1 family were intramolecularly crosslinked making them inaccessible to peroxide bleach. This also suggests that since only a relatively small fraction of the cysteines were oxidized, these cysteine residues were probably more exposed in the tertiary structure of the proteins and hence were also more likely to form intermolecular crosslinks with keratin proteins [38].

Studies involving sequential extraction and labelling of cysteine residues followed by mass spectrometry analysis of wool proteins provided a better understanding of the accessibility of cysteine residues within the wool fiber structure. Stepwise increase in reducing agents and chaotropes were used to systematically open the fibre molecular structure. At each step, unique alkylating agents were used to identify accessible cysteines. An initial step involved wool fibre that had not been exposed to any reducing agents or chaotropes. Alkylation was performed on the wool fibre with an aim to label easily accessible free cysteines or cysteines in a constant state of flux. Statistical significance of the results from this step could not be obtained indicating that there are no specific cysteine residues that exist consistently as free thiols or as very weak disulfide bridges that are highly accessible.

Fibres subjected to a low concentration of a reducing agent in an attempt to break relatively accessible disulfide bonds, followed by alkylation of the free thiols that were generated, revealed labelling of mainly KAP proteins. Extraction of KAPs only in the presence of a reducing agent without any chaotropic disruption of the fibre, suggest that KAP-KAP and KAP-keratin interactions occur in the fibre mainly through the formation of relatively easily accessible disulfide bridges. These included HSPs, UHSPs, HGTPs and uncategorized KAPs (e.g., KAP 27.1). Most of the labelled cysteines in HSPs and UHSPs were determined to be in close proximity to a proline residue. In KAPs, proline residues are responsible for forming closed β-turn conformation of cysteine rich pentapeptide sequences. This β-turn conformation is stabilized by the formation of a disulfide bridge between cysteine residues that are five residues apart. This pentapeptide repeat can take several forms: A1, C-C-Q-P-X or A2, C-C-R-P-X or B, C-C-X-S/T-- S/T, and is often repeated contiguously within the KAP protein sequence. Presence of a proline residue in the 4th position of the sequence favours a closed loop conformation with the formation of disulfide bridges between the 1st and the 2nd position stabilizing the structure (Chap. 7, Sect. 2). These contiguous closed loop conformations also support the hypothesis that disulfide bonds are not randomly formed between cysteines within these proteins but that these loops are topologically linked by specific disulfides, which are not part of the loop, spanning the polypeptide backbone [39]. These proline containing pentapeptides are also thought to confer stability to KAPs by assisting in intermolecular KAP-KAP and KAP-keratin interactions. A high incidence of labelled cysteines that were in close proximity to proline residues indicate that proline induced conformation changes probably increases cysteine accessibility within the KAP proteins.

Of particular interest are the proteins belonging to the KAP2 family and their interactions with trichocyte keratins.]. It is highly likely that these interactions are confined to the head and tail regions of the keratin dimers mainly due to their less structured nature compared to the highly structured rod domains. However, the degree of influence that these KAPs have on regulating the helical arrangement of the intermediate filaments is not confirmed [28]. Labelling studies also show that the majority of these accessible cysteines are located near the termini of the KAP proteins (Fig. 12.1). This is indicative of KAP proteins where the terminal regions are not buried within the tertiary structure but are available for protein-protein interactions to help stabilize the fibre [7].

An interesting feature of some of the labelled cysteines in the HGTP family of KAPs is their presence in glycine rich regions of these proteins. These regions are similar to the glycine-loop regions found on some epithelial keratins [40], where the glycine residues form flexible loops. Presence of cysteines in these flexible loops are therefore thought to be more exposed and in a better position to interact with other proteins (Figs. 12.2 and 12.3).

Analysis of amino acid sequences indicate that semiregular disulfide bridges are present in KAPs providing these proteins with partial conformation [34]. Also many of these disulfide bridges are predicted to be intramolecular in nature. KAPs have been shown to preferentially swell in water, which is possible if disulfides are intramolecular in nature, as intermolecular bridges will severely limit this process [1]. It is possible to solubilize KAPs with only reducing agents in the absence of any chaotropes which also imply that KAP-KAP and KAP-keratin interactions are possibly mainly due to disulfide bridges that are easily accessible within the fibre.

Labelled cysteines were also identified in trichocyte keratin proteins from fibres extracted only with reducing agents and without chaotropes. Some of these cysteines are found to be present in the rod domains of Type I and II keratins. Unambiguous assignment of these labelled cysteines to any one keratin is not possible due to the high sequence homology present in this family of proteins. Presence of labelled cysteines in the rod domains are indicative of these cysteine residues appearing on the outer surface of the rod, opposite the residues forming the hydrophobic stripe, making them readily accessible [7]. Within the heptad repeat (a-b-c-d-e-f-g)$_n$ of the α-helical rod domains of keratin proteins, cysteine residues in position 'c' only are

```
KAP2.3
TGSCCGPTFS SLSCGGGCLQ PRYYRDPCCC RPVSCQTVSR PVTFVPRCTR PICEPCRRPV
CCDPCSLQEG CCRPITCCPT SCQAVVCRPC CWATTCCQPV SVQCPCCRPT SCQPAPCSRT
TCRTFRTSPC C

KAP3.2
ACCAPRCCSV RTGPATTICS SDKFCRCGVC LPSTCPHNIS LLQPTCCDNS PVPCYVPDTY
VPTCFLLNSS HPTPGLSGIN LTTFIQPGCE NVCEPRC

KAP3.4
ACCARLCCSV PTSPATTICS SDKFCRCGVC LPSTCPHTVW FLQPTCCCDN RPPPCHIPQP
SVPTCFLLNS SQPTPGLESI NLTTYTQPSC EPCIPSCC

KAP4.3
VSSCCGSVCS DQSCGRSLCQ ETCCRPSCCQ TTCCRTTCYR PSCGVSSCCR PVCCQPTCPR
PTCYISSCSR PSCCVSSCGS SCYRPTGCIS SCYRPQCCQP VCCQPTCPRP TCCISSCRPR
CCQPVCCQPT CPRPTCCISS CYRPSSCGSS CGSSCYRPTG CISSCYRPQC CQPVCCQPTC
SRPTCCISSC YRPQCCQPVC CQPTCPRPTC CISSCYRPSS CGSSCGSSCY RPTCCISSCR
PRCCQPVCCQ PSCPRISSCC RPSCYSSSCC RPSCCLRPVC GRVSCHTTCY RPTCVISTCP
RPVSCPSSCC

KAP4.8
VSSCCGSVCS DQSCGRSLCQ ETCCRPSCCQ TTCCRTTCYR PSCGVSSCCR PVCCQPTCPR
PTCCISSCSR PSCCVSSCGS SCYRPTGCIS SCYRPQCCQP VCCQPTCPRP TCCISSCRPR
CCQPVCCQPT CPRPTCCISS CYRPSSCGSS CGSSCCRPTC CISSCRPRCC QPVCCQPSCP
RISSCCRPSC YSSSCCRPSC CLRPVCGRVS CHTTCYRPTC VISTCPRPVC CPSSCC

KAP4.19
VSSCCGSVCS DQSCGRSLCQ ETCCRPSCCQ TTCCRTTCYR PSCGVSSCCR PVCCQPTCPR
PTCCISSCYR PSCCVTRCGS SCYRPTGCIS SCRPQCCQPV CCQPTCPPPT CCISSCYRPS
SCGSSCGSSC CRPTCCISSC CRPQCCQPVC CQPTCPRPTC CISSCYRPSS CGSSCGSSCC
RPTCCISSCR PRCCQSVCCQ PSCPRISSCC RPSCYSSSCC RPSCCLRPVC GRVSCHTTCY
RPTCVISTWP RPVSCPSSCC

KAP6.1
CGYYGNYYGG LGCGSYSYGG LGCGYGSCYG SGFRRLGCGY GCGYGYGSRS LCGSGYGYGS
RSLCGSGYGC GSGYGSGFGY YY

KAP6.3
CGYYGNYYGG LGCGSYGYGG LGCGYGSCYG SGFRRLGCGY GCGYGYGSRS LCGSGYGCGS
RPLYGCGYGC GSGYGSGFGY YY

KAP9.2
THSCCSPCCQ PTCCESSCCR PCCPPTCYQT SEHTCCRTTC SKPTCVTTCC QPACGGSSCC
QPCCRPISCQ TTCCRTTCLK PTCVTTCCQP TCCESSCCRP CCPPTCYQTS ENTCCRTTCS
KPTCVTTCCQ PACGGSSCCQ PCCRPISCQT TCCRTTCLKP VCATTCCQPA CCESSCSQPS
CPQTCCQITE TTCCKPTCVT SCCQPTCCGS SSCGQPCGGS NCCQPASCAP VYCHRTCYHP
TCCCLPGCQA QSCGSSCCQP CSRPVCCQTT CCRTTRCRPS CVSSCCQPSC C
```

Fig. 12.1 KAPs with identified sequences from the mapping the accessibility of the disulfide crosslink network in the wool fibre cortex study. (Reproduced from Proteins 2014 [7] with permission from Wiley Periodicals).

```
KAP10-like
QPVSYKATIC EPACPVSSCA QPVSCEATIC EPSCSVSSCA QPVSYKATIC EPSCSVSSCA
QPVSYKATIC EPTCPVSSCA QPVSCEATIC EPSCSVSSCA QPVSYKATIC EPACPVSSCA
QPVCCEVPPG QRVFCVPSSC QPILCKPSYC QPVICEPSCY QPVSSGVRCC PSVCSVANSC
QSACCDSSPC EPSCSEPSIC QSATRVSLVC EPICVRPVCC VSSPCEPPCV SSTCQEPSCC
VSSICQPICS EPSPCLPSVC VPRPCQPTCY VVKRSRSISC EPLSCRPLSG RPGSSASAVC
QPTCSRTFYI PSSCKQPCTT STSYRPICRP ICSGPITYRQ PYLTSISYRP ACYRPFYSIL
RRPACIASVP YRSVCSRLPC ADSCKRDCKK STSSQPDCAD STPCKTEVSE ASPCQPTEAK
PTSPTTREAA VSQPAATKPT NC

KAP11.1
SYSCSTRNCS SRRIGGEYTV PVVTVSSPDA DCLSGIYLPS SFQTGSWLLD HCQETCCEPT
VCQSTCYQPT PCVSSPVRVT SRQTTCVSSP CSTTCSRPLT FISSGCQPLS GVSTVCKPVR
SISTVCQPVG GVSTICQPTC GVSRTYQQSC VSSCRRIC

KAP12.2-like
MCHTSFSSGC QAACVPSSCQ PSCSTSSPCQ PSCLPVSCRP AVYVTPSCQS SVRLPVSCRP
AVYAVLVAPS CQSSGCYQPS RPTLVYRPVS CSTPSCL

KAP13.1
SYNCCSGNFS SLSLRDHLRY SGSSCGSSFP SNLVYRTDLY SPSSCQLGSS LYSQETCCEP
IRTQTVVSSP CQTSCYRPRT STFFSPCQTT CSGSLGFGSS NLQSAGHVFP SLGFGSGGFQ
SVGHSPNIFS SLSCRSSFYR PTFFSSRSGQ SLSFQPTCGS GFY

KAP16.1-like
MSGNCCSRKC PSVPAISLCS TEVSCRGPVC LPSSCRSQTW QLVTCQDSCG SSSCDPQCCE
PSCSASSCVQ PVCCETTICE PACPVSSRAQ PVSCEATICE PACPVSNCAQ PVCYKATICE
PACPVSSCAQ PVYCEATIYE PACPVSSCAQ PVCYKATICE PACPISSCAQ PVCCEATICE
PSCSVSSYAQ PVSYKATICE SACPVSSCAQ PVSRAQPVSC EAPICEPACP VSNCAQPVCY
KATICEPACP VISCAQTVCY ETTICEPACP VSSCAQPVYC EATIYEPACP VSNCAQPVCY
KATICEPACP VISCAQTVCY ETTICEPACP VSSCAQPVYC EATICEPSCS VSSCAQPVCC
EATICEPSCS VSSYAQPVSY KATICESACP VSSCAQPVYY KATICEPSCP VSSCAQPVSC
EATICEPSCS VSSCAQPVSY KATICEPSCS VSSCAQPVSY KATICEPACP VSSCAQPVSC
EATICEPSCS PILCKPSYCQ PVICEPSCYQ PVSSGVRCCP SVCSVANSCQ SACCDSSPCE
PSCSEPSICQ SATRVSLVCE PICVRPVCCV SSPCEPPCVS STCQEPSCCV SSICQPICSE
PSPCLPSVCV PRPCQPTCYV VKRSRSISCE PLSCRPLSGR PGSSASAVCQ PTCSRTFYIP
SSCKQPCTTS TSYRPICRPI CSGPITYRQP YLTSISYRPA CYRPFYSILR RPACIASVPY
RSVCSRLPCA DSCKRDCKKS TSSQPDCADS TPCKTEVSEA SPCQPTEAKP TSPTTREAAV
SQPAATKPTN C
```

Fig. 12.1 (continued)

capable of forming interchain disulfide linkages outside the keratin heterodimer. Intracoil disulfide linkages are not possible due to distance and steric hindrance [41]. A number of cysteines were also labelled from the head and tail domains of Type I and II keratins. Identification of these labelled cysteines indicate that these are not buried deep within the fibre structure.

Some of the cysteines were observed in the rod domain of Type I keratins only when fibres were treated with both reducing agents and chaotropes, indicating that these were less accessible in the tertiary structure of the keratin protein family and possibly oriented towards the centre of the IF. A few of these labelled cysteines are present within or in close proximity to linker and stutter regions. A stutter is thought to maintain the structural rearrangement of the apolar amino acid residues of the coiled-coil rope of the keratin proteins in

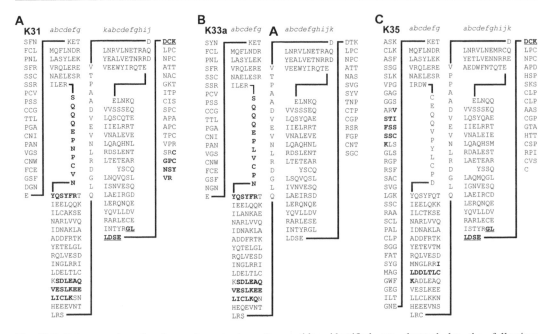

Fig. 12.2 Primary and predicted secondary structure of Type I keratin K31 (**A**), K33a (**B**) and K35 (**C**) from the mapping the accessibility of the disulfide crosslink network in the wool fibre cortex study [7]. The tryptic peptides identified are denoted by the following: **Bold** = structurally accessible in the presence of a reductant only; **Bold underlined** = accessible only in the presence of a reductant + a chaotrope

the formation of IFs [22]. It is a highly conserved area and apparently results from insertion of four or deletion of three residues in the heptad sequence. It is possible that a rearrangement of the coiled coil structure results from the presence of the stutter and is confined to one or two turns of the α-helix on either side of it. This structural rearrangement possibly makes some of the cysteines in the region more accessible. It has been reported in earlier studies that the head domain in Type II keratins has the potential to fold back over part of Coil 1 and hence cysteines present in Coil 1 may not be accessible for forming disulfide bridges with KAPs [42]. The L1 linker region in Type II keratins is hypothesized to act as a flexible hinge that allows the head domain to fold back and interact with the coil region [42]. A number of labelled cysteines present in the non-helical terminal regions of both the Type I and II keratins suggest that the orientation of these regions is possibly towards the outside in the coiled-coil structure and hence capable of forming disulfide links with KAPs.

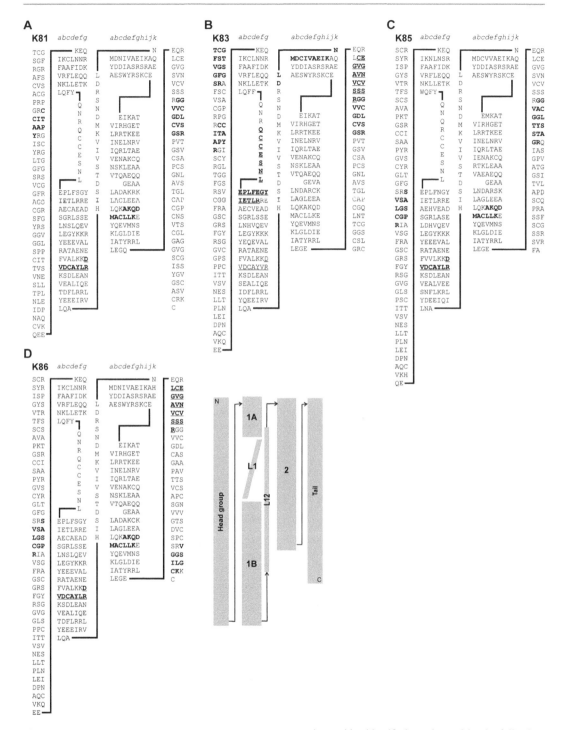

Fig. 12.3 Primary and predicted secondary structure of Type II keratin K81 (**A**), K83 (**B**), K85 (**C**) and K86 (**D**) from the mapping the accessibility of the disulfide crosslink network in the wool fibre cortex study [7]. The tryptic peptides identified are denoted by the following: **Bold** = structurally accessible in the presence of a reductant only; **Bold underlined** = accessible only in the presence of a reductant + a chaotrope

References

1. Fraser, R. D. B., MacRae, T.P., & Rogers, G. E. (1972). *Keratins: Their composition, structure and biosynthesis* (The Bannerstone division of American lectures in living chemistry, p. 320). Springfield: Charles C Thomas Publisher, Ltd.
2. Langbein, L., et al. (1999). The catalog of human hair keratins. I. Expression of the nine type I members in the hair follicle. *Journal of Biological Chemistry, 274*(28), 19874–19884.
3. Lee, Y. J., Rice, R. H., & Lee, Y. M. (2006). Proteome analysis of human hair shaft – From protein identification to posttranslational modification. *Molecular & Cellular Proteomics, 5*(5), 789–800.
4. Orwin, D. F. G. (1979). The cytology and cytochemistry of the wool follicle. *International Review of Cytology, 60*, 331–374.
5. Hill, P., Brantley, H., & Van Dyke, M. (2010). Some properties of keratin biomaterials: Kerateines. *Biomaterials, 31*(4), 585–593.
6. Alexander, P., & Earland, C. (1950). Structure of wool fibres: Isolation of an α- and β-protein in wool. *Nature, 166*(4218), 396–397.
7. Deb-Choudhury, S., et al. (2015). Mapping the accessibility of the disulfide crosslink network in the wool fiber cortex. *Proteins: Structure, Function, and Bioinformatics, 83*(2), 224–234.
8. Arai, K., et al. (1996). Crosslinking structure of keratin. VI. Number, type, and location of disulfide cross-linkages in low-sulfur protein of wool fiber and their relation to permanent set. *Journal of Applied Polymer Science, 60*(2), 169–179.
9. Stappenbeck, T. S., et al. (1993). Functional analysis of desmoplakin domains: Specification of the interaction with keratin versus vimentin intermediate filament networks. *The Journal of Cell Biology, 123*(3), 691–705.
10. Getsios, S., et al. (2004). Coordinated expression of desmoglein 1 and desmocollin 1 regulates intercellular adhesion. *Differentiation, 72*(8), 419–433.
11. Syed, S. E., et al. (2002). Molecular interactions between desmosomal cadherins. *Biochemical Journal, 362*(Pt 2), 317–327.
12. Orwin, D. F. G., & Thomson, R. W. (1973). Plasma membrane differentiations of keratinizing cells of the wool follicle (4. Further membrane differentiations). *Journal of Ultrastructure Research, 45*(1), 41–49.
13. Plowman, J. E., & Deb-choudhury, S. (2016). Wool proteomics. In S. G. H. (Ed.), *Animal proteomics* (pp. 201–213). Springer.
14. Steinert, P. M., & Marekov, L. N. (1995). The proteins elafin, filaggrin, keratin intermediate filaments, loricrin, and small proline-rich proteins 1 and 2 are isodipeptide cross-linked components of the human epidermal cornified cell envelope. *Journal of Biological Chemistry, 270*(30), 17702–17711.
15. Harding, H. W., & Rogers, G. E. (1971). (γ-glutamyl) lysine cross-linkage in citrulline-containing protein fractions from hair. *Biochemistry, 10*, 624–630.
16. Rogers, G. E. (2004). Hair follicle differentiation and regulation. *International Journal of Developmental Biology, 48*(2-3), 163–170.
17. Rogers, G. E. (2006). Biology of the wool follicle: An excursion into a unique tissue interaction system waiting to be re-discovered. *Experimental Dermatology, 15*(12), 931–949.
18. Zahn, H. (1980). Wool is not keratin only. In *Proceedings of the 6th International Wool Textile Research*. Pretoria: Deutsches Wollforschungsinstitut.
19. Nienhaus, M. & Föhles, J. (1980). Zur chemie von humanhaar. In *6th quinquennial international wool textile Research conference* (pp. 487–495). Pretoria: South African Wool and Textile Reseach Institute.
20. Harding, H. W. J., & Rogers, G. E. (1972). The occurrence of the (γ-glutamyl)lysine cross-link in the medulla of hair and quill. *Biochimica et Biophysica Acta, 257*(1), 37–39.
21. Rogers, G. E. (1989). Special biochemical features of the hair follicle. In G. E. Rogers, P. J. Reis, K. A. Ward, & R. C. Marshall (Eds.), *The biology of wool and hair* (pp. 69–85). London/New York: Chapman and Hall.
22. Parry, D. A. D., & Steinert, P. M. (1995). *Intermediate filament structure*. Heidelberg: Springer.
23. Steinert, P. M. (1993). Structure, function and dynamics of keratin intermediate filaments. *Journal of Investigative Dermatology, 100*(6), 729–734.
24. Gillespie, J. M. (1972). Proteins rich in glycine and tyrosine from keratins. *Comparative Biochemistry and Physiology B.Biochemistry and Molecular Biology, 41B*, 723–734.
25. Rafik, M. E., Doucet, J., & Briki, F. (2004). The intermediate filament architecture as determined by X-ray diffraction modeling of hard α-keratin. *Biophysical Journal, 86*, 3893–3904.
26. Fraser, R. D. B., MacRae, T. P., & Suzuki, E. (1976). Structure of the α-keratin microfibril. *Journal of Molecular Biology, 108*, 435–452.
27. Parry, D. A., et al. (2007). Towards a molecular description of intermediate filament structure and assembly. *Experimental Cell Research, 313*(10), 2204–2216.
28. Matsunaga, R., et al. (2013). Bidirectional binding property of high glycine-tyrosine keratin-associated protein contributes to the mechanical strength and shape of hair. *Journal of Structural Biology, 183*(3), 484–494.
29. Fujikawa, H., et al. (2012). Characterization of the human hair keratin-associated protein 2 (KRTAP2) gene family. *Journal of Investigative Dermatology, 132*(7), 1806–1813.
30. Parry, D. A. D. (1995). Hard α-keratin IF: A structural model lacking a head-to-tail molecular overlap but having hybrid features characteristic of both epi-

dermal keratin and vimentin IF. *Proteins: Structure, Function, and Bioinformatics, 22*(3), 267–272.
31. Parry, D. A. D. (1996). Hard α-keratin intermediate filaments: An alternative interpretation of the low-angle equatorial X-ray diffraction pattern, and the axial disposition of putative disulphide bonds in the intra- and inter-protofilamentous networks. *International Journal of Biological Macromolecules, 19*(1), 45–50.
32. Wang, H., et al. (2000). In vitro assembly and structure of trichocyte keratin intermediate filaments: A novel role for stabilization by disulfide bonding. *Journal of Cell Biology, 151*(7), 1459–1468.
33. Whiteley, K. J., Balasubramaniam, E., & Armstrong, L. D. (1970). The swelling and supercontraction of sulphur-enriched wool fibers. *Textile Research Journal, 40,* 1047–1048.
34. Parry, D. A. D., et al. (2006). Human hair keratin-associated proteins: Sequence regularities and structural implications. *Journal of Structural Biology, 155*(2), 361–369.
35. Wu, D.-D., Irwin, D. M., & Zhang, Y.-P. (2008). Molecular evolution of the keratin associated protein gene family in mammals, role in the evolution of mammalian hair. *BMC Evolutionary Biology, 25*(8), 241–255.
36. Fujimoto, S., et al. (2013). Krtap11-1, a hair keratin-associated protein, as a possible crucial element for the physical properties of hair shafts. *Journal of Dermatological Science, 74*(1), 39–47.
37. Langbein, L., et al. (2001). The catalog of human hair keratins. II. Expression of the six type II members in the hair follicle and the combined catalog of human type I and II keratins. *Journal of Biological Chemistry, 276*(37), 35123–35132.
38. Plowman, J. E., et al. (2003). The effect of oxidation or alkylation on the separation of wool keratin proteins by two-dimensional gel electrophoresis. *Proteomics, 3*(6), 942–950.
39. Benham, C. J., & Saleet Jafri, M. (1993). Disulfide bonding patterns and protein topologies. *Protein Science, 2*(1), 41–54.
40. Steinert, P. M., et al. (1991). Glycine loops in proteins: Their occurrence in certain intermediate filament proteins, loricrins and single-stranded RNA binding proteins. *International Journal of Biological Macromolecules, 13*(3), 130–139.
41. Fraser, R. D. B., et al. (1988). Disulfide bonding in α-keratin. *International Journal of Biological Macromolecules, 10,* 106–112.
42. Parry, D. A. D., et al. (2002). A role for the 1A and L1 rod domain segments in head domain organisation and function of intermediate filaments: Structural analysis of trichocyte keratin. *Journal of Structural Biology, 137,* 97–108.

The Thermodynamics of Trichocyte Keratins

13

Crisan Popescu

Contents

13.1	General Notions of Thermodynamics	186
13.2	Assembly and Stabilization of Intermediate Filaments	189
13.3	Thermodynamics of Keratin denaturation and Effect of Ligands	192
13.4	Thermodynamics of Alpha-Helix to Beta Sheet Transition in Keratins	195
13.5	Glass Transition of Keratins	195
13.6	Moisture Sorption and Desorption by Keratins	197
13.7	Thermodynamics of Keratin Mechanics	198
13.8	Conclusions	201
	References	201

Abstract

This chapter is an attempt at an excursion into the world of keratins with the help of thermodynamics.

After briefly introducing some of the thermodynamic concepts involved in deciphering the behaviour of keratins, we will use them to look into the process of aggregation of keratin molecules into intermediate filaments, and keratin fibres, and then for analysing how keratin materials react to mechanical, thermal and moisture stresses, respectively.

In most of the cases entropy appears to be the major driving force of the response occurring in keratins under environmental assault. This fact points to the important role played for keratins by temperature, which, aside from influencing the kinetics of the processes (accelerating or decelerating the rates of the rates), helps increase or decrease the entropic contribution to the Gibbs free energy and, thus, allows thermodynamically the occurrence of the observed behaviour of keratins.

C. Popescu (✉)
KAO Germany GmbH, Darmstadt, Germany
e-mail: crisan.popescu@kao.com

Keywords

KIF assembling · Keratin denaturation · Alpha helix-beta sheet transition · Glass transition · Moisture sorption–desorption · Keratin fibre mechanics

13.1 General Notions of Thermodynamics

Thermodynamics is the branch of science that deals with heat and its relation to work and energy.

The objects of thermodynamics are called thermodynamic systems. A thermodynamic system is defined as a region of the Universe surrounded by an arbitrary border, considered in such a way that the ensemble of bodies and fields within it may be taken as a whole. The border may, or may not, allow exchange with the surroundings of matter ('open' and 'closed' systems, respectively) and/or energy ('adiabatic' and 'non-adiabatic' systems, respectively).

The state of the system is characterized by thermodynamic parameters, which are physical variables describing the whole characteristics of a thermodynamic system and the relations of the system with the surrounding ones. They may be internal, noted "X_i", and external variables, respectively, noted "a_i". The variables relate to the amount of elements composing the system (number of particles i, N_i, mass of system, m, or volume of system, V) and to their interaction (temperature, T, pressure, p, entropy, S). According to how the variable depends on the size of the system they are also classified as intensive variables (whose magnitude is independent of the size of the system) or extensive variables, respectively (whose magnitude is dependent of the size of the system).

The results of thermodynamics rely on the existence of idealized states of thermodynamic equilibrium.

The macroscopic parameters are considered to be subjected to certain constraints known as the laws of thermodynamics. The first law of thermodynamics, sometimes named the general law of thermodynamics, introduces the concept thermodynamic equilibrium by stating that:

(i) equilibrium exists when all macroscopic variables are constant in time (their values do not fluctuate any more) and the total fluxes of matter/energy through the system border are nil (over a period of time as much as goes out also comes in), and
(ii) that thermodynamic equilibrium has the property of transitivity. This is to say that if system A is in equilibrium with system B and system B is in thermodynamic equilibrium with C, then system A is also in thermodynamic equilibrium with system C.

The other three laws of thermodynamics, historically numbered 1–3, introduce the concept of conservation of energy and relation of work with heat, and the concept of "entropy", a parameter related to the order within the system and the way it changes with a process.

Thermodynamics use equations of state to describe equilibrium states and make use of potentials for representing these states. The potentials are functions of thermodynamic variables and give account of the capacity of the system to perform work.

The most used potentials in discussing chemical and chemical-related processes (chemical thermodynamics of reactions) are the energy potentials: internal energy, U, Gibbs free energy, G, enthalpy, H, and Helmholtz free energy, A.

One defines "mechanical work", W, as the variation of internal energy of a thermodynamic system, U, due only to external parameters, a_i. Considering a_i as generalized coordinates and introducing Y_i as generalized forces one may write:

$$\delta W = \sum Y_i da_i \quad (13.1)$$

By convention, if work is done on the system, it is a positive quantity, $W > 0$, and if it is done by the system it is a negative quantity, $W < 0$.

Similarly, "heat", Q, is defined as the variation of internal energy of the thermodynamic system due only to internal parameters, X_i. According to second law of thermodynamics:

$$\delta Q = TdS \quad (13.2)$$

and it is conventionally considered that the heat absorbed by the system, as in the *endothermic* processes, is positive, $Q > 0$, and those released by the system, as in *exothermic* processes, is negative, $Q < 0$.

There are commonly two types of heat considered in thermodynamics, namely: '*latent heat*', the heat released or absorbed by the system during a change of its state but without change of temperature; and '*sensible heat*', the heat released or absorbed by the system which reflects only in change of temperature. Consequently the latent heat is associated with phase transitions like melting-crystallisation, or boiling-condensation, whilst sensible heat is used for the chemical reactions, or physical changes of the systems.

Heat capacity, C, is derived from sensible heat, being defined as the ratio of heat change of a system to the corresponding change of temperature:

$$C = \frac{\partial Q}{\partial T} = \frac{Q}{\Delta T} \quad (13.3)$$

As defined by Eq. (13.3) heat capacity is an extensive property; for easy comparison of systems, it is transformed in its intensive form by dividing it by the mass of the system. The specific heat capacity defined this way is the amount of heat required to change the temperature of a mass, m, of 1 kg with 1 kelvin:

$$c = \frac{Q}{m \cdot \Delta T} \quad (13.4)$$

For chemistry purposes the definition is modified into the amount of heat required to change the temperature of a mole of substance, of mass M, with 1 kelvin, which is the molar heat capacity:

$$C_M = \frac{Q}{M \cdot \Delta T} \quad (13.5)$$

The fluxes of matter and chemical reactions occurring within systems are accounted for by considering the contribution of various particles comprising the system, N_i, through their chemical potentials, μ_i (the energy transferred during the chemical reaction of particles of species "i"): $\mu_i dN_i$.

The fundamental equations of thermodynamics, from which one derives the description of the systems and their transformations, are Legendre transforms of the law of conservation of energy (first law of thermodynamics) which states that any variation of the internal energy is due to transfer of heat, performance of work and transfer of matter:

$$dU = \delta Q - \delta W + \sum \mu_i dN_i = TdS - \sum Y_i da_i + \sum \mu_i dN_i \quad (13.6)$$

For most of the systems pressure, p, is the generalized force, volume, V, is the generalized coordinate, and Eq. (13.6) becomes:

$$dU = TdS - pdV + \sum \mu_i dN_i \quad (13.7)$$

Legendre transforms applied to this form of variation of internal energy results in the expressions of the other potentials as:

Enthalpy, H: $\quad dH = TdS + Vdp + \sum \mu_i dN_i$
$$(13.8)$$

Gibbs free energy, G: $dG = -SdT + Vdp + \sum \mu_i dN_i$
$$= dH - d(TS)$$
$$(13.9)$$

Helmholtz free energy, A: $dA = -SdT - pdV$
$$+ \sum \mu_i dN_i$$
$$= dU - d(TS)$$
$$(13.10)$$

During the study of the system evolution the investigation maintains constant two thermodynamic variables, be they pressure (isobar), volume (isochore), and/or temperature (isotherm), describing the change of energy makes use of one of the potential forms which suits the variable of the experiment. One calculates

$$\Delta G = \Delta H - T\Delta S \text{ for } T, p = \text{cons} \quad (13.11)$$

as a measurement of the ability of the system to perform non-mechanical work, under isothermal and isobaric conditions, and

$$\Delta H = Q_p \text{ for } p = \text{const} \quad (13.12)$$

as a measurement of heat changed by the system under isobaric conditions.

$$\Delta A = \Delta U - T\Delta S \text{ for } T, V = \text{const} \quad (13.13)$$

indicating the part of internal energy of the system which can be converted in work (useful mechanical work) under isothermal and isochoric conditions.

Under these circumstances heat capacity, as defined by Eq. (13.3), is re-defined according to the way it is determined: under isochore, C_V, or isobaric, C_p, conditions:

$$C_V = \partial U / \partial T \mid_{V=\text{const}} \quad (13.14)$$

and

$$C_p = \partial H / \partial T \mid_{p=\text{const}} \quad (13.15)$$

It is worth to noting that heat capacity measured under isobaric conditions provides information on enthalpy for a closed-system process.

Similarly with mechanical systems, for which processes occur driven by decreasing activity, any spontaneous and irreversible process occurring in an isolated thermodynamic system is driven either by increasing entropy, $\Delta S > 0$ for adiabatic processes, or by decreasing the free energy, $\Delta G < 0$ for isotherm-isobaric processes, or $\Delta A < 0$ for isotherm-isochoric processes, respectively. For chemical processes this statement is usually expressed as 'the processes which are thermodynamically allowed are those for which the Gibbs free energy decreases, $\Delta G < 0$'. Using the definition of Gibbs free energy as in Eq. (13.11) we may write:

$$\Delta G = \Delta H - T\Delta S < 0 \quad (13.16)$$

from which we derive the thermodynamic conditions (inequalities) for a process to be allowed:

$$\Delta H \langle 0 \text{ and } \Delta S \rangle 0 \text{ process allowed}$$
$$\text{at any temperature } T \quad (13.16a)$$

$$\Delta H > 0 \text{ and } \Delta S > \text{ process allowed}$$
$$\text{from temperature } T > \Delta H / \Delta S \quad (13.16b)$$

For systems in which chemical reactions occur, the driving force is the chemical potential of reactants, X_i, and products, Y_i, of the general chemical reaction:

$$x_i X_i \to y_j Y_j \quad (13.17)$$

where x_i, y_j are the stoichiometric coefficients of reactants and products, respectively.

The rates of chemical reactions and the pathways used by systems to go from a state of equilibrium to another is the object of the field of "kinetics of processes".

The rate at which the species in general reaction (13.17) react, v, is given by the law of mass action, for reactants:

$$v_r = k_r \prod_i [X_i]^{x_i} \quad (13.18)$$

and, respectively for products:

$$v_p = k_p \prod_j [Y_j]^{y_j} \quad (13.19)$$

where constants $k_{r,p}$ used in expressing the proportionality of rates to the product (Π_j) of concentrations of chemical species (indicated by square brackets) are named '*rate constants*'.

At equilibrium the two rates are equal, $v_r = v_p$, which allows the definition of the reaction equilibrium constant, K:

$$K = \frac{k_p}{k_r} \cdot \frac{\prod_j [Y_j]^{y_j}}{\prod_i [X_i]^{x_i}} \quad (13.20)$$

With the equilibrium constant, K, defined, the Gibbs free energy of the system is:

$$\Delta G = -RT \ln K \quad (13.21)$$

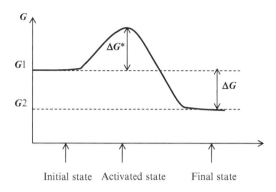

Fig. 13.1 Evolution of Gibbs free energy during a process

Equations (13.20) and (13.11) allow evaluating the variation of the equilibrium constant with temperature, namely the van't Hoff equation:

$$\frac{d(\ln K)}{dT} = \frac{\Delta H}{R \cdot T^2} \quad (13.22)$$

The dependence of equilibrium constant on temperature, as derived from above:

$$K = \exp(-\Delta H / RT) \quad (13.23)$$

serves to understand that the pathway of a system from the initial state to its final state needs to go through an intermediate state, named 'activated state'.

A diagram showing the evolution of Gibbs free energy, G, during the pathway of the system from the initial to the final state is shown in Fig. 13.1. ΔG^* is understood to be the energy required to initiate the process and the rate of process occurring is proportional to the probability of the system to reach this activated state.

Figure 13.1 shows that a process, even if thermodynamically allowed to unfold from the initial state to the final one ($\Delta G < 0$), cannot occur until it receives an amount of energy, defined as *activation energy*, for crossing the barrier of the activated state. This concept lies at the base of kinetics, the discipline which deals with the rates at which processes occur, and the pathway the systems take for going from one state to another.

13.2 Assembly and Stabilization of Intermediate Filaments

The protein material, synthesized in the follicle bulb by living cells, is pushed out of the follicle and organised into the hair shaft. Along the way certain proteins aggregate to form the keratin intermediate filaments (KIF), which are anchored by surrounding proteins (keratin associated proteins, KAPs) and, later on, the disulfide bonds freeze the material in the fibre shaft.

The keratin fibres which result from the process which begins in the follicle and ends with the protruding of keratin fibre out of the skin, are, simply composite materials made of rod-like KIFs embedded by the KAPs [1].

The organization of the trichocyte intermediate filament protein into the KIF is a complex process occurring via self-assembling of α-helices building units, first laterally, to form dimer, tetramer, up to 32mer and along this pathway there is also assembling end-to-end to form the end-product [2].

A simplified schema is shown in Fig. 13.2, redrawn after [1]:

Keratins are generally poorly soluble, reacting only at very low ionic strength and high pH. Various studies of the whole assembly process showed that it is initiated by raising the ionic strength with divalent cations such as Ca^{2+}, Mg^{2+}, Cu^{2+} or Mn^{2+} being more efficient than monovalent ions such as K^+, Na^+ and Li^+ [3, 4]. The assembly of keratins is also sensitive to pH and the correct process of formation runs over a quite narrow range of pHs [5]. The kinetics of the assembling process is thought to have an activation energy of about 125 kJ·mol^{-1} as has been determined for keratin assembly in neurons [6]. As suggested by XRD micro-beam investigations, and supported by AFM measurements, the length of assembly pathway is about 600 to 800 μm [7, 8], after which apparently the keratinization begins freezing the structure. At a rate of hair growth from a follicle of about 3.5 nm s^{-1} [9], one may calculate that it takes about 55 hours to complete the process of KIF assembly to the keratinization point. The activation energy allows for the effect of temperature on this assembly

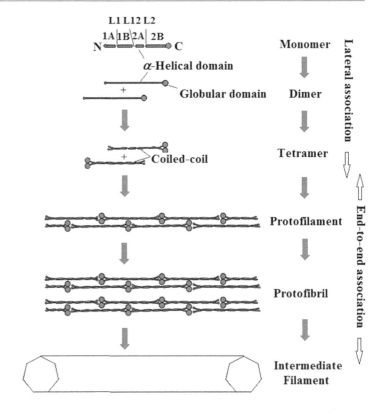

Fig. 13.2 Assembly of keratins into keratin intermediate filaments

process. Assuming that the process runs to completion at 37°C, a change of temperature of +3°C will shorten the process by about 37%, whilst a change of −3°C will lengthen the association time by 60%. This brief evaluation shows how important the control of temperature is for the whole process, if the assembly process is not yet finished when keratinization starts (case of −3°C) the keratin material is weakened.

Behind each step forming the assembly process, as per Fig. 13.2, thermodynamics requires the existence of a driving force and the fulfilling of thermodynamic conditions given by the relationship in Eq. 13.16.

The first step in formation of the KIFs is the building up of the structural unit, α-helix, out of the amino acids in the cell. The ribosomes assemble amino acids in the sequences commanded by the mRNA-transported instructions and the long chain so-formed in the crowded environment of the cell adopts the ordered, helical conformations due to entropy constraint [10]. Considering the polypeptide chain as an impenetrable tube surrounded by hydrophobic forces and other chain-chain interactions of radius of action r, forming a colloid-like environment, the entropy of the total interactions increases as the excluded volume decreases by the chain folding into a helix arrangement and formation of 'overlapping' volume between helix spirals. The proposed physical model indicates that if the ratio of force acting radius to the width of polypeptide chain is small, the helices which form have a pitch to spiral radius ratio ranging from 2.5122 to 2, retrieving data available from other simulations [11]. When the ratio of radius of acting forces to width of chain increases the helix spiral radius tends to infinity, which means that the chain bends by a certain length, producing the beta-sheet arrangement.

The step of assembling two α-helices into a dimer (double-helix) is by far the most studied one, starting with the early use of X-ray diffraction analysis [12, 13]. The precise sequences of amino acid residues of some members of the two α-keratin families (Type I and Type II, acid and neutral-basic, respectively) are known [14]; the central rod domain largely consists of three α-helical segments (1A, 1B, 2A) separated by two amorphous linker segments (L1 and L12). The repeating heptads in the α-helical segments favour the formation of both the α-helical structure of the segments and the parallel, super helical dimer structure of two IF monomers [15, 16]. The current view is of Type I and Type II keratin heterodimers held in a rod-like conformation by heptad and hendecad coiled-coil segments interrupted by a flexible linker [17]. Head and tail segments are random coil, but play a part in stable filament assembly [18].

The dimerization process is the combination of a keratin of Type I with a Type II; it appears, therefore, that the process is mainly driven by the acid-base interaction, being a neutralization process. When a strong acid reacts with a strong base the enthalpy change of the reaction (neutralization) is -55.9 kJ mol^{-1}, meaning that the process releases heat; it is exothermic. For weak acids and bases the value is smaller, as they do not fully dissociate and some energy is consumed for the dissociation process, reducing thus the enthalpy change and increasing the importance of the entropic factor, as resulted from inequality (Eq. 13.16b).

A study of dimerization reactions of two keratins, namely K35 (Type I, acidic) and K85 (Type II, basic), by using the isothermal titration calorimetry of one of the keratins with the other, in a solution of 8 M urea, at 25°C, allowed the direct evaluation of the contribution of the enthalpy and entropy factors to the formation of heterodimer (coiled-coil) as being [19]:

$\Delta H = 2546$ kJ·mol^{-1}
$T\Delta S = 2585$ kJ·mol^{-1}
$\Delta G = -39$ kJ·mol^{-1} for temperature of titration at 298 K (25°C)

It appears that the first assembly process, of dimerisation (coiled-coil formation) is an endothermic process and the contribution of the entropy factor is decisive in allowing it to occur. The experimentally obtained values above allow calculation of the variation of Gibbs free energy at 37°C, as being of -143 kJ·mol^{-1}; it also permit estimating the minimum temperature at which the dimerization process is still allowed thermodynamically ($\Delta G = 0$) as being of 20.5°C, which means that below 20°C the coiled-coil cannot form anymore.

The increase in entropy, physically, is assumed to be due to the release of water molecules, which surrounded the individual keratins (and have low entropy) into the bulk solution (high entropy), during the assembling process.

The formed dimers associate exothermally [19] to form tetramers composed of antiparallel dimers, which stack longitudinally [20]. Eight tetramers organize laterally, slightly randomly [21], to form an annular 32mer [22], the so-called Unit Length Filaments, ULF. All these steps of assembling appear, from direct measurements, to be entropically driven [19].

The elongation of the ULFs into KIFs and the assembly in macrofibrils of KIFs with KAPs stabilising the fibrous structure, has been suggested to be a solution process of polymerisation of ULFs into KIFs, followed by a phase separation process of KIFs assembled with KAPs [23].

The concepts of polymerisation at equilibrium for forming filaments of some 230 nm from monomers (ULFs) with a length of about 50–60 nm and width of 10–16 nm, along with those of isotropic solution of polydisperse rods (ideal geometry for the ULFs) allows for the qualitative description of the formation of filaments in macrofibrils of the cells as being dictated by initial concentration of ULFs in the isotropic phase, which leads to segregation of long rods (KIFs) and formation of the small size phase-separated regions, named '*tactoids*', by aggregation of rods of various lengths [24]. Support for the role of concentration of ULFs according to the type of cell in which macrofibrils form is the observation that the fibril-matrix ratio is highest in orthocortical cells, indicating a high concentration of the segregated macrofibrils [25], and is low in paracortical cells, with mesocortical cells being intermediate [26].

The slight differences of concentration of ULFs existing in different cell types can produce tactoids of different shape and length, which in turn give rise to the observed structural variations in cell types [23]. The KIFs are likely to self-organize in macrofibrils in a double-twist (blue-phase) cylindrical structure. The intensity of the double-twist thus gives the differences between cells, with the orthocortical cells having a high intensity double-twist structure, and the paracortical cells having rather a nematic structure than a very low intensity double-twist [23, 27].

The process of forming the macrofibrils by the assembly (polymerisation) of ULFs into KIFs and of stabilization by the association of KIFs with KAPs is exclusively a self-assembly process, involving protein and solvent, without the involvement of any motor proteins, or chaperones. This has been shown by mixing *in vitro* callus reconstituted intermediate filaments (IFs) and KAPs separated previously from wool, nails and callus in a ratio of 2:1 IFs-to-IFAPs and leaving the mixture at room temperature for 15 min. The process led in all cases to the formation of callus macrofibrils, relatively independent from the original nature of IFAPs used [28].

The thermodynamic force behind the process is entropy, namely the variation of side-chain conformational entropy with head group characteristics, and the conservation of head group entropy after polymerization, and the increase in water entropy as a result of the arrangement of hydrophobic regions in the middle of ULF, for the polymerisation process, and the stabilization of rigid-rod molecules (KIFs) as an emulsion by side-chain entropy gain, a 'steric stabilization'.

The enthalpic contribution to the process is evaluated from the theory of polymers solution, according to which the change of Gibbs free energy of mixing, ΔG_m, is [29, 30]:

$$\Delta G_m = RT\left[n_1 \cdot \ln\phi_1 + n_2 \cdot \ln\phi_2 - n_1 \cdot \phi_2 \cdot \chi_{12}\right]$$
(13.24)

Equation (13.24) shows that the change of free energy of mixing the proteins (ULFs, KIFs) with water, ΔG_m, is a function of the number of moles, n_1, n_2, and volume fractions ϕ_1, ϕ_2, of each component of the mixture. The parameter χ_{12}, named 'Flory-Huggins parameter', is used for taking into account the energy of inter-dispersing protein and solvent molecules, R being the gas constant and T the absolute temperature) and allows calculating the enthalpic contribution of the mixing of polymer monomers (ULFs, volume fraction ϕ_2) with water molecules, n_1, as [29, 30]:

$$\Delta H_m = R \cdot T \cdot n_1 \cdot \phi_2 \cdot \chi_{12}$$
(13.25)

The Flory-Huggins parameter in Eq. (13.25) was not evaluated directly for ULFs but for those of keratin fibres interaction with water. Its value is around 1.0, which indicates that the change of enthalpy at this mixing is likely to be positive, hence an endothermic process [31].

This means that, for the process of polymerizing ULFs and macrofibrils formation, the thermodynamic force which allows the self-assembling is the entropy, the enthalpy change being positive. This implies, also, that the process is sensitive to temperature change, since $\Delta G < 0$ stands only for a certain temperature range.

13.3 Thermodynamics of Keratin denaturation and Effect of Ligands

The assembly process described previously generally occurs for proteins: the linear polypeptides arrange into a 3D structure named '*folded state*', or '*native state*'. Under the effect of various parameters this structure may go back into the '*unfolded state*', and chemically the process is described as a reversible reaction:

$$\text{Unfolded State} \rightleftharpoons \text{Folded State}(\text{Native}) \quad \text{(I)}$$

The folded state is that in which the protein fulfils its biological functions; the unfolded state is an altered structure, which does not allow the protein to realize its specific functions. Consequently the study of the reaction (I) is of importance.

Thermodynamically the unfolded state has high entropy, due to a disorganized arrangement of the chain. This state also possesses high enthalpy, due to the amount of hydrogen bonds,

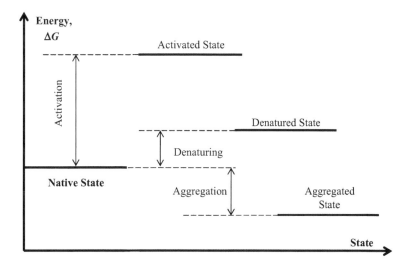

Fig. 13.3 Energy of various states of a protein. (Drawn after [33])

which are built by water molecules surrounding the unfolded protein. The folded state has small entropy, being much more ordered than the unfolded one; it has a smaller enthalpy too, but overall the huge cost of loss of entropy by ordering is only slightly compensated by the new stabilizing interactions and the stability of the folded state is not too far from those of the unfolded one.

The process of losing the ordered structure of Native state is called 'protein denaturation' and the parameters which may contribute to it are temperature, mechanical stress, pH, ionic stress, or surface interaction. For many small size proteins this process is reversible according to reaction (I), but for large ones, keratins included, the process of denaturing is irreversible. The process of denaturing is accompanied by aggregation of protein to proteins, coagulation and jellification [32]. A general scheme of these states and the relative free energy corresponding to each step is graphed in Fig. 13.3.

The kinetic barrier of activation energy in Fig. 13.3 determines the temperature-time dependence of the denaturing process. The change of enthalpy throughout the processes described above is measureable by calorimetry. As is observed, often the denatured state is energetically above the native state (is a metastable state), but the final state (aggregated state, after coagulation, aggregation and/or jellification occurred) has always a lower energy than those of native state,

fulfilling the thermodynamic condition for reaching a stable state [33].

The kinetics of the denaturing process pictured by Fig. 13.3 is traditionally described by Lumry-Eyring model [34]:

$$N \underset{k_1}{\overset{k_{-1}}{\rightleftharpoons}} U \overset{k_2}{\rightarrow} D \quad \text{(II)}$$

N, U and D being the Native, Partial denatured and Denatured states of the protein, respectively, and k_1, k_{-1}, k_2 are the rate constants of the corresponding reactions. This model assumes that the irreversible denaturing of the protein involves two stages, namely: (a) a reversible denaturing of the native protein, N; (b) the irreversible altering of the partly denatured protein, U, towards the final, aggregated state, D, from which the protein cannot recover the initial native state.

The movement of a protein along the pathway from a state to another may be induced by changing temperature, pressure, pH, concentration of ligands, or of the solvent around the protein. The move is allowed only if the native, N, and denatured, D, state, respectively, has different affinity toward the used parameter. The temperature induced denaturation is thus due to the difference between the enthalpies of the two states; the pressure-induced denaturing comes if the two states have different partial molar volumes; chemically-induced denaturing (pH, ligand, solvent) is due to differences in binding the chemical to protein in native and in denatured state.

The chemical products, which interfere with the denaturing process, protecting or sensitizing it, are named *'modifiers of denaturing process'* [33]. Such agents acting at thermal-induced denaturing are the pH, or solvents like methanol, ethanol, propanol and butanol [35, 36]. When a ligand binds preferentially to the folded protein it stabilizes the native state and the denaturation process is less favoured with increasing ligand concentration; when the ligand binds preferentially to the denatured protein, the native state is destabilized [37].

Supposing that ligand molecules, L, bind specifically only to folded protein, N, one may write the reactions at equilibrium and corresponding equilibrium constants as [37]:

Ligand binding : $N + L \rightleftharpoons NL$; $K_{L,N} = [N][L]/[NL]$
Unfolding : $N \rightleftharpoons U$; $K_0 = [U]/[N]$

(III)

$K_{L,N}$ being the dissociation constant for binding ligand to native protein and K_0 is the equilibrium constant for non-bonded protein. The effective equilibrium constant, K_{unf}, is:

$$K_{unf} = \frac{[U]}{[N]+[NL]} = \frac{K_0}{1+[L]/K_{L,N}} \approx \frac{K_0 \cdot K_{L,N}}{[L]}$$

(13.26)

Eqn (13.27) shows the decrease of equilibrium constant K_{unf} and the increase of the stability of native form with increasing ligand concentration.

The change of Gibbs free energy is:

$$\Delta G_{unf} = -RT \cdot \ln(K_{unf}) = \Delta G_{unf,0} + RT \cdot \ln\left(\frac{1+[L]}{K_{L,N}}\right) \approx$$
$$\approx \Delta G_{unf,0} + \Delta G^0_{diss,N} + RT \cdot \ln[L]$$

(13.27)

with $\Delta G_{unf,0}$ being the Gibbs free energy of denaturing the non-bonded protein and $\Delta G^0_{diss,N} = -RT\ln(K_{L,N})$ is standard Gibbs free energy for dissociating the ligand from the binding site of native protein.

Gibbs free energy may be separated in the enthalpic and entropic contributions:

$$\Delta H_{unf} \approx \Delta H_{unf,0} + \Delta H^0_{diss,N}$$
$$\Delta S_{unf} \approx \Delta S_{unf,0} + \Delta S^0_{diss,N} - R \cdot \ln[L]$$

(13.28)

The stabilizing effect of the ligand is, thus, understood as being due to additional free energy required to remove ligand molecules prior to denaturation and the supplementary contribution ($R\ln[L]$) of mixing entropy of free ligand in the solvent [37].

The effect of ligand on the temperature of thermal denaturing may be thus evaluated as being [37, 38]:

$$\frac{\Delta T_m}{T_m} = \pm \frac{n \cdot RT_0}{\Delta H_{unf,0}} \cdot \ln \frac{1+[L]}{K_L}$$

(13.29)

with $\Delta T_m = T_m - T_0$ being the shift of denaturing temperature, n stands for the number of binding sites of ligand molecules to protein, and the sign +, or − indicates the stabilizing action of the ligand: of native form, or of the denatured state, respectively.

For low concentrations of ligands ($[L]/K_L < 1$), Eq. (13.29) turns into the linear approximation, which describes the initial part of the logarithmic curve:

$$\frac{\Delta T_m}{T_m} \approx \pm \frac{n \cdot RT_0 \cdot [L]}{K_L \cdot \Delta H_{unf,0}}$$

(13.30)

For a ligand which binds strongly to the native protein and less to the denaturing protein, the denaturation temperature, T_m, shifts to higher values with increasing the concentration of ligand, [L], until both states, N and D, are bonded completely and the denaturing temperature reaches a plateau [39].

A particular case of the ligand binding is the case of protons (H⁺), hence the effect of low pH on the thermal stability of keratin, which was shown by thermal analysis investigations on human hair to shift the temperature of keratin denaturing towards higher values [40].

13.4 Thermodynamics of Alpha-Helix to Beta Sheet Transition in Keratins

As mentioned in Sect. 13.2 of this chapter, the α-helix arrangement of amino acids drives the building units of keratin chains. Around the time the keratin structure was first deciphered by XRD investigations of hair fibres, it was observed that, on stretching the hair fibre the pattern of α-helix vanishes and a new one, named latter β-sheet, evolves [41, 42]. The transition proved to be reversible, the pattern of α -helix reappearing when the stress is suppressed and the fibre is left to relax [43]; it appeared to be also quasi-independent of alkali, or other chemical treatments of the fibre, as long as these do not alter the α-helix crystalline amount [42].

Because this occurred on stretching the transformation of α-helix into β-sheet and back plays an important role in the understanding the mechanics of the keratin fibre. Following the stress-strain diagram the XRD investigations of gradual disappearance of meridional reflection at 5.1 Å, corresponding to α-helix, and the appearance of 4.65 Å equatorial reflection, corresponding to β-sheet, supported the idea that the transition begins during the first 5% of strain [43]. Recent studies indicate that the process requires a minimum 45% relative humidity to occur [44], and it is not a one-step transformation α → β evolving linearly with increasing strain, but a multi-step process beginning at 5% strain with the unravelling of α-helical domains and the unfolding of α-helices, followed, after 20% strain, by their folding into β structure in a centrifugal propagation, due, probably, to the strain gradient across the fibre cross-section, from core towards periphery [45].

The enantiotropic transformation of a crystalline phase into the other is a reversible reaction:

$$\alpha - \text{Helix State} \rightleftharpoons \beta - \text{Sheet State} \quad \text{(IV)}$$

with α form appearing to be thermodynamically more stable than β-sheet one.

Since the transformation is very much influenced by the stress applied, σ, the Clausius-Clapeyron equation, characterizing a discontinuous phase transition between two phases of a single constituent, at constant pressure and temperature, can be written as:

$$\frac{\partial \left(\sigma/T \right)}{\partial \left(1/T \right)} \bigg|_p = \frac{\Delta H}{\Delta l} \quad (13.31)$$

where Δl is the variation of fibre length due to the stress. Using this formalism and experimentally measured stress-strain at isothermal different temperatures (values of stress recorded at beginning of yielding region of stress-strain diagram) yielded from calculations the enthalpy of the α → β transformation as being about 10 kJ·mol^{-1} of peptide unit, for an average molecular weight of 110 Da, and the temperature of transformation, in absence of any stress, of about 180°C [46]. The temperature estimated this way is in line with Differential Scanning Calorimetry measurements, DSC, which record the unfolding of α-helices at around 150°C in swollen fibres [47].

13.5 Glass Transition of Keratins

The concepts of glassy state and the corresponding glass transition temperature are largely used for describing the behaviour of macromolecules in semi-crystalline polymers. The glassy state is understood as a metastable state, a state for which small isothermal changes of the thermodynamic variables result in sudden increase of the free energy of the system [48]. The typical example is the supercooled liquid, which is in metastable state with respect to its crystalline phase.

Although keratin fibres do not form by the cooling of a melted phase, they are similar to semi-crystalline polymers by having around 20 to 40% keratin chains arranged in ordered manner as KIFs and the rest packed amorphously [49]. The macromolecular chains packed in the random manner are those supposed to be in a metastable state, with respect to a relaxed state; as a consequence these chains would try to move towards the equilibrium positions, with their mobility being hindered by packing or increased suddenly after a certain temperature, the glass transition temperature.

Imagining the chains move as a viscous Newtonian flow, the rate, k_0, at which the chains jump from one equilibrium "lattice position" in the liquid to another is [50]:

$$k_0 \approx R \cdot T / V \cdot \eta \qquad (13.32)$$

where R is the gas constant, T is the absolute temperature, V is the mole volume and η is the viscosity. Assuming for the keratins that the mole volume is 90 cm^3 (ratio of their molecular weight, 110 Da, and density of 1.3 g·cm^{-3}) and using the viscosity value of ~ 10^{12} Pa·s, just at the limit between liquids and solids, it appears that k_0 indicates about one molecular chain jump at ~10 h at room temperature (27°C).

The mobility of chains, as estimated by Eq. (13.32), is strongly dependent on temperature because the viscous flow is a thermally activated process:

$$\eta = Z \exp(E_v / R \cdot T) \qquad (13.33)$$

Z being a scaling constant, similar to a pre-exponential factor and also temperature dependent ($Z = Z_0 T$) and E_v is the activation energy of viscous flow.

The viscosity of keratins is also affected by the solvents, water being one of the strongest agents able to change it. The effect of water (moisture) on glass transition temperature of keratins was shown to follow Fox's law [51]:

$$w_1 / T_{g1} + w_2 / T_{g2} = 1 / T_g \qquad (13.34)$$

with w_1, w_2 being the weight percentage of the keratin and water, respectively, hence: $w_1 + w_2 = 1$. The glass transition of water is known, T_{g1} = 125 K (−148°C as measured by Kalichevsky et al. [52]) and this along with Eq. (13.34) allows for the estimation of the glass transition temperature of a particular keratin from experimental measurements of various keratin-to-moisture ratio mixtures. Several works used measurements from fibre mechanics, which show an abrupt change at the glass transition temperature [53, 54], as well as the fibre specific heat, determined by DSC, which records a step-wise change occurring at the glass transition [55–58] allowed claims that, depending on the fibre source, the glass transition temperature of wool fibres is around 175°C and that of Caucasian human hair around 145°C. Using Eq. (13.34) and the values of glass transition temperature for water, and for wool, respectively hair, given above, one easily calculates that at a moisture content of keratin fibres of 18% (a moderate value of moisture content of a keratin at a relative humidity of the environment of 65–70%) the glass transition temperature of wool reaches 33°C, and of human hair falls at 21°C.

The concept of glass transition and viscous flow in physics of polymers explains the process of fibre aging (the decay of mechanical properties over time) as being due to the fact that the thermodynamically glassy state is not at equilibrium and the entangled chains tend to reach it. In other words, below the glass transition temperature the macromolecular chains still move (jump), at a very low frequency, towards their equilibrium position. The process depends on the degree of packing or the '*free-volume*' existing in their arrangement [59]. The free-volume is related, hence, to segmental mobility of the chains, which allows changes in the chain configuration.

As aging is thermo-reversible and occurs below glass-transition temperature, the system can be de-aged (history erased) if moved above that temperature [60]. In terms of keratins, with the strong influence of water molecules on the T_g, the humidification of the fibres brings this temperature below the room temperature, therefore de-aging of keratins occurs readily at their wetting. This plays an important role at fibre setting.

Aging of materials starts at the glass transition temperature, T_g, and occurs over a wide range of temperatures down to the first secondary transition. This low-temperature transition is named, for keratins, the '*gamma-transition*', T_γ, and has been measured, for Corriedale wool, to be of around −150°C [61, 62]. Both T_g and T_γ are related to the mobility of the polymer chains, which means that they occur in the amorphous regions of keratin materials: the glass transition is considered to be the moment when the system changes from a viscoelastic into a glassy state, and the movement of the backbone of polymer chains is frozen in, and the gamma-transition indicates the moment when even the movement of protein side chains is frozen in.

13.6 Moisture Sorption and Desorption by Keratins

The interaction of water with keratins is of fundamental interest for understanding the behaviour and performance of keratin materials.

Water absorption results in swelling, the hair fibre showing considerable swelling anisotropy. An increase in the amount of absorbed moisture from 0% to 33%, corresponds roughly to change of relative humidity from 0 to 100%, and leads to a longitudinal swelling of ~ 2% and a radial swelling of ~16% at wool and hair fibres [1]. The extent of swelling depends on the pH value, with a minimum at around isoelectric point.

Moisture sorption by keratins is an exothermic process, the evolved heat decreasing with the amount of moisture adsorbed from 10 to about 0.9 kJ·mol^{-1}, as has been measured calorimetrically by Hedges [63].

As mentioned previously (Sect. 13.1) the state of a system is fully described by a surface in the space of the state parameters pressure, temperature and number of molecules. For the purpose of describing keratin-water systems the parameters to be used are the relative pressure of the water vapour, or the relative humidity, RH, the amount of water in keratin material at equilibrium, ewc (equilibrium water content as defined in [64]) and the temperature, T:

$$f(RH, ewc, T) = 0 \qquad (13.35)$$

The isothermal recording of sorption and desorption curves of moisture in keratin materials produces characteristic sigmoidal isotherms, as plotted in Fig. 13.4. The area between sorption and desorption curves is the hysteresis area, which shrinks with increasing temperature of measurement [65].

Finding the analytical expression of the relationship (13.35) in terms of ewc versus RH for any temperature is still a field of investigation, most of the proposed models either being empirical, without a sound physical background

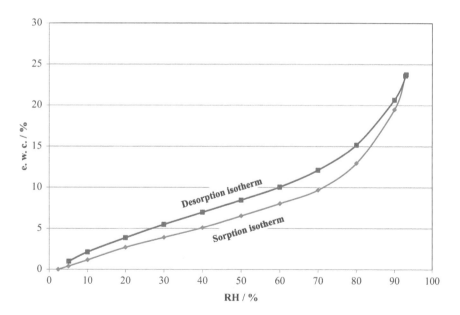

Fig. 13.4 Sorption and desorption isotherms recorded for human hair at 30°C

of the constants, or addressing only part aspects of the keratin-water system.

The shrinking of the hysteresis area with the increasing temperature of measuring the isotherms is what is expected from the behaviour of glassy polymer-water systems, the size of the area being considered as proportional to the amount of free-volume (see discussion on T_g in previous section). This fact led to the proposition of the models, which use Flory-Huggins relationship for polymer-water solutions [29, 30]. Rosenbaum's model [31] is among the first ones to use the Flory-Huggins treatment, adding a correction factor depending on the structural changes occurring within the system when water molecules enter:

$$\ln(RH) = \ln(v_1) + v_2 + \chi_{12} \cdot v_2^2 - \frac{1}{2 \cdot \beta \cdot R \cdot T} \cdot V_d$$

(13.36)

v_1 and v_2 being the volume fractions of water (index 1) and keratin (index 2), respectively ($v_1 + v_2 = 1$), χ_{12} is the Flory-Huggins parameter of water-keratin interaction, $\chi_{12} \sim 1$, β is the keratin compressibility (Pa^{-1}) and V_d is the differential change of water volume (cm$^3 \cdot$g^{-1}).

A similar model, Vrentas-Vrentas, was developed from the Flory-Huggins equation for polymer-solvent systems by considering the changes of polymer structure with the increasing amount of diffusing solvent into the polymer structure [66–68]. This model has been adjusted for the keratin-water system by Pierlot [69]:

$$\ln(RH) = \ln(v_1) + v_2 + c_{12} \cdot v_2^2 + k \cdot F \quad (13.37)$$

The correction factor $k \cdot F$ in Eq. (13.37) includes Rosenbaum's approach and makes explicit use of the glass transition temperature, T_g, as F becomes nil for $T > T_g$.

The state of water molecules inside the keratin material is another line followed in the study of sorption-desorption isotherms. Presuming three types of water, namely strongly bonded, weakly bonded and freely, D'Arcy and Watt [70] produced a three terms model, which fits the experimental data well. Analysis of the model constants casts doubts on its validity [71] and the evidence from various physical measurements (thermal analysis, proton nuclear magnetic resonance and electrical conductivity measurements) point to no difference existing among the water molecules inside the keratin-water system annuls the working hypothesis.

Assuming that the water molecules are energetically equivalent inside the keratin-water system, and for a heterogeneous inside surface of the keratins, this allows the use of the formalism of Jura and Harkins [72], who demonstrated, thermodynamically, that for molecules adsorbed on a surface the following general relationship holds (adapted for keratin-water system):

$$\ln(RH) = B - A / (ewc)^2 \quad (13.38)$$

A and B being two constants, A relating to the square of the inner surface area of the keratin. The plot of experimental data ln(RH) *versus* 1/(ewc)^2 leads to two straight lines for data below and above around 50% relative humidity, indicating a change of the probably monolayer film of water molecules formed at low values of relative humidity into a plurimolecular film at high values of *RH*, accompanied by an opening of the keratin material structure [73]. This approach allows for the estimation the size of the inner surface of the keratins, giving, for the data below 50% *RH*, similar values as those obtained by using BET equation [74]. Moreover, the opening of the structure indicated by the application of the model corresponds to the fibre swelling as a result of increasing diffused water molecules in system.

As a result we need to combine models like those given by Eqs. (13.37) and (13.38) to understand the complexity of water-keratin interaction.

13.7 Thermodynamics of Keratin Mechanics

The various aspects discussed so far aim at understanding the peculiarities of keratin mechanics.

The change of Gibbs free energy ΔG of a fibre subjected to a uniform tensile force F, acting in

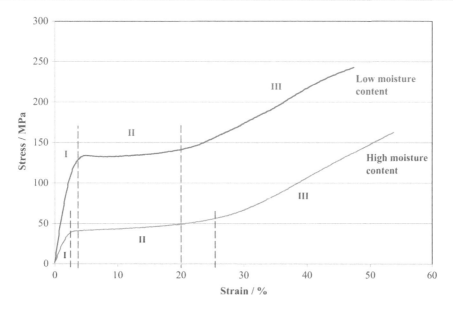

Fig. 13.5 Stress-strain curve of a keratin fibre

the direction of the fibre length L may be written [75]:

$$\Delta G = \Delta H - T \cdot \Delta S - F \cdot \Delta L \qquad (13.39)$$

The tensile test of a keratin fibre, measured as stress versus strain, produces the general sigmoidal shapes illustrated in Fig. 13.5. With the increasing moisture content of the fibre (increasing relative humidity) the curve elongates, as also shown in Fig. 13.5. The diagram is generally discussed as being composed of three regions, namely: "*elastic*", "*yield*" and "*post-yield*" region, respectively, numbered on diagram as I to III.

The curve is understood to result from the contribution of hydrogen bonds, ionic bonds and weak van der Waals interactions over the first region, and mainly from the contribution of disulfide bonds for the other two regions. As discussed, the α-to-β transition also occurs along this diagram, providing the amount of moisture in fibre is higher than 6% (relative humidity higher than 45% as shown by Krepak et al. [44]) for decreasing the viscosity of material and allowing the KIFs to unravel and α-helices to unfold, as described in Section 13.4.

The contribution of the various types of bonds to the tensile diagram is through a two-stage model: a fast opening of hydrogen, ionic and van der Waals bonds, followed by a slower scission of disulfide bonds [76, 77]. The approach developed by Flory [75] allowed evaluating the activation enthalpy and entropy for the diffusion of hydroxyl ions, OH^-, into keratin fibres (human hair) as being of 10.76 kcal·mol^{-1} (45.0 kJ·mol^{-1}) and 10.5 cal·K^{-1}·mol^{-1} (43.9 J·K^{-1}·mol^{-1}) respectively. These values suggested that "hair in the swollen state acts like a partially oriented membrane with water in many of the cavities in the fibre" [75]. The activation energy of the disulfide interchange reaction, evaluated from the stress-relaxation kinetic study [76], is of 22…24 kcal·mol^{-1} (92…100 kJ·mol^{-1}), and is close to the value of 18 kcal·mol^{-1} (75.2 kJ·mol^{-1}) reported for the whole stress-relaxation process [78]. The comparison of all these values for activation energy (enthalpy) suggests that disulfide bonds play the most important role in the overall mechanical behaviour of keratins.

Due to the readiness to break and the low energy involved, the hydrogen bonds contribute mainly to the behaviour of keratins during the first region, the elastic (Hookean) one. Modelling this, showed that the slope of the curve in that region, the Young's modulus (modulus of elasticity), E, is proportional to the amount of *effective* hydrogen bonds per unit volume, N as [79]:

$$E = k \cdot N^{1/3} \qquad (13.40)$$

k being a constant of proportionality. The term "effective bonds" is used generally to stress that not all bonds contribute to the mechanics of the keratin, some of them being too stereo-chemically hindered to contribute to the stress on the material.

Understanding the visco-elastic properties of keratins requires consideration of not only the contribution of the hydrogen and disulfide bonds, but also the knowledge that the material is a composite of interconnected crystalline and amorphous zones. Yet, the theory of elasticity developed for rubber, says that entropy decreases with increasing strain (and aligning the chains) and that this accounts for the increasing stress required for further extension, does not work well with keratins because entropy plays a minor role in their elastic recovery [80], and the melting of crystalline zones of rubber is not tantamount to the α-β transition.

In taking into account the α-β transformation it is assumed that when the material is under stress the folded α-keratin chains of average cross-section s transforms into β-sheet and that the cross-links existing within fibre keep the folded chains apart a distance 2λ through various bonds. Taking into consideration that at each moment of strain, ε, there is an equilibrium between unfolded and refolded chains, therefore for that strain [81]:

$$\epsilon = \frac{1}{2} \cdot \varepsilon_\infty \cdot \left[\tanh\left(\frac{\lambda \cdot N_A \cdot s}{R \cdot T} \cdot f - \frac{\Delta G_0}{2 \cdot R \cdot T}\right) + \tanh\frac{\Delta G_0}{2 \cdot R \cdot T} \right]$$

(13.41)

ε_∞ being the extension at which all chains are fully unfolded (transformed into β-sheets), N_A is the Avogadro number, f is the stress applied per chain, and *tanh* stands for hyperbolic tangent operator.

The stress-strain experiments analysed by this equation allows for the evaluation of the enthalpy and entropy of 1.74 kcal·mol⁻¹ (7.27 kJ·mol⁻¹) and of 6.0 cal·K⁻¹·mol⁻¹ (25.1 J·K⁻¹·mol⁻¹) [81].

Obtaining an understanding of molecular events in the keratins during the stretching of the fibre was also approached using molecular dynamics simulations of the stress-strain diagram, taking into consideration that under stretching there are two states in equilibrium, α and β, with a partition function for each state of:

$$q_i = \exp(S_i / k_B) \cdot \exp(-E_i / k_B T) \qquad (13.42)$$

S_i and E_i being the entropy and energy, respectively, of state "i", and k_B stands for Boltzmann constant [82]. This leads to the expression of average force as:

$$\langle F \rangle = 1/\Delta x \cdot \left[\Delta E - T \cdot \Delta S + k_B \cdot T \cdot \ln(n/(N-n)) \right]$$

(13.43)

N being the number of chains out of which n are in β-state and Δx is the change of length of a unit α and/or β with applied force. Eqn (13.43) is, practically, the thermodynamic relationship (13.39) expressed at molecular level for the equilibrium (i.e. for $\Delta G = 0$), the energy change, ΔE being equivalent with enthalpy change, ΔH.

This model predicts that, when half of the units are in α and half in β state (i.e. $n = N/2$), the force depends linearly on temperature:

$$\langle F \rangle = \Delta E / \Delta x - T \Delta S / \Delta x \qquad (13.44)$$

This moment is assumed to be equivalent to the yield point of the diagram (Fig. 13.5), namely the point between regions I and II. The experimental stress-strain diagrams at various temperatures, translated into the stress on each α-helix unit, allow for the evaluation of the enthalpy and entropy of the α-β transformation during stretching. Assuming a change in length of α-unit at complete unfolding, Δx, of 1 Angstrom, and the cross-section area of the same unit of ~1 nm², the variation of enthalpy is determined to be 5.1 kJ·mol⁻¹ and entropy of 11 J·K⁻¹·mol⁻¹, which is in very good agreement with the results above [82].

Moreover, the model and the experimental data used for calculation allows for an estimation the temperature at which F becomes nil, hence the temperature of complete α-β unfolding is around 200°C, which is in the range of data from DSC [47]. The enthalpy of 5.1 kJ·mol^{-1} as evaluated above, for a molecular weight of 110 Da coupled with the number of α-helices per keratin fibre of ~25%, gives the enthalpy of unfolding of 11 J·g^{-1}, which is also close to the results measured by Popescu and Gummer [47].

13.8 Conclusions

Summing up, in this excursion using thermodynamics to decipher the behaviour of keratins it is noticeable that, from the aggregation of keratin molecules in the formation of microfibrils to the response of keratin materials under mechanical, thermal or moisture stress, the entropic component of the free energy Gibbs is the driving factor of most of the processes. This points to the important role played by temperature for the keratins, which, along with contributions to the process kinetics (accelerating or decelerating the process rate), helps increase or decrease the entropic contribution and, thus, allows thermodynamically the occurrence of any process.

References

1. Popescu, C., & Höcker, H. (2007). Hair – The most sophisticated biological composite material. *Chemical Society Reviews, 36*(8), 1282–1291.
2. Herrmann, H., & Aebi, U. (2004). Intermediate filament assembly: Molecular structure, assembly mechanism, and integration into functionally distinct intracellular scaffolds. *Annual Review of Biochemistry, 73*, 749–789.
3. Yang, Z. W., & Babitch, J. A. (1988). Factors modulating filament formation by bovine glial fibrillary acidic protein, the intermediate filament component of astroglial cells. *Biochemistry, 27*, 7038–7045.
4. Stromer, M. H., Ritter, M. A., Pang, Y. Y., & Robson R. M. (1987). Effect of cations and temperature on kinetics of desmin assembly. *Biochemical Journal, 246*, 75–81.
5. Aebi, U., et al. (1988). Unifying principles in intermediate filament (IF) structure and assembly. *Protoplasma, 145*, 73–81.
6. Angelides, K. J., Smith, K. E., & Takeda M. (1989). Assembly and exchange of intermediate filament proteins of neurons: Neurofilaments are dynamic structures. *The Journal of Cell Biology, 108*, 1495–1506.
7. Rafik, M. E., et al. (2006). In vivo formation steps of the hard α-keratin intermediate filament along a hair follicle: Evidence for structural polymorphism. *Journal of Structural Biology, 154*(1), 79–88.
8. Bornschlögl, T., et al. (2016). Keratin network modifications lead to the mechanical stiffening of the hair follicle fiber. *Proceedings of the National Academy of Sciences, 113*(21), 5940–5945.
9. Popescu, C., & Höcker, H. (2009). *Chapter 4 cytomechanics of hair: Basics of the mechanical stability*. In W. J. Kwang (Ed.), *International review of cell and molecular biology* (pp. 137–156). Academic Press.
10. Snir, Y., & Kamien, R. D. (2015). Entropically driven helix formation. *Science, 307*, 1067.
11. Maritan, A., et al. (2000). *Nature, 406*, 287–290.
12. Crick, F. H. C. (1952). Is α-keratin a coiled coil? *Nature, 170*, 882–883.
13. Crick, F. H. C., & Kendrew, J. C. (1957). X-ray analysis and protein structure. *Advances in Protein Chemistry, 12*, 133–214.
14. Steinert, P. M., et al. (1983). Complete amino acid sequence of a mouse epidermal keratin subunit and implications for the structure of intermediate filaments. *Nature, 302*, 794–800.
15. Steinert, P. M., Torchia, D. R., & Mack, J. W. (1988). In G. E. Rogers et al. (Eds.), *The biology of wool and hair*. London: Chapman & Hall.
16. Parry, D. A. D., & Steinert, P. M. (1999). Intermediate filaments: Molecular architecture, assembly, dynamics and polymorphism. *Quarterly Reviews of Biophysics, 32*(2), 99–187.
17. Chernyatina, A. A., Guzenko, D., & Strelkov, S. V. (2015). Intermediate filament structure: The bottom-up approach. *Current Opinion in Cell Biology, 32*, 65–72.
18. Parry, D. A. D., & Fraser, R. D. B. (1985). Intermediate filament structure. 1. Analysis of IF protein sequence data. *International Journal of Biological Macromolecules, 7*, 203–213.
19. Ishii, D., et al. (2011). Stepwise characterization of the thermodynamics of trichocyte intermediate filament protein supramolecular assembly. *Journal of Molecular Biology, 408*(5), 832–838.
20. Fraser, R. D., & Parry, D. A. (2005). The three-dimensional structure of trichocyte (hard α-) keratin intermediate filaments: Features of the molecular packing deduced from the sites of induced crosslinks. *Journal of Structural Biology, 151*(2), 171–181.
21. Rafik, M. E., Doucet, J., & Briki, F. (2004). The intermediate filament architecture as determined by X-ray

diffraction modeling of hard α-keratin. *Biophysical Journal, 86,* 3893–3904.
22. Parry, D. A., et al. (2007). Towards a molecular description of intermediate filament structure and assembly. *Experimental Cell Research, 313*(10), 2204–2216.
23. McKinnon, J., & Harland, D. P. (2011). A concerted polymerization-mesophase separation model for formation of trichocyte intermediate filaments and macrofibril templates 1: Relating phase separation to structural development. *Journal of Structural Biology, 173*(2), 229–240.
24. Flory, P. J., & Frost, R. S. (1978). Statistical thermodynamics of mixtures of rodlike particles. 3. The most probable distribution. *Macromolecules, 11*(6), 1126–1133.
25. Plowman, J. E., Paton, L. N., & Bryson, W. G. (2007). The differential expression of proteins in the cortical cells of wool and hair fibres. *Experimental Dermatology, 16*(9), 707–714.
26. Harland, D. P., et al. (2011). Arrangement of trichokeratin intermediate filaments and matrix in the cortex of merino wool. *Journal of Structural Biology, 173*(1), 29–37.
27. Harland, D. P., et al. (2014). Three-dimensional architecture of macrofibrils in the human scalp hair cortex. *Journal of Structural Biology, 185*(3), 397–404.
28. Kueppers, B., & Hoecker, H. (1990). Cross-reaction of keratin filaments and intermediate filament-associated proteins from various tissues: Assembly of macrofibrils. In *Proceedings of the 8th international wool textile research conference* 1990. Christchurch: Wool Research Organisation of New Zealand.
29. Huggins, M. L. (1941). Solutions of long chain compounds. *The Journal of Chemical Physics, 9,* 440.
30. Flory, P. J. (1941). Thermodynamics of high polymer solutions. *The Journal of Chemical Physics, 9,* 660–661.
31. Rosenbaum, S. (1970). Solution of water in polymers: The keratin–water isotherm. *Journal of Polymer Science, Part C: Polymer Symposia, 31,* 45–55.
32. Gossett, P., Rizvi, S., & Baker, R. (1984). Symposium: Gelation in food protein systems quantitative analysis of gelation in egg protein systems. *Food Technology, 38,* 67.
33. Bischof, J. C., & He, X. (2005). Thermal stability of proteins. *Annals of New York Academy of Sciences, 1066,* 12–33.
34. Lumry, R., & Eyring, H. (1954). Conformation changes of proteins. *Journal of Physical Chemistry, 58,* 110–120.
35. Joly, M. (1965). *A physico-chemical approach to the denaturation of proteins* (pp. 153–170). London: Academic Press.
36. Eyring, H. (1974). Temperature. In F. H. Johnson, H. Eyring, & B. J. Stoner (Eds.), *The theory of rate processes in biology and medicine.* New York: Wiley.
37. Cooper, A. (1999). Thermodynamics of protein folding and stability. In G. Allen (Ed.), Protein: A comprehensive treatise (Vol. 2). Stamford: JAI Press Inc.
38. Cooper, A., & McAuley-Hecht, K. E. (1993). Microcalorimetry and the molecular recognition of peptides and proteins. *Philosophical Transactions of Royal Society London A, 345,* 23–35.
39. Sanchez-Ruiz, J. M. (1992). Theoretical analysis of Lumry-Eyring models in differential scanning calorimetry. *Biophysical Journal, 61,* 921.
40. Istrate, D., et al. (2013). The effect of pH on the thermal stability of fibrous hard alpha-keratins. *Polymer Degradation and Stability, 98,* 542–549.
41. Astbury, W.T., & Street, A. (1932). X-ray studies of the structure of hair, wool, and related fibres. I General. *Philosophical Transactions of Royal Society London A, 230,* 75–101.
42. Astbury, W. T., & Woods, H. J. (1934). X ray studies of the structure of hair, wool, and related fibres. II. The molecular structure and elastic properties of hair keratin. *Philosophical Transactions of Royal Society London A, 232,* 333–394.
43. Bendit, E. G. (1960). A quantitative x-ray diffraction study of the alpha-beta transformation in wool keratin. *Textile Research Journal, 30,* 547–555.
44. Kreplak, L., et al. (2002). A new deformation model of hard α-keratin at the nanometer scale: Implications for hard α-keratin intermediate filament mechanical properties. *Biophysical Journal, 82,* 2265–2274.
45. Kreplak, L., et al. (2004). New aspects of the a-helix to b-sheet transition in stretched hard a-keratin fibres. *Biophysical Journal, 87,* 640–647.
46. Ciferri, A. (1963). The alpha-beta transformation in keratin. *Transaction Faraday Society, 59,* 562–569.
47. Popescu, C., & Gummer, C. (2016). DSC of human hair: A tool for claim support or incorrect data analysis? *International Journal of Cosmetic Science, 38,* 433–439.
48. Kauzmann, W. (1948). The nature of the glassy state and the behavior of liquids at low temperatures. *Chemical Reviews, 43,* 219–256.
49. Feughelman, M., & Note, A. (1989). On the water-impenetrable component of α-keratin Fibres. *Textile Research Journal, 59,* 739–742.
50. Eyring, H. (1936). Viscosity, plasticity, and diffusion as examples of absolute reaction rates. *The Journal of Chemical Physics, 4,* 283–292.
51. Fox, T. G. (1956). Influence of diluent and of copolymer composition on the glass temperature of a polymer system. *Bulletin of the American Physical Society, 1,* 123.
52. Kalichevsky, M. T., Jaroszkiewics, E. M., & Blanchard, J. M. V. (1992). Glass transition of gluten. 1: Gluten and gluten-sugar mixtures. *International Journal of Biological Macromolecules, 14,* 257–266.
53. Menefee, E., & Yee, G. (1965). Thermally-induced structural changes in wool. *Textile Research Journal, 35,* 801–812.

54. Wortmann, F. J., Rigby, B. J., & Phillips, D. G. (1984). Glass transition temperature of wool as a function of regain. *Textile Research Journal, 54*, 6–8.
55. Phillips, D. G. (1985). Detecting a glass transition in wool by differential scanning calorimetry. *Textile Research Journal, 55*, 171–174.
56. Huson, M. G. (1991). DSC investigation of the physical ageing and deageing of wool. *Polymer International, 26*, 157–161.
57. Kure, J. M., et al. (1997). The glass transition of wool: An improved determination using DSC. *Textile Research Journal, 67*, 18–22.
58. Wortmann, F. J., et al. (2006). The effect of water on the glass transition of human hair. *Biopolymers, 81*, 371–375.
59. Struik, L. C. E. (1978) *Physical aging in amorphous polymers and other materials*. Amsterdam/Oxford/New York: Elsevier Scientific Publishing Company.
60. Struik, L. C. E. (1977). Physical aging in plastics and other glassy materials. *Polymer Engineering & Science, 17*, 165–173.
61. Druhala, M., & Feughelman, M. (1971). Mechanical properties of keratin fibres between −196 °C and 20 °C. *Kolloid-Zeitschrift und Zeitschrift für Polymere, 248*, 1032–1033.
62. Druhala, M., & Feughelman, M. (1974). Dynamic mechanical loss in keratin at low temperatures. *Colloid Polymer Science, 252*, 381–391.
63. Hedges, J. J. (1926). The absorption of water by colloidal fibres. *Transactions of the Faraday Society, 22*, 178–193.
64. Watt, I. C., & D'Arcy, R. L. (1979). Water-vapour adsorption isotherms of wool. *Journal of Textile Institute, 70*, 298–307.
65. Wortmann, F. J., Hullmann, A., & Popescu, C. (2007). In *IFSCC Magazine* (pp. 317–320).
66. Vrentas, J. S., & Vrentas, C. M. (1991). Sorption in glassy polymers. *Macromolecules, 24*, 2404–2412.
67. Vrentas, J. S., & Vrentas, C. M. (1994). Evaluation of a sorption equation for polymer-solvent systems. *Journal of Applied Polymer Science, 51*, 1791–1795.
68. Vrentas, J. S., & Vrentas, C. M. (1996). Hysteresis effects for sorption in glassy polymers. *Macromolecules, 29*, 4391–4396.
69. Pierlot, A. P. (1999). Water in Wool. *Textile Research Journal, 69*(2), 97–103.
70. D'Arcy, R. L., & Watt, I. C. (1970). Analysis of sorption isotherms of non-homogeneous sorbents. *Transactions of Faraday Society, 66*, 1236–1245.
71. Wortmann, F. J., Augustin, P., & Popescu, C. (2001). Temperature dependence of the water-sorption isotherms of wool. *Journal of Applied Polymer Science, 79*, 1054–1061.
72. Jura, G., & Harkins, W. D. (1946). Surfaces of solids. XIV. A unitary thermodynamic theory of the adsorption of vapors on solids and of insoluble films on liquid subphases. *Journal of the American Chemical Society, 68*, 1941–1952.
73. Popescu, C., & Wortmann, F. J. (2002). Water vapour sorption and desorption by wool. *Wool Technology and Sheep Breeding, 50*(1), 52–63.
74. Bull, H. B. (1944). Adsorption of water vapor by proteins. *Journal of the American Chemical Society, 66*, 1499–1507.
75. Flory, P. J. (1956). Phase equilibria in solutions of rod-like particles. *Proceedings of the Royal Society of London. Series A. Mathematical and Physical Sciences, 234*(1196), 73.
76. Reese, C. E., & Eyring, H. (1950). Mechanical properties and the structure of hair. *Textile Research Journal, 20*, 743–753.
77. Weigmann, H.-D., & Rebenfeld, L. (1966). Reduction of wool with dithiothreitol. *Textile Research Journal, 36*(2), 202–203.
78. Feughelman, M. (1965). *III-ieme Congres International de la Recherche Textile Lainière*. Section 1: p. 413.
79. Nissan, A. H. (1976). H-bond dissociation in hydrogen bond dominated solids. *Macromolecules, 9*(5), 840–850.
80. Bull, H. B. (1945). Elasticity of keratin fibers. II. Influence of temperature. *Journal of the American Chemical Society, 67*, 533–536.
81. Peters, L., & Speakman, J. B. (1948). The viscoelastic properties of wool fibres. *Textile Research Journal, 18*, 511–518.
82. Akkermans, R. L. C., & Warren, P. B. (2004). Multiscale modelling of human hair. *Philosophical Transactions of Royal Society London A, 362*.

Oxidative Modification of Trichocyte Keratins

14

Jolon M. Dyer

Contents

14.1	**Introduction & Overview**	206
14.1.1	Oxidative Modification – Why Important?	206
14.1.1.1	The Effects of Keratin Oxidation	206
14.2	**Oxidative Insults**	208
14.2.1	Photo-Oxidative Damage	208
14.2.1.1	Photo-Oxidative Pathways	208
14.2.1.2	Influence of Metal Ions	208
14.2.2	Hydrothermal Modification	208
14.2.3	Bleaching & Relaxation	209
14.3	**Profiling Molecular Modification**	209
14.3.1	Modification Types	209
14.3.1.1	Residue Specific Side-Chain Modification	209
14.3.1.2	Carbonylation	210
14.3.1.3	Side-Chain Scission & Backbone Cleavage	210
14.3.1.4	Protein-Protein Crosslinking	210
14.4	**Evaluation of Keratin Oxidative Modification**	211
14.4.1	Characterisation	211
14.4.2	Modification Profiling	212
14.4.3	Damage Hierarchies	212
14.4.4	Redox Scoring	212
14.4.5	Evaluating Carbonylation	214
14.4.6	Characterising Protein Crosslinking	214
14.5	**Conclusions & Future Trends**	215
	References	215

Abstract

Oxidation of keratin results in a range of deleterious effects, including discolouration and compromised physical and mechanical properties. Keratin oxidative degradation is driven by molecular-level events, with accumulation of modifications at the protein primary level resulting directly in changes to secondary, tertiary and quaternary structure, as well as eventually changes in the observable physical and chemical properties. Advances in proteomic analysis techniques provide an increasingly

J. M. Dyer (✉)
AgResearch Ltd., Lincoln, New Zealand
e-mail: Jolon.Dyer@agresearch.co.nz

clearer insight into the cascade of molecular modification underpinning keratin oxidation and how this translates through to higher order changes in properties. This chapter summarises the effects of oxidation on keratin-based materials, the types of molecular modification associated with this, and advances in techniques and approaches for characterising this modification.

Keywords

Keratin · Protein · Oxidation · Redox · Modification

14.1 Introduction & Overview

14.1.1 Oxidative Modification – Why Important?

For proteins in general, oxidation is generally a degradative process. The process is driven by oxidative events at the molecular level, with accumulation of oxidative modifications at the protein primary level resulting directly in changes to secondary, tertiary and quaternary structure, as well as eventually changes in the observable physical and chemical properties of the protein material [1] Deleterious effects of protein oxidation are wide-ranging, including in a food context, for example, loss of nutritional value and altered digestibility.

Typically understanding of protein oxidation has been largely at the holistic levels, in terms of overall changes in the properties of the protein material, while the underpinning molecular mechanisms have typically been poorly understood. This can be attributed to the highly complex cascade of oxidative products formed at the molecular level, with any given specific modification (type and location) often only present in low relative abundance. The more recent advent and progressive development of highly sensitive and specific analytical techniques, however, notably mass spectrometry, have begun to facilitate a clearer insight into protein oxidation at the molecular level, and how this translates through to higher order changes in properties. This is also true in the case of understanding keratin oxidation.

Before expanding further on the application of these developing approaches, the next section will summarise some of the key keratin-based materials and the specific deleterious effects of oxidation for these materials.

14.1.1.1 The Effects of Keratin Oxidation

14.1.1.1.1 Hair

Human hair is made up of central inner cortical cells, largely comprised of proteins, enclosed by overlapping layers of cuticle cells [2]. The predominant protein class in human hair are keratins, which also comprise the key structural component of the fibre. The keratins are organised into filamentous bundles which are embedded within a protein matrix comprised largely of keratin-associated proteins (KAP) [3]. As for all mammalian fibres, human hair is characterised by a very high sulfur content, resultant from high levels of cysteine and an extensive network of disulfide cross-links. The cuticle also has high levels of protein-protein crosslinking, characterised by the presence of isopeptide crosslinks in addition to disulfide crosslinks [4]. It is noteworthy that cysteine is particularly sensitive to oxidative modification and, in general, oxidation of human hair proteins therefore, has progressively profound effects on the fibre properties.

Human scalp and facial hair in particular have important social value, being a significant factor in the perception of appearance and with modification of this hair therefore being an important cosmetic consideration. Oxidation of human hair proteins affects a wide-range of cosmetic-important properties, including fibre colour, lustre, moisture retention, brittleness and tactile properties [5, 6]. Typically high levels of exposure to sunlight means that photo-oxidation is a major source of oxidative insult for scalp and facial hair. In the case of photo-oxidation, the levels of melanin present in the fibre influence the degree of protein oxidation, with the melanin able to act as both a protective and pro-oxidant agent [6]. In addition, common cosmetic treatments applied to human hair, such as bleaching and dyeing, can result in significant levels of protein oxidative damage [7].

14.1.1.1.2 Wool

Wool has an analogous anatomy to human hair, with a central cortex comprised mostly of IFP within a KAP matrix, enveloped by an outer cuticle region, and also characterised by high levels of sulfur-containing amino acid residues and high levels of protein-protein crosslinking. One difference is that while human hair typically has many layers of cuticle cells, some breeds of sheep have wool with few layers, notably merino wool, which has only one layer of cuticle cells [8].

Traditionally, wool fibres have been used in apparel and interior textile applications. For both these uses, appearance retention has come to be a key factor in the quality and value of these products. As keratins comprise the key structural component of wool, modification of these keratins can have a profound effect on the physicomechanical properties, and therefore appearance retention of wool products. For example, "phototendering" is the process by which fibrous proteins gradually lose their structural integrity through progressive photo-induced oxidation, leading to an increase in brittleness, lowered strength, and therefore ultimately wool products which wear out faster [9].

Another notable deleterious effect of oxidation in wool keratins is discolouration. In the case of photo-oxidation, the colour of the wool fibre is affected via two competing processes, photobleaching which leads to loss of colour through destruction of chromophores, and photoyellowing which leads to increasing colouration through formation of chomophores [10]. Chomophores, visible light absorbing moieties, can be produced through the attack of reactive oxygen (ROS) species on protein residues sides chains, in particular the aromatic and cyclic amino acid residues [11, 12]. For wool-based apparel exposed to UVB outdoors, photoyellowing can result directly in discolouration via progressive yellowing of the fibre and also decreased brightness, which is a particularly noticeable and therefore negative effect for pale or pastel shaded fabrics. Additives which increase fabric brightness, known as fluorescent whitening agents or optical brightening agents, are now commonly added to laundry detergents. However, these additives have been shown to excacerbate photoyellowing for natural protein based fibres exposed to UVB, via photosensitisation and consequent increased generation of ROS [9, 13]. Bleaching of wool fibres, which is a routine part of wool processing from greasy to scoured wool has also been shown to exacerbate subsequent discolouration. For wool carpets, which are typically indoors behind glass windows which block out UVB but still allow exposure to UVA, photobleaching is the dominant process, but with potentially equally deleterious effect through uneven loss of colour [9].

14.1.1.1.3 Skin

There are three major protein classes present in mammalian skin, namely collagens, keratins and elastins. For skin proteins, the most significant form of oxidative damage is via exposure to UV, particularly of concern with increasing levels of UV reaching the earth's surface as the ozone layer is depleted. Exposure of skin proteins to UV results in ROS-mediated oxidation, associated with photoageing and also implicated in the development of skin disorders, including notably skin cancers [14].

Skin is comprised of three main layers, namely the outer epidermis, the dermis and the hypodermis [15]. Within the epidermis, the outermost sub-layer is the stratum corneum contains packets of keratin, which have an important function preventing loss of moisture and therefore maintaining skin hydration. In terms of photodamage, UVB is absorbed primarily in the epidermis, where it can generate ROS and cause oxidative modification, such as carbonylation, to stratum corneum proteins [16, 17]. Accumulation of protein oxidation has been observed in photodamaged skin [18]. Such progressive protein oxidation is believed to play a contributing role to loss of structural integrity associated with cosmetic skin damage, such as the formation of wrinkles and lines. [19].

This chapter will next consider keratin oxidative damage from the viewpoint of the major types of oxidative insult. It will then examine what modifications occurring at the molecular level underpin keratin oxidation. And finally, approaches and techniques for evaluating and tracking keratin oxidation will be considered.

14.2 Oxidative Insults

This section summarises the major forms of oxidative insult of relevance to keratins.

14.2.1 Photo-Oxidative Damage

Ultraviolet light (UV) induced photo-oxidation of proteins generally has been linked to a wide range of deleterious effects [20]. For keratin-based substrates, notably scalp hair, wool and skin, outdoor exposure to UV is an important consideration This is increasing in concern, with depletion of the earth's ozone layer and consequent rising levels of UVB intensity [21]. For keratin-based fibres, exposure to UV is strongly associated with discolouration, increased brittleness, harsher handle and loss of tensile strength, while for skin exposure is associated with general skin damage and negative cosmetic changes.

Photo-oxidative modification of keratins results from generation and attack of ROS at amino acid residues side chains or at the protein backbone itself. The main ROS associated with keratin oxidative modification are the hydroxyl radical, particularly dominant in aqueous systems, and singlet oxygen. The main targets of attack are amino acid residues with high electron density, namely the sulfur-containing residues and the aromatic side-chain residues. Modification of aromatic residues, in particular, is associated with keratin photodamage. When phenylalanine or tyrosine residues are proximal to tryptophan, absorption of electromagnetic radiation and energy transfer can take place, with fluorescence quenching [22–24]. Due to this energy transfer, tryptophan is particularly susceptible to photo-oxidation [12].

14.2.1.1 Photo-Oxidative Pathways
Keratins, as with other proteins, do not have significant absorbance in the UV range, and therefore photo-oxidation is mediated by the photosensitised generation of ROS [9]. Tryptophan and tyrosine amino acid residues in keratin can absorb light at wavelengths over 290 nm [25, 26]. These can then generate ROS, such as singlet oxygen and superoxide, when molecular oxygen is present [27]. Generally speaking, both Type I (electron transfer and hydrogen abstraction) and Type II (energy transfer with molecular oxygen) photochemical reactions can occur on exposure to UV. For either of these reaction types, however, ultimately it is the direct attack of resultant ROS, notably singlet oxygen and the hydroxyl radical, that leads to protein photo-oxidation products [28]. Additionally, photo-oxidative products themselves, such as N-formylkynurenine, can act as photosensitisers [29, 30].

14.2.1.2 Influence of Metal Ions
Trace metal ions, particularly Fe (II) and Cu (II) ions, can have an exacerbating influence on photo-oxidation [31–35]. For instance, Cu (II) ions are very commonly present in human hair, as it is deposited during washing with tap water. . Recent studies have shown that the level of Cu (II) ions in hair directly affects the observed levels of keratin photo-oxidative products, with increasing levels of Cu (II) ions correlating to increasing levels of keratin photo-oxidation [36, 37]. Fe (II) ions also promote keratin photo-oxidation through the generation of hydroxyl radicals via Fenton chemistry [38]. Disruption of the binding of metal ions to the proteins, therefore, such as via chelation, has been demonstrated in some cases to have a protective effect for keratins against photo-oxidation [39, 40].

14.2.2 Hydrothermal Modification

Wool, human hair and skin are all subjected at times to elevated temperatures in aqueous conditions and consequent hydrothermal insult. Typically, though, molecular changes associated with hydrothermal insult to keratins has not been well understood, with most evaluation approaches focused on holistic evaluation [41, 42]. Recent studies have discovered a range of oxidative modification induced by hydrothermal treatment, and these are generally the same range of ROS-induced modications observed for photo-oxidation, although in differing ratios and relative abundance [43]. There are some modifications more specifically-related to thermal insult also observed though, such as the generation of dehydroalanine from cysteine or serine. Dehydroalanine in turn can react with the side-chains of basic amino acid residues, to form protein-protein crosslinks, such as lysinoalanine in the case of reaction with lysine [44]. Lanthione, a derivative of cystine, is the other particularly noteworthy crosslink type associated with hydrothermal treatment of keratins [10, 45, 46].

14.2.3 Bleaching & Relaxation

Bleaching and relaxation are common consumer treatments applied to human hair for the purposes of modifying the cosmetic properties. Hydrogen peroxide-based bleaching of human hair was found in recent studies to lead to significantly elevated levels of keratin oxidative modification [47]. Once again, the general range of observed oxidative products are similar to those observed for photo-oxidation, implying the role of ROS in this type of insult also, with oxidation of cysteine to cysteic acid being a predominant oxidative product. Relaxation treatment of human hair with sodium hydroxide resulted in significant formation of dehydroalanine and other amino acid residue side-chain dehydration products, together with elevated levels of oxidation to aromatic and sulfur-containing amino acid residues.

14.3 Profiling Molecular Modification

14.3.1 Modification Types

There are a range of different types of protein oxidative modification initiated at the primary structural level. Oxidative modification is induced typically through (1) reaction of an amino acid residue side chain [48], (2) scission of an amino acid residue side-chain [49], (3) reaction of either the *N*- or *C*-terminus of the protein or peptide [50], or (4) protein backbone cleavage [51]. After initiation through reaction of an amino acid residue side chain, another class of modification can occur as a secondary reaction, namely (5) protein-protein crosslinking. It is important to note that although in a biological system, i.e. within a living organism, oxidised proteins can essentially be repaired, though identification, disposal and recycling, this is not the case for *ex vivo* oxidative modification. For materials such as human hair, wool and the outer surface layers of skin, accumulation of modifications can progressively alter appearance and properties of the material.

We will now consider in summary form the major classes of protein oxidative modification. It is important to consider each and all of these modification types when evaluating keratin oxidation at the molecular level.

14.3.1.1 Residue Specific Side-Chain Modification

ROS are able to attack any of the amino acid residues [52]. However, there are differing levels of sensitivity to oxidation. The amino acid residues most sensitive to oxidation are the sulfur-containing residues, with cysteine being the most sensitive, followed by methionine. The second tier of sensitive residues are the aromatic amino acid residues, tryptophan, tyrosine, phenylalanine and histidine, in order of susceptibility [53]. The cyclic amino acid residue proline is also relatively susceptible to oxidation.

14.3.1.1.1 Modification to Sulfur-Containing Residues

Of all the amino acid residue side-chain types, the sulfur-containing residues are the most susceptible to oxidation. Keratins contain a high relative abundance of cysteine, found in the either the free thiol form or as part of the disulfide protein-protein crosslink network (where it is termed cystine). As an integral part of this crosslink network, cysteine plays a critical role in the structural and mechanical properties of the fibre [54–56]. The relative amounts of cysteine and cystine within the keratins is a factor in the degree of damage suffered by fibres under oxidative insult, as these two forms are redox interconvertible. However, once oxidised to cysteic acid, the modification is not reversible. Another important cystine modification is conversion to lanthionine, which is a crosslinked modification that is not redox convertible back to cysteine and therefore locks the crosslink in place. Lanthionine is formed with exposure of disulfide crosslinks to heat and/or alkali conditions [57, 58]. Oxidation of the amino acid residue methionine results in formation of methionine sulfoxide [59].

14.3.1.1.2 Modification to Aromatic Residues

The aromatic amino acids most susceptible to oxidative modification are in order of sensitivity tryptophan, tyrosine, phenylalanine and histidine [60]. A broad range of primary and secondary oxidation products can potentially be formed from these residues. The hydroxyl radical ROS has been noted to have a particular affinity for aromatic structures [61–64], and can react via hydroxylation to form primary oxidation products such as hydroxytryptophan from native tryptophan residues, dihydroxyphenylalnnine (DOPA) from native tyrosine residues. These initial products in turn can be progressively further oxidised to products such as N-formylkynurenine, kynurenine and hydroxykynurenine, in the case of tryptophan [12]. For histidine, hydroxylation is also the major observed route of oxidation, with subsequent potential degradation to asparagine or aspartic acid residues [65].

14.3.1.1.3 Modification to Other Residues

Other notable oxidative modifications to amino acid residue side chains include hydroxylation of proline to form hydroxyproline and dihydroxyproline [66, 67]. Proline residues can also be modified to glutamic semialdehyde and pyroglutamic acid. Deamidation of asparagine to aspartic acid, and glutamine to glutamic acid is another common oxidative modification [68]. Deamidation can also occur to the N-terminus of the peptide or protein. Decarboxylation of terminal amino acid residues can also occur.

As mentioned previously, ROS can attack any of the amino acid residue side-chains, and so there are multiple other potential oxidation products possible [38]. However, the side-chain oxidative modification types specifically noted here are considered the ones predominantly influencing the overall protein oxidation in terms of relative abundance and effect on overall properties.

14.3.1.2 Carbonylation

Exposure of keratins to ROS results in the formation of carbonyl groups, a modification type referred to as carbonylation [69, 70]. This is generally regarded as an irreversible modification type. Introduction of carbonyl moieties can occur on any amino acid residue side-chain, but the most commonly reported are additions to arginine, lysine, proline and threonine [71].

14.3.1.3 Side-Chain Scission & Backbone Cleavage

Carbon-centred radicals can be formed within keratins under oxidative insult via ROS-mediated extraction of the α-hydrogen [72]. These radicals can subsequently generate peroxy radicals through reaction with molecular oxygen, and progressively be modified to peroxides and alkoxyl groups. The introduction of these moieties into proteins can in turn lead to cleavage of the protein or peptide backbone. Backbone cleavage in keratins is likely to play an important role in modifying the physical properties of the material under oxidative insult, as this is irreversible and results in changes at the protein secondary and tertiary structural levels and beyond.

Scission of the amino acid residue side chain can also occur under oxidative insult, particularly under exposure to ionising radiation. Cleavage occurs at the β-position [1, 49]. This type of scission is one route for the introduction of carbonyl groups as side-chain derivatives, and is also another route for the generation of carbon-centred radicals into the keratin backbone.

14.3.1.4 Protein-Protein Crosslinking

Both intra- and intermolecular crosslinking can occur in keratins under oxidative insult [53, 73]. As is the case for other types of modification, there is a wide range of different crosslinks that can be generated. One characteristic crosslinked oxidation product generated in keratins is dityrosine. This is formed via ROS attack on tyrosine residues to initially generate tyrosyl radicals, which in turn react with proximal tyrosine residues to form dityrosine [74]. Dityrosine itself can then be further progressively modified, such as by hydroxylation to form hydroxydityrosine. Dityrosine formation contributes to colouration changes in the keratins under oxidative insult, and, in the case of photo-oxidation, contributes to secondary photochemical reactions via photosensitisation [75]. Another important crosslink associated with oxidative modification of keratins is lanthionine, a modification of cystine associated with thermal and/or alkali insult.

14.4 Evaluation of Keratin Oxidative Modification

Keratin oxidation, via any kind of oxidative insult, can result in a highly complex pattern of modifications. This complexity is attributable to: (1) the broad range of specific modifications possible, (2) the typically low relative abundance of any given specific modification (type and location in the protein), and combining these factors, (3) the fact that any given specific protein location can have multiple modification types present at differing relative abundances, together with the native unmodified variant.

Any approach to evaluation of keratin oxidation at the molecular level is also severely confounded by the fact that keratins exist biologically in highly protein-protein crosslinked networks (for both mammalian fibres and skin), and additional protein-protein crosslinks are induced by the oxidative insult itself. In order to analyse the proteins effectively, it is typically necessary to disrupt the native disulfide crosslink network via reduction. The oxidative-induced crosslinks, such as dityrosine or lanthionine, are not broken under these conditions and undoubtedly play a role in lowering the protein extractability observed typically with keratin oxidation. Information from non-extracted protein and peptides is therefore often lost.

Due to these complicating factors, evaluation of keratin oxidation by the wool or personal care industries are therefore largely using fairly holistic techniques, such as fluorescence analysis or protein extractability assays. However, the advent of highly sensitive mass spectrometric instrumentation and associated proteomic analysis techniques have allowed advances in our understanding of keratin oxidation and tools to evaluate it. The following sections will summarise these advances.

14.4.1 Characterisation

Accurate and sensitive characterisation of low relative abundance modifications is the key underpinning factor in evaluating keratin oxidation at the molecular level. Mass spectrometry is the tool currently best suited to this type of analysis, as it is able to evaluate proteins in a system and profile modifications to these proteins relatively rapidly and, importantly, without the need for individual peptides or proteins to be purified. To summarise a proteomic evaluation approach generally, the keratin-based substrate would typically be reduced to break the disulfide bonds and alkylated to prevent these bonds re-forming. The sample is then enzymatically or chemically digested to produce a mixture of peptides. For fibres, as mentioned previously, widespread crosslinking can confound extraction. One approach that has been used to overcome this problem is the utilisation of tandem enzymatic/chemical sequential digestion. One example of the successful application of this is the use of 2-nitro 5-thiocyanobenzoic acid (NTCB), which cleaves proteins at cysteine residue N-termini, following by subsequent trypsin digestion [4].

After digestion, fractionation of resultant peptides, generally by one- or multidimensional liquid chromatography, allows the best chance of observing and characterising low abundance modified peptides [76]. Mass spectrometric analysis of these peptides, generally utilising electrospray ionisation (ESI) or matrix-assisted laser desorption ionisation (MALDI) mass spectrometry, distinguishes peptides based on their mass to change (m/z) ratio. This typically generates large amounts of data which can be bioinformatically mined to sequence the peptides observed and hopefully characterise modifications present [77, 78].

14.4.2 Modification Profiling

To understand the molecular processes occurring during keratin oxidation and to design damage mitigations strategies it is imperative to profile the widest possible range of modifications and track these through the oxidative process. The utilization of keratin marker peptides is one useful approach to doing this, with subsequent tracking of this set of peptides.

Selection of marker peptides should be based around two key criteria. Firstly, each keratin-derived peptide needs to be consistently and reproducibly observed at high relative abundance in the material of interest after enzymatic digestion. Secondly, to be useful the selected peptide needs to contain at least one oxidatively sensitive residue [43]. In addition, it is advisable for the overall set of marker peptides to contain a range of differing oxidatively sensitive residues, as opposed to all being one type. A notable example of a useful peptide marker for tracking keratin oxidation in wool is the peptide with amino acid residue sequence DVEEWYIR, which is observed consistently in proteomic wool analyses and contains an oxidatively sensitive tryptophan (W) residue [79, 80].

Where specific keratin sites are of interest, profiling can also be expanded to include tracking the relative degradation of targeted amino acid residues through direct comparison of mass spectrometric ion abundance levels between the modified and unmodified native peptide of interest [81]. Variation in this relative abundance can be tracked over time or correlated to changes in other parameters such as degree of UV exposure, temperature and pH. Stable isotope labelling approaches can also be utilized to assess with high sensitivity modifications at specific sites and changes in the relative levels of differing modifications at this site [82, 83]. Ultimately, measuring the absolute abundance, as opposed to relative abundance, of specific keratin modifications would provide the best way of evaluating keratin oxidation and it is envisaged development of this area will provide future advancements in our understandings.

14.4.3 Damage Hierarchies

Oxidative modification to keratins is not a series of discrete conversions from a native unmodified amino acid residue to a modified one, but rather a progressive cascade and accumulation of modifications, beginning with primary oxidation events but with modified residues then being further modified. This can be described for any given amino acid residues by way of what could be termed "modification hierarchies" outlining the progression of modification [11, 84]. The most well elucidated of these to date is for tryptophan residues, for which a very broad range of oxidative modification products have been characterised. Two predominant oxidative pathways have been observed for tryptophan (W) (Fig. 14.1). The first is via initial oxidation to hydroxytryptophan (W + O) via attack by an ROS such as the hydroxyl radical, then further oxidation to dihydroxytryptophan (W + 2O), from which various dione-related products can be formed. The second is via initial formation of N-formylkynurenine (W + 2O) through attack of an ROS such as singlet oxygen, which can then subsequently be further modified to products such as kynurenine (W + O-C) and hydroxykynurenine (W + 2O-C), or to hydroxyformylkynurenine (W + 3O) and dihydroxyformylkynurenine (W + 4O) [85]. To fully understand and evaluate keratin oxidation, characterising and profiling the relative abundance of each and all of these modifications is important.

14.4.4 Redox Scoring

An emerging tool in the evaluation of keratin oxidation, that makes use of the concept of modification hierarchies, is the use of redox

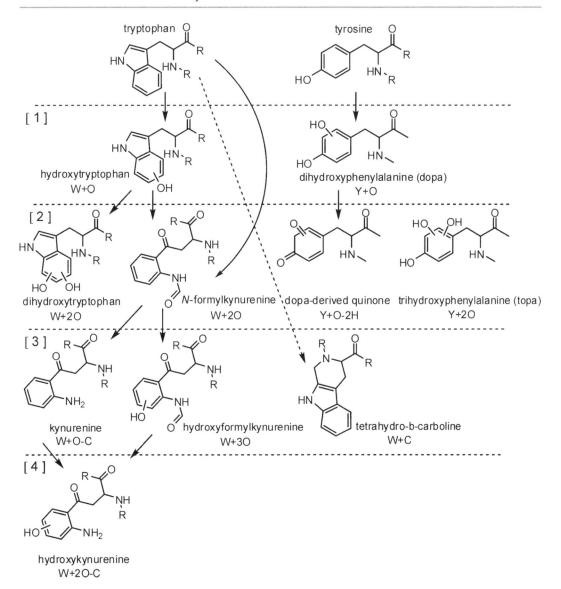

Fig. 14.1 Classification of photo-oxidative damage to tryptophan and tyrosine residues based on the degree of photomodification reprinted from Journal of Photochemistry and Photobiology B: Biology 2010 [85] with permission from Elsevier

scoring systems. The fundamental concept behind this approach is to assign a weighted score to each qualitatively observed modification. The weighting is related to the degree of oxidation relative to the native unmodified amino acid residue. To exemplify, observation of hydroxytryptophan at a specific protein location would be assigned a score of 1 as a primary oxidation event, whereas observation of kynurenine at a specific location would be assigned a score of 3, since it is a tertiary modification (Fig. 14.1) (Fig. 14.1) [85]. Scores are only assigned where the modification has been characterised by tandem mass spectrometric evaluation (MS/MS), comparing observed fragment ions with theoretical fragment ions, above a certain identification threshold, to avoid false identifications. The total redox score is comprised of the addition of each individual modification score. This total redox score can then also be broken down and presented by specific kinds of modification class, such as oxidative modification to sulfur-containing amino acid residues. One particularly useful class of oxidative modification to score separately as a sub-set of the total modification is deamidation, which is strongly associated with changes in protein secondary structure. Deamidation is also generally a good marker for assessing oxidative degradation occurring in historical and archaeological keratinous samples, such as ancient textiles or human remains [80, 86].

Redox scores can be used to track the relative degrees of modification and modification types occurring during the course of oxidative insult, such as during hair bleaching or the heat treatment of wool. They can also be used to objectively assess the effect of damage protection, mitigation or repair treatments or strategies at the keratin molecular level, such as the use of UV-blockers for skin protection.

A further, more recent, adaption of the redox scoring approach has been to include an additional weighting designed to help account for difference in keratin extraction and digestion between samples [85, 87]. This additional weighting normalises the score between samples by utilising the ratio of the number of observed oxidative modifications for any specific amino acid residue to the total number of observed native, non-modified instances of that residue. This allows for differences in extraction as samples from which the proteins have been extracted efficiently to a high level also have a similarly higher level of observable native residues present.

14.4.5 Evaluating Carbonylation

The introduction of carbonyl moieties has long been recognised as characteristic of protein oxidation. The established method for assessing overall oxidation levels holistically is via derivatisation with 2,4-dinitrophenylhydrazine (DNPH), which reacts with carbonyl groups, followed by either immune- or spectrophotometric evaluation [76]. In more advanced approaches applied to other non-keratinous substrates, affinity purification can be combined with tandem mass spectrometry (LC-MS/MS) to characterise specific carbonylation sites. For affinity purification, carbonyl groups react with biocytin hydrazide resulting in hydrazine conjugate generation [76]. Streptavidin immobilised on agarose is then used to pull out these modified peptides. This approach offers promise for further understanding carbonylation in keratins.

14.4.6 Characterising Protein Crosslinking

Protein-protein crosslinking, put simply, is the formation of molecular bridges either within (intramolecular) or between (intermolecular) proteins. Although all forms of oxidative insult can produce protein-protein crosslinking, accurate characterisation and evaluation of these oxidatively-induced crosslinks is very difficult. Peptides which contain crosslinks are not easily identifiable with common mass spectrometric/bioinformatics workflows as the fragmentation patterns differ significantly in complexity from non-crosslinked peptides. In addition, as mentioned previously, crosslink formation hinders protein extractability and any given crosslinked peptide is likely present is very low relative abundance. Of course, any extraction protocol utilised must also ensure that the crosslinks of interest remain intact during the extraction, if information on the specific location of the crosslinks in the keratins is to be maintained.

Until now, there has only been highly limited progress in the proteomic characterisation and detailed profiling of oxidatively-induced protein-protein crosslinks in keratins [12]. However, novel approaches applied to other non-keratinous systems offer some promise for application to keratins. Potentially the most significant of these approaches is the utilisation of partial digestion followed by isotopic labelling [88, 89]. Water isotopically enriched with ^{18}O is used to add ^{18}O into the C-termini of peptides during enzymatic digestion [90]. In contrast to non-crosslinked peptides, peptides containing a crosslink have at least two C-termini, and so two ^{18}O atoms are added to these. When combined with parallel samples digested in regular ^{16}O water, shifts in the mass to charge ratio can be used to identify those peptides that contain a crosslink [89].

14.5 Conclusions & Future Trends

Keratin oxidation plays a critical role in the retention of appearance and functionality in human hair, wool, and skin. Advances in technologies and approaches for mapping and tracking this oxidation at the molecular level have significantly increased our understanding. For further advances in the future, several key technical areas require development. Firstly, reliable and sensitive techniques for measuring the absolute abundance of modifications. Secondly, effective techniques for evaluating keratin oxidatively-induced crosslinking. In addition, further research which links and correlates the molecular level profiling with higher order structural changes and holistic material properties and performance will be important. Ultimately, increasingly accurate mathematical models based on these correlations which can take molecular level information and material property information and make two-way predictions is an ongoing goal. This two way predictive modelling will facilitate a new level in highly functional personal care treatments as well as the accurate matching of natural fibre specification and production to specific consumer performance demands.

References

1. Dean, R. T., et al. (1997). Biochemistry and pathology of radical-mediated protein oxidation. *Biochemical Journal, 324*(Pt 1), 1–18.
2. Hashimoto, K. (1988). The structure of human hair. *Clinics in Dermatology, 6*(4), 7–21.
3. Popescu, C., & Höcker, H. (2007). Hair – The most sophisticated biological composite material. *Chemical Society Reviews, 36*(8), 1282–1291.
4. Bringans, S. D., et al. (2007). Characterization of the exocuticle a-layer proteins of wool. *Experimental Dermatology, 16*(11), 951–960.
5. Nogueira, A. C. S., et al. (2007). Photo yellowing of human hair. *Journal of Photochemistry and Photobiology B: Biology, 88*(2-3), 119–125.
6. Nogueira, A. C. S., Dicelio, L. E., & Joekes, I. (2006). About photodamage of human hair. *Photochemical and Photobiological Sciences, 5*, 165–169.
7. Dawber, R. (1996). Hair: Its structure and response to cosmetic preparations. *Clinics in Dermatology, 14*(1), 105–112.
8. MacKinnon, P. J., Powell, B. C., & Rogers, G. E. (1990). Structure and expression of genes for a class of cysteine-rich proteins of the cuticle layers of differentiating wool and hair follicles. *Journal of Cell Biology, 111*(6), 2587–2600.
9. Millington, K. R. (2006). Photoyellowing of wool. Part 1: Factors affecting photoyellowing and experimental techniques. *Coloration Technology, 122*(4), 169–186.
10. Duffield, P. A., & Lewis, D. M. (1985). The yellowing and bleaching of wool. *Review of Progress in Coloration, 15*, 38–51.
11. Dyer, J. M., Bringans, S. D., & Bryson, W. G. (2006). Determination of photo-oxidation products within photoyellowed bleached wool proteins. *Photochemistry and Photobiology, 82*(2), 551–557.
12. Dyer, J. M., Bringans, S. D., & Bryson, W. G. (2006). Characterisation of photo-oxidation products within photoyellowed wool proteins: Tryptophan and tyrosine derived chromophores. *Photochemical and Photobiological Sciences, 5*(7), 698–706.
13. Dyer, J. M., et al. (2008). The photoyellowing of stilbene-derived fluorescent whitening agents—Mass spectrometric characterization of yellow photoproducts. *Photochemistry and Photobiology, 84*(1), 145–153.
14. Giacomoni, P. U. (2007). Biophysical and physiological effects of solar radiation on human skin. In D. P. J. Häder & Giulio (Eds.), *Comprehensive series in photochemical and photobiological sciences* (Vol. 10, p. 341). Cambridge: RSC Publishing.
15. Song, H. K., Wehrli, F. W., & Ma, J. (1997). In vivo MR microscopy of the human skin. *Magnetic Resonance in Medicine, 37*(2), 185–191.
16. Duval, C., Regnier, M., & Schmidt, R. (2001). Distinct melanogenic response of human melanocytes in mono-culture, in co-culture with keratinocytes and in reconstructed epidermis, to UV exposure. *Pigment Cell Research, 14*(5), 348–355.
17. Thiele, J. J., et al. (2001). The antioxidant network of the stratum corneum. *Current Problems in Dermatology, 29*, 26–42.
18. Sander, C. S., et al. (2002). Photoaging is associated with protein oxidation in human skin in vivo. *Journal of Investigative Dermatology, 118*(4), 618–625.
19. Gilchrest, B. A. (1996). A review of skin ageing and its medical therapy. *British Journal of Dermatology, 135*, 867–875.
20. Kochevar, I. E. (1999). Molecular and cellular effects of UV radiation relevant to chronic photodamage. In B. A. Gilchrest (Ed.), *Photodamage* (pp. 51–67).
21. Frederick, J. E., & Lubin, D. (1988). Possible long-term changes in biologically active ultraviolet

radiation reaching the ground. *Photochemistry and Photobiology, 47*, 571–578.
22. Teale, W. W. J. (1960). The ultraviolet fluorescence of proteins in neutral solution. *Biochemical Journal, 76*, 381–388.
23. Chiu, H. C., & Bersohn, R. (1977). Electronic energy transfer between tyrosine and tryptophan in the peptides trp-(pro)$_n$-Tyr. *Biopolymers, 16*(2), 277–288.
24. Simpson, W. S. (1995). Origins of variation in the fluorescence patterns of wool and wool proteins. In *Proceedings of the 9th International Wool Textile Research Conference*. Biella, Italy.
25. Creed, D. (1984). The photophysics and photochemistry of the near-UV absorbing amino acids – II. Tyrosine and its simple derivatives. *Photochemistry and Photobiology, 39*(4), 563–575.
26. Creed, D. (1984). The photophysics and photochemistry of the near-UV absorbing amino acids – I. Tryptophan and its simple derivatives. *Photochemistry and Photobiology, 39*(4), 537–562.
27. Gracanin, M., et al. (2009). Singlet-oxygen-mediated amino acid and protein oxidation: Formation of tryptophan peroxides and decomposition products. *Free Radical Biology and Medicine, 47*, 92–102.
28. Kerwin, B. A., & Remmele, R. L. J. (2007). Protect from light: Photodegradation and protein biologics. *Journal of Pharmaceutical Sciences, 96*(6), 1468–1479.
29. Mizdrak, J., et al. (2008). Tryptophan-derived ultraviolet filter compounds covalently bound to lens proteins are photosensitizers of oxidative damage. *Free Radical Biology and Medicine, 44*(6), 1108–1119.
30. Davies, M. J. (2004). Reactive species formed on proteins exposed to singlet oxygen. *Photochemical and Photobiological Sciences, 3*(1), 17–25.
31. Evans, A. O., Marsh, J. M., & Wickett, R. R. (2011). The uptake of water hardness metals by human hair. *Journal of Cosmetic Science, 62*(4), 383–391.
32. Marsh, J. M., et al. (2014). Role of copper in photochemical damage to hair. *International Journal of Cosmetic Science, 36*(1), 32–38.
33. Marsh, J. M., et al. (2007). Hair coloring systems delivering color with reduced fiber damage. *Journal of Cosmetic Science, 58*(5), 495–503.
34. Zhang, H., et al. (2013). The influence of copper (II) ions on wool photostability in the dry state. *Coloration Technology, 29*(5), 323–329.
35. Naqvi, K. R., et al. (2013). The role of chelants in controlling Cu(II)-induced radical chemistry in oxidative hair colouring products. *International Journal of Cosmetic Science, 35*(1), 41–49.
36. Grosvenor, A. J., et al. (2016). Oxidative modification in human hair: The effect of the levels of Cu (II) Ions, UV exposure and hair pigmentation. *Photochemistry and Photobiology, 92*(1), 144–149.
37. Tang, Y., et al. (2016). Trace metal ions in hair from frequent hair dyers in China and the associated effects on photo-oxidative damage. *Journal of Photochemistry and Photobiology B: Biology, 156*, 35–40.
38. Stadtman, E. R., & Oliver, C. N. (1991). Metal-catalyzed oxidation of proteins. Physiological consequences. *Journal of Biological Chemistry, 266*(4), 2005–2008.
39. Linxiang, L., et al. (2007). Iron-chelating agents never suppress Fenton reaction but participate in quenching spin-trapped radicals. *Analytica Chimica Acta, 599*, 315–319.
40. Pande, C. M., & Jachowicz, J. (1993). Hair photodamage – Measurement and prevention. *Journal of the Society of Cosmetic Chemists, 44*(2), 109–122.
41. Lin, M.-G., et al. (2006). Evaluation of dermal thermal damage by multiphoton autofluorescence and second-harmonic-generation microscopy. *Journal of Biomedical Optics, 11*, 064006.
42. Reutsch, S. B., & Kamat, Y. K. (2004). Effects of thermal treatments with a curling iron on hair fiber. *Journal of Cosmetic Science, 55*, 13–27.
43. Grosvenor, A. J., Morton, J. D., & Dyer, J. M. (2011). Proteomic characterisation of hydrothermal redox damage. *Journal of the Science of Food and Agriculture, 91*(15), 2806–2813.
44. Schwass, D. E., & Finley, J. W. (1984). Heat and alkaline damage to proteins: Racemization and lysinoalanine formation. *Journal of Agricultural and Food Chemistry, 32*(6), 1377–1382.
45. Sweetman, B. J. (1967). The hydrothermal degradation of wool keratin. Part I: Chemical changes associated with the treatment of wool keratin with water at 50 – 100°C. *Textile Research Journal, 37*(10), 834–844.
46. Brack, N., et al. (1999). Effect of water at elevated temperatures on the wool fibre surface. *Surface and Interface Analysis, 27*(12), 1050–1054.
47. Dyer, J. M., et al. (2013). Redox proteomic evaluation of bleaching and alkali damage in human hair. *International Journal of Cosmetic Science, 35*(6), 555–561.
48. Bender, D. A., & Barrett, G. C. (1985). *Chemistry and biochemistry of the amino acids* (pp. 169–171). London: Chapman & Hall.
49. Stadtman, E. R., & Levine, R. L. (2006). Chemical modification of proteins by reactive oxygen species. In I. S. Dalle-Donne, A. Scaloni, & D. A. Butterfield (Eds.), *Redox proteomics – From protein modifications to cellular disfunctions and diseases* (pp. 3–23). Hoboken: Wiley.
50. Gill, A. C., et al. (2000). Post-translational hydroxylation at the N-terminus of the prion protein reveals presence of PPII structure in vivo. *EMBO Journal, 19*(20), 5324–5331.
51. Davies, M. J. (2003). Singlet oxygen-mediated damage to proteins and its consequences. *Biochemical and Biophysical Research Communications, 305*(3), 761–770.
52. Goshe, M. B., Chen, Y. H., & Anderson, V. E. (2000). Identification of the sites of hydroxyl radical reaction

with peptides by hydrogen/deuterium exchange: Prevalence of reactions with the side chains. *Biochemistry, 39*, 1761–1770.
53. Berlett, B. S., & Stadtman, E. R. (1997). Protein oxidation in aging, disease, and oxidative stress. *Journal of Biological Chemistry, 272*(33), 20313–20316.
54. Hearle, J. W. (2000). A critical review of the structural mechanics of wool and hair fibres. *International Journal of Biological Macromolecules, 27*(2), 123–138.
55. Parbhu, A. N., Bryson, W. G., & Lal, R. (1999). Disulfide bonds in the outer layer of keratin fibers confer higher mechanical rigidity: Correlative nanoindentation and elasticity measurement with an AFM. *Biochemistry, 38*(36), 11755–11761.
56. Katsumi, A., et al. (2000). Localization of disulfide bonds in the cystine knot domain of human von Willebrand factor. *Journal of Biological Chemistry, 275*(33), 25585–25594.
57. Earland, C., & Raven, D. J. (1961). Lanthionine formation in keratin. *Nature, 191*(4786), 384–384.
58. Bessems, G. J., Rennen, H. J., & Hoenders, H. J. (1987). Lanthionine, a protein cross-link in cataractous human lenses. *Experimental Eye Research, 44*(5), 691–695.
59. Garner, M. H., & Spector, A. (1980). Selective oxidation of cysteine and methionine in normal and senile cataractous lenses. *Proceedings of the National Academy of Sciences of the United States of America, 77*, 1274–1277.
60. Davies, M. J., et al. (1999). Stable markers of oxidant damage to proteins and their application in the study of human disease. *Free Radical Biology and Medicine, 27*, 1151–1163.
61. Asquith, R. S., & Rivett, D. E. (1971). Studies on the photooxidation of tryptophan. *Biochimica et Biophysica Acta, 252*(1), 111–116.
62. Simat, T. J., & Steinhart, H. (1998). Oxidation of free tryptophan and tryptophan residues in peptides and proteins. *Journal of Agricultural and Food Chemistry, 46*(2), 490–498.
63. Maskos, J., Rush, J. D., & Koppenol, W. H. (1992). The hydroxylation of tryptophan. *Archives of Biochemistry and Biophysics, 296*(2), 514–520.
64. Żegota, H., et al. (2005). o-Tyrosine hydroxylation by OH. radicals.2,3-DOPA and 2,5-DOPA formation in γ-irradiated aqueous solution. *Radiation Physics and Chemistry, 72*(1), 25–33.
65. Wei, C., et al. (2007). Luminescence and Raman spectroscopic studies on the damage of tryptophan, histidine and carnosine by singlet oxygen. *Journal of Photochemistry and Photobiology A: Chemistry, 189*, 39–45.
66. Sionkowska, A., & Kaminska, A. (1999). Thermal helix-coil transition in UV irradiated collagen from rat tail tendon. *International Journal of Biological Macromolecules, 24*(4), 337–340.
67. Dyer, J. M., et al. (2009). Photoproducts formed in the photoyellowing of collagen in the presence of a fluorescent whitening agent. *Photochemistry and Photobiology, 85*(6), 1314–1321.
68. Stadtman, E. R., & Berlett, B. S. (1999). Fenton chemistry. Amino acid oxidation. *Journal of Biological Chemistry, 266*, 17201–17211.
69. Holt, L. A., & Milligan, B. (1977). The formation of carbonyl groups during irradiation of wool and its relevance to photoyellowing. *Textile Research Journal, 47*, 620–624.
70. Stadtman, E. R. (2006). Protein oxidation and aging. *Free Radical Research, 40*(12), 1250–1258.
71. Scaloni, A. (2006). Mass spectrometry approaches for the molecular characterisation of oxidatively/nitrosatively modified proteins. In I. S. Dalle-Donne, S. Andrea, D. M. Desiderio, & N. M. Nibbering (Eds.), *Redox proteomics*. Hoboken: Wiley.
72. Dalle-Donne, I., Scaloni, A., & Butterfield, D. A. (Eds.). (2006). *Redox proteomics: From protein modifications to cellular dysfunction and diseases* (Wiley-Interscience series on mass spectrometry, D. M. N. Desiderio & M. Nico, Eds.). Wiley: Hoboken.
73. Gerhardt, K. E., Wilson, M. I., & Greenberg, B. M. (1999). Tryptophan photolysis leads to a UVB-induced 66 kDa photoproduct of ribulose-1,5-bisphosphate carboxylase/oxygenase (rubisco) in vitro and in vivo. *Photochemistry and Photobiology, 70*(1), 49–56.
74. Amadò, R., & Neukom, H. (1976). Formation of dityrosine cross-links in proteins by oxidation of tyrosine residues. *Biochimica et Biophysica Acta, 439*(2), 292–301.
75. Vazquez, S., et al. (2002). Novel protein modification by kynurenine in human lenses. *Journal of Biological Chemistry, 277*(7), 4867–4873.
76. Thomas, S. N., et al. (2006). MudPIT (multidimensional protein identification technology) for identification of post-translational protein modifications in complex biological mixtures. In I. S. Dalle-Donne, Andrea, & D. A. Butterfield (Eds.), *Redox proteomics – From protein modifications to cellular dysfunctions and diseases* (pp. 233–252). Hoboken: Wiley.
77. Domingues, M. R. M., et al. (2003). Identification of oxidation products and free radicals of tryptophan by mass spectrometry. *Journal of the American Society for Mass Spectrometry, 14*(4), 406–416.
78. Abello, N., et al. (2009). Protein tyrosine nitration: Selectivity, physicochemical and biological consequences, denitration and proteomics methods for the identification of tyrosine-nitrated proteins. *Journal of Proteome Research, 8*(7), 3222–3238.
79. Dyer, J. M., et al. (2014). Molecular marker approaches for tracking redox damage and protection in keratins. *Journal of Cosmetic Science, 65*(1), 25–36.
80. Solazzo, C., et al. (2013). Modeling deamidation in sheep α-keratin peptides and application to archeological wool textiles. *Analytical Chemistry, 86*(1), 567–575.

81. Grosvenor, A. J., Morton, J. D., & Dyer, J. M. (2010). Profiling of residue-level photo-oxidative damage in peptides. *Amino Acids, 39*(1), 285–296.
82. Wiese, S., et al. (2007). Protein labeling by iTRAQ: A new tool for quantitative mass spectrometry in proteome research. *Proteomics, 7*(3), 340–350.
83. Wu, W. W., et al. (2006). Comparative study of three proteomic quantitative methods, DIGE, cICAT, and iTRAQ, using 2D gel- or LC-MALDI TOF/TOF. *Journal of Proteome Research, 5*(3), 651–658.
84. Deb-Choudhury, S., et al. (2014). Effect of cooking on meat proteins: Mapping hydrothermal protein modification as a potential indicator of bioavailability. *Journal of Agricultural and Food Chemistry, 62*(32), 8187–8196.
85. Dyer, J. M., et al. (2010). Proteomic evaluation and location of UVB-induced photomodification in wool. *Photochemistry and Photobiology B: Biology, 98*(2), 118–127.
86. Solazzo, C., et al. (2013). Proteomic evaluation of the biodegradation of wool fabrics in experimental burials. *International Biodeterioration & Biodegradation, 80,* 48–59.
87. Lassé, M., et al. (2015). The impact of pH, salt concentration and heat on digestibility and amino acid modification in egg white protein. *Journal of Food Composition and Analysis, 38*(0), 42–48.
88. Chen, X., Chen, Y. H., & Anderson, V. E. (1999). Protein cross-links: Universal isolation and characterization by isotopic derivatization and electrospray ionization mass spectrometry. *Analytical Biochemistry, 273*(2), 192–203.
89. Back, J. W., et al. (2002). Identification of cross-linked peptides for protein interaction studies using mass spectrometry and 18O labeling. *Analytical Chemistry, 74*(17), 4417–4422.
90. Mirza, S. P., Greene, A. S., & Olivier, M. (2008). ^{18}O labeling over a coffee break: A rapid strategy for quantitative proteomics. *Journal of Proteome Research, 7*(7), 3042–3048.

Index

7 + 1, 25, 65, 66, 174
8 + 0, 65, 66, 160

A
A_{11}, 60, 62–65, 67, 162–164
A_{12}, 60, 62, 63, 65, 67, 162–164
A_{22}, 60, 62–65, 67, 162–164
Acid phosphatase, 103, 127, 129, 140, 141, 145
Actin, 119, 137
Activation energy, 189, 193, 196, 199
Adamson's fringe, 92, 94, 124, 144
Adherens junction, 114, 137, 140
Adiabatic, 186, 188
    See also Non-adiabatic
Adipocyte, 103, 104
Allergies, 148
Alpaca, 10, 54, 156
Alpha-helices, 35
Alpha-keratin, see Keratin
Amphibian, 37, 39, 40
Anagen, 15–17, 89–92, 97–106, 111–115, 117, 145–147
Androgen, 98, 101, 106
    See also Hormone
Androgenic alopecia, 101, 106
    See also Balding
Angora, 54
Ape, 25, 53
    See also Primate
Apoptosis, 15, 127, 129, 131
Arginine to citrulline, 12, 132
Armadillo, 54
Arrector pili muscle, 16, 92, 98, 103, 105–106, 147
Artiodactyl, 40
Atomic force microscopy, 90, 123
Auber's critical level, see Critical level
Auchene, 42
Avian, 35
Awl, 42

B
Bacteria, 148, 158
Balding, 101, 106
Baleen, 58, 73
Basement membrane, 17, 92, 94, 98, 101, 103–106, 111, 113–116, 118, 131, 132
β-Catenin, 117, 137
Beard, 111, 121, 124, 131, 133, 137, 138
Beta-keratin, 35, 195
Bird, 34, 58, 82
    See also Feather
Blood
    blood capillaries, 91, 101
    blood vessels, 101, 103
Bony fish, 37
Bovine, 73, 130
β Sheet, 61, 81, 190, 195, 200
β Strand, 29, 81

C
Cadherin, 114, 126
Capillaries, see Blood
Carbonylation, 207, 210, 214
Carboxy-terminal motif (CTM), 38
Cartilaginous fish, 37
Cashmere, 54, 55
    See also Goat
Cat, 73, 100, 148
Catagen, 15–17, 117
Caudate, 40
Ceiling temperature, 164–166
Cell membrane complex (CMC), 11, 12, 96, 126, 127, 130, 140–142, 145, 148
Cell migration, 106, 113, 114, 117, 119
Cell reshaping, 92, 94, 117–121, 125, 126, 129, 140
Cellular junctions, 91, 92, 94, 118, 119, 125–127, 135–138, 140, 141, 143, 147
Cellulose, 22, 23, 162
Cetacean, 42
Chaotropes, 7, 176, 177, 179–181
Chaperones, 192
Chicken, 48, 49
Chimeric genes, 34
Chimpanzee, 43
Chitin, 161
cIF, see Cytoplasmic IF (cIF)

Claw, 4, 34, 39, 40, 42, 58, 59, 67, 73
Club hair, 16, 17
CMC, *see* Cell membrane complex (CMC)
Coated vesicles, 130
Coiled-coil, 60–64, 67, 82, 179, 180, 191
Collagen, 91, 98, 99, 101, 102, 105, 161, 207
Companion layer, 41, 42, 91, 92, 94, 96, 103, 104, 111, 116, 124, 131, 143, 144, 146–148
Configurational and mixing entropy, 162, 163
Connective tissue, 98, 99, 101
Consensus sequence, 38
Co-operative nucleated supramolecular polymerisation, 164
Corneous beta proteins, 34
Cornification, 35, 42, 95, 120, 122, 125, 143, 144, 146, 147
  See also Consolidation; Hardening
Cortex
  cortical cells, 4, 7, 10, 12, 75, 90, 143, 156
  mesocortex, 10, 11, 54, 158
  orthocortex, 9–11, 54, 58, 59, 72, 74, 75, 82, 85, 120, 156, 160, 166, 175
  paracortex, 9–11, 54, 58, 59, 74, 75, 85, 123, 156–158, 166, 175
Cortical cells, *see* Cortex
Crewther, W.G., 59, 61
Critical level, 94, 115, 138, 166
Crystallisation, 61, 103, 123, 187
CTM, *see* Carboxy-terminal motif (CTM)
Curvature, 43, 90, 119, 157, 174
Cuticle
  cuticle cells, 4, 6, 92, 94, 119, 135–142, 145, 148, 174, 206, 207
  endocuticle, 4, 7, 140
  epicuticle (*see* Fibre cuticle surface membrane (FSCM))
  exocuticle, 4, 6, 7, 94, 95, 139–142, 174
  exocuticle a-layer, 4, 6, 7, 95, 139, 140
  fibre cuticle surface membrane (FSCM), 5, 6, 142
Cysteic acid, 176, 209, 210
Cysteine, 6, 22–30, 34, 38–40, 42, 48, 49, 55, 60, 61, 63–65, 67, 72–74, 77, 79, 80, 82, 84, 85, 117, 126, 132, 139, 175–180, 206, 209–211
  See also Cysteic acid; Keratin associated protein (KAP)
Cytochrome c oxidase, 55, 131
Cytoplasmic IF (cIF), 35, 143, 147
Cytoplasmic remnant, 8, 12, 125, 129, 130, 135, 140
Cytoskeleton, 26, 48, 59, 90, 91, 94, 117, 119, 126, 137, 138, 144, 147, 158, 165

D
Decapeptide, 27, 28, 76, 77, 79
Deer, 10, 156, 167
Defleecing, 73
Dermal papilla (DP), 16, 17, 90–94, 98, 101–103, 111–118, 120, 121, 124, 129, 131, 132, 135, 139, 140, 146, 175
Dermal plaque, 101, 102

Dermal sheath, 90, 91, 101, 104, 105, 116, 118
Dermis, 17, 98, 99, 101, 105, 111, 113, 207
Desmoglein, 17, 126, 127, 146, 174
Desmoplakin, 125–127, 174, 175
Desmosome, 91, 120, 121, 125–127, 137, 138, 143–147, 158, 165, 174
Differential scanning calorimetry (DSC), 195, 196, 201
Dimerization, 82, 140, 191
Disease, 72, 84, 142
Disulfide, 11, 34, 38–40, 48, 54, 60–65, 67, 72, 73, 79, 80, 84, 93, 95, 122–125, 131, 139, 143, 174–181, 189, 199, 200, 206, 209–211
Disulfo-succinimidyl-tartrate (DST), 58, 62–65
DNA, 16, 34, 49, 53, 93, 127, 129
Dog, 25, 49, 50
Dolphin, 42, 48, 53
Double-twist, *see* Macrofibril
DP, *see* Dermal papilla (DP)
DSC, *see* Differential scanning calorimetry (DSC)
DST, *see* Disulfo-succinimidyl-tartrate (DST)

E
Echidna, 54, 73, 83
Ectodermal dysplasia, 30, 84
Egg case, 82
Eisenberg, A., 164
Electrospray ionisation (ESI), 211
Elongation zone, 94
  See also Follicle Zone B
Endocuticle, *see* Cuticle
Endonuclease, 129
Endoplasmic reticulum, 114, 120, 158
Energy, 93, 164–166, 186–189, 191–196, 198–201, 208
ε-(N-γ-glutamyl)-lysine, 7, 122, 135
Enthalpy, 165, 186–188, 191–193, 195, 199–201
Entropic, 191, 194, 201
Epidermis, 17, 37, 38, 41, 42, 53, 94, 98, 105, 111, 113, 117, 125, 126, 144, 148, 207
  See also Skin; Stratum corneum
Epithelia
  corneal epithelium, 39
  epithelial cells, 16, 17, 23, 48, 111, 126
  epithelial keratins, 22, 38, 59, 60, 131, 158, 176, 177
  oral epithelium, 39
  simple epithelia, 38
  stratified epithelia, 38
ESI, *see* Electrospray ionisation (ESI)
Exocuticle, *see* Cuticle
Exocytosis, 133
Exogen, 17
Exothermic, 165, 175, 187, 191, 197
Extracellular matrix (ECM), 16, 113, 115, 126–127

F
Feather, 58
Fenton chemistry, 208
Fibre cuticle surface membrane (FSCM), 5, 6, 142
Fibre diameter, 6, 11, 54, 75, 90, 119, 143

Fibroblast, 16, 17, 91, 98, 101, 114
Filament-matrix, 58, 72, 73, 82, 122
Filament network, 91, 120, 143, 163
Fingernail, 83
 See also Nail
Flory-Huggins, 161, 192, 198
 See also Vrentas-Vrentas model
Flory, P.J., 161, 164–166, 192, 198, 199
Flügelzellen, 143, 144, 147
Follicle
 bulb, 101, 111, 112, 116, 117, 189
 bulb volume, 112
 bulge, 105, 106
 groups, 98, 99, 105, 106, 131
 miniaturisation, 106 (*see also* Vellus hair)
 permanent portion, 106
 primary hair follicle, 55
 secondary hair follicle, 55
 transient portion, 106
 Zone A, 92, 94, 111–116, 126, 132, 134, 137, 145, 146
 Zone B, 92–94, 111, 116–122, 125–127, 129, 130, 132, 135–139, 142–144, 146, 147
 Zone C, 93, 94, 120–123, 126, 127, 129, 131, 133, 134, 137–144, 146, 147
 Zone E, 94, 95, 122–125, 127, 129, 134, 135, 140, 141, 144, 146, 147
 Zone F, 94, 96, 141, 147, 148
 Zone G, 94, 96
Follicle cycle
 Anagen (*see* Anagen)
 Catagen (*see* Catagen)
 Telogen (*see* Telogen)
Follicle Zone B, 94
Fourier transform, 60
Frog, 40
Fungi, 148

G

Gap junction, 114, 126, 127, 137, 138, 143, 174
GC nucleotide, 54
Gene duplication, 34, 35, 37, 41, 42, 50–52
Gene family, 48–54
Gene marker, 54, 55
Genome sequencing, 48, 51
Geometric artefact, 157
Germative matrix, 72, 75, 94, 101, 111–117
Gibbs free energy, 186–189, 191, 192, 194, 198
Glass transition, 195–198
Glycine, 22, 23, 25–30, 38, 48, 49, 55, 59, 72–79, 81–85, 121, 175, 177
 See also Keratin associated protein (KAP)
Glycogen, 91
Goat, 10, 25, 29, 54, 55, 156
Golgi apparatus, 114, 120, 130, 131, 140, 158
Gorilla, 43
Guard hair, 42, 98, 100, 106, 114, 115, 117, 146
Guinea pig, 73, 137, 140

H

Hair matrix, *see* Elongation zone
Hair shaft, 17, 28, 89, 117, 120, 129, 189
 hair internal lipids, 148
 See also Cortex; Cuticle; Medulla
Hardening, 90, 92–94, 120, 121, 125, 129, 133, 140, 141, 143, 144, 146
 See also Consolidation; Cornification; Keratinization
Hardness, 11, 82, 90–94, 120, 121, 123, 125, 129, 133, 140, 141, 143, 144
Head domain, 25, 29, 37, 38, 60–63, 65, 76–78, 82, 174–176, 179, 180
Helix initiation motif (HIM), 58, 60
Helix termination motif (HTM), 58, 60
Helmholtz free energy, 186, 187
Hemidesmosome, 111, 131
Hendecad, 60, 61, 67, 191
Heptad, 60, 61, 67, 178, 180, 191
Heterodimer, 39, 60–63, 120, 131, 139, 158, 174, 179, 191
HGTP, *see* Keratin associated protein (KAP)
High glycine-tyrosine (HGT), 22–25, 29, 30, 48, 49, 54, 55, 59, 72, 74, 75, 78–84
 See also Keratin associated protein (KAP)
High sulfur protein, *see* Keratin associated protein (KAP)
High/ultrahigh cysteine, *see* Keratin associated protein (KAP)
HIM, *see* Helix initiation motif (HIM)
Homodimer, 60, 158, 163
Hooves, 39, 58
Hormone, 101, 103, 148
Horn, 40, 58, 73, 83, 148
HS-KRTAP, *see* Keratin associated protein (KAP)
HSP, *see* Keratin associated protein (KAP)
HTM, *see* Helix termination motif (HTM)
Hydroxyl radical, 208, 210, 212

I

IF, *see* Intermediate filament (IF)
IFP, *see* Intermediate filament protein (IFP)
Immunohistochemistry, 90, 129
Infundibulum, 90, 92, 94, 147
Inner root sheath (IRS), 17, 39, 41–43, 90–96, 111, 113, 116, 118–120, 122, 125, 127, 129–133, 135–148, 163
 Henle's layer, 91–94, 103, 110, 117, 120, 121, 124, 136, 140–147
 Huxley's layer, 91, 124, 136, 140, 142–145, 147, 148
 IRS breakdown, 91, 147, 148
 IRS cuticle, 91, 94, 124, 135–146, 148
Intercellular space, 118, 127, 133, 141
Intermacrofibrillar material, 125, 174
Intermediate filament (IF), 9, 36–38, 57–67, 71–75, 81, 82, 84, 85, 94, 122, 124, 125, 132, 137, 143, 145–147, 156–160, 162–166, 173, 174, 179, 191

Intermediate filament protein (IFP), 22, 33, 35, 59, 189, 207
   type III, 35–38
   type IV, 35
   type VI, 35
   See also Keratin
Ionic, 189, 193, 199
IRS, *see* Inner root sheath (IRS)
Isobar, 187, 188
Isochore, 187, 188
Isopeptide bond, 7, 11, 135, 174
   ε-amino-(γ-glutamyl)lysine, 7, 122, 135 (*see also* Transglutaminase)
   See also Trichohyalin
Isothermal titration calorimetry, 165, 191
Isthmus, 90, 94, 147, 148

J
Jones, L.N., 6, 59, 142, 158, 159

K
KAP, *see* Keratin associated protein (KAP)
Keratin
   corneal epithelium
      K3, 23, 36, 38, 164
      K12, 23, 36, 38
   epithelial
      K1, 22, 23, 36, 38, 41, 193
      K2, 23, 36, 41, 193
      K5, 23, 36, 41, 131, 164
      K8, 22, 23, 35–38, 40
      K10, 23, 36, 38, 41
      K14, 16, 23, 36, 41, 131
      K18, 23, 35–37, 40
      K77, 23, 36, 41
   K23, 23, 36, 40, 41, 163
   K80, 23, 36, 40, 41
   keratin genes
      KRT7, 23, 35
      KRT8, 23, 35, 37
      KRT9, 23, 37
      KRT18, 23, 35, 37
      KRT19, 23, 37
   keratin head domain (*see* Head domain)
   keratin intermediate filaments (*see* Intermediate filament (IF))
   keratins across species, 73
   keratin tail domain (*see* Tail domain)
   keratogenous region, 122, 123, 129, 143, 147
   kinetics, 185, 188, 189, 193, 199, 201
   oral epithelium
      K4, 23, 36, 39
      K13, 23, 36, 39, 41
   type I keratins
      K31, 23–26, 36, 41, 42, 120, 129, 131, 158, 163, 165, 166, 175, 180
      K32, 7, 23, 24, 36, 41, 139
      K34, 23–26, 36, 41, 42, 82, 121, 163, 166, 176
      K35, 7, 23–26, 36, 41, 120, 138, 139, 158, 163, 165, 166, 180, 191
      K36, 23, 24, 36, 38, 41, 42, 121, 131
      K37, 24, 36, 41, 42, 121, 131
      K38, 23–26, 36, 41, 42, 120, 131, 163, 166
      K39, 23, 24, 26, 36, 38, 41, 140, 175
      K40, 23, 24, 26, 36, 41, 138, 140
      K33a, 23–26, 36, 41, 42, 121, 180
      K33b, 23–26, 36, 41, 42, 121
   type II keratins
      K81, 23, 26, 36, 41, 42, 121, 131, 181
      K82, 7, 23, 26, 36, 41, 139
      K83, 23, 26, 36, 41, 42, 121, 181
      K84, 23, 26, 36, 38, 41, 42, 139
      K85, 7, 23, 26, 30, 36, 41, 82, 120, 131, 138, 139, 158, 165, 175, 176, 181, 191
      K86, 23, 26, 36, 41, 42, 82, 121, 131, 176, 181
      K87, 26
Keratin associated protein (KAP)
   high glycine tyrosine proteins (HGT, HGTP or HGT-KRTAP)
      KAP6, 25, 29, 54, 75, 78, 79, 81
      KAP7, 25, 29, 54, 75
      KAP8, 25, 29, 54, 59, 74, 75, 81
      KAP16, 25, 29
      KAP19, 25, 29, 74
      KAP20, 25, 75, 78, 79, 81
      KAP21, 25, 75, 78, 79, 81
      KAP22, 25
   high sulfur proteins (HS, HSP or HS-KRTAP)
      KAP1, 24–28, 75–79, 81, 82, 176
      KAP2, 25, 26, 28, 74–79, 82, 176, 177
      KAP3, 25, 26, 28, 75
      KAP10, 25, 27, 74, 75, 140
      KAP13, 25, 75
      KAP15, 25, 27, 75
      KAP24, 25
      KAP26, 25
      KAP29, 24
      KAP31, 24
   KAP9.2, 29, 55
   ultra-high sulfur proteins (UHS or UHSP)
      KAP4, 24, 25, 28, 74, 75, 121
      KAP5, 7, 25, 28, 29, 75, 140
      KAP28, 24
      KAP30, 24
      KAP32, 24
   See also High glycine tyrosine (HGT-KRTAP)
Keratinization, 28, 54, 91–95, 110, 122–125, 127, 140, 144, 145, 147, 166, 189, 190
Keratinocytes, 17, 35, 113, 117, 129, 137
Keratogenous zone, 28, 92
Knockout mutant, 126

L
Lamin, 35, 59, 61
Landmark, 90–92, 140, 141
Lanolin, 148
Lattice model, 161
Legendre transforms, 187
Ligand, 51, 53, 161, 192–194
Linker, 60, 62, 67, 76, 77, 79, 179, 180, 191

Lipid, 4–6, 11, 96, 103, 104, 110, 141, 148
Liquid chromatography, 62, 211
Liquid crystal, 9, 155, 156, 159–161, 166
Loose anagen hair syndrome, 145
Loricrin, 122
Low sulfur proteins, 22, 72
　　See also Keratin
Lumry-Eyring model, 193
Lung fish, 37
Lyotropic mesophase, 160, 161

M

Macrofibril
　　double-twist macrofibril, 10, 157
　　intensity, 66, 157, 166, 176, 192, 208
　　macrofibril nucleation, 94, 121, 160
Mammal, v, 4, 22, 34, 35, 37, 39, 40, 42, 48, 49, 51–55, 72, 74, 75, 90, 98, 100, 105, 106, 116, 133, 135, 145, 146, 163, 206, 207, 211
Maple syrup urine disease, 142
Mass spectrometry, 6, 176, 206, 211, 214
　　See also Electrospray ionisation (ESI); Matrix-assisted laser desorption ionisation (MALDI)
Mast cell, 91, 114
Matrix, 9, 16, 17, 22, 24, 39, 41, 48, 54, 58–60, 72, 73, 75, 78, 81, 82, 84, 90, 93, 94, 101, 111–117, 120, 122–124, 126, 127, 145, 156, 174, 175, 191, 206, 211
　　See also Germative matrix; Keratin associated protein (KAP)
Matrix-assisted laser desorption ionisation (MALDI), 211
Matrix proteins, see Keratin associated protein (KAP)
MEA, see Methyleicosanonic acid (MEA)
Medulla
　　medullary cortex cells, 133
　　medullary granules, 132
　　uni-serial ladder, 133
Melanin
　　eumelanin, 117
　　pheomelanin, 117
Melanin granule, 4, 93, 118, 134, 135, 148
Melanocytes, 93, 116–118
Melanosome, 116, 117
Mercer, E.H., 74, 92, 93
Mesenchymal, 101
Mesocortex, see Cortex
Mesogen, 160–163, 166
Mesophase, 160–167
18-methyleicosanoic acid, see Methyleicosanonic acid (MEA)
Methyleicosanonic acid (MEA), 6, 141, 142
　　See also Cuticle
Microfibril, 59, 123, 201
　　See also Intermediate filament (IF)
Micro-niche, 92, 113, 114, 146
Microtubule, 39, 59, 117, 119

Mitochondria
　　mitochondrial DNA, 129
　　mitochondrial membrane potential, 130
Mitosis, 94, 113
Modification, 37, 40–42, 98, 135, 205–215
Mohair, 54
　　See also Goat
Monkey, 25, 73
Mouse, 16, 25, 48, 53, 54, 73, 103, 106, 111, 117, 120, 126, 129, 131, 133, 140, 145, 148
Moustache, 115, 146
Muscle, see Arrector pili
Muskox, 54
Myosin, 119, 137

N

NADH dehydrogenase, 55
NADH-ubiquinone oxidoreductase, 55
Nail, 30, 39, 42, 83, 192
Nematic, 156, 157, 166, 192
Nerve, 103–106
　　See also Neuron
Neurofilament, 35, 59
Neuron, 91, 101, 104, 105, 189
Newtonian flow, 196
NMR, 6, 79
Non-adiabatic, 186
Normarski, 92
Nuclei
　　nuclear pocket, 129
　　nuclear remnant (see Cytoplasmic remnant)
　　nucleophagy, 144
Nutrition, 72, 175, 206

O

Olfactory receptors, 51–53
Opossum, 25, 49, 53
ORS, see Outer root sheath (ORS)
Orthocortex, see Cortex
Orwin, D.F.G., v, 92, 99, 119, 123, 126–128, 136–138, 143
Outer root sheath (ORS), 17, 41, 42, 51, 91–94, 104, 118, 146–148
Ovine, see Sheep
Oxidation, 66, 175, 176, 206–215
Oxidizing, 62, 67, 72

P

Papilla, see Dermal papilla
Paracortex, see Cortex
Pelage, 42, 92, 98, 106, 117, 119, 126, 131, 133, 135, 146–148
Pentapeptide repeats, 77–81, 177
Peptides, 12, 62, 195, 209, 210, 212, 214
Peptidylargenine deiminase, 132
Peri-nuclear basket, 129

Peripherin, 35, 59
Photobleaching, 207
Photodamage, 207, 208
Photo-oxidation, 206–209, 211
   See also Photobleaching; Photodamage; Photosensitisation; Phototendering
Photosensitisation, 207, 211
Phototendering, 207
Pilary canal, 141, 142, 147, 148
Plakoglobin, 125, 127
Plasma membrane, 12, 114, 123, 126, 127, 139–142, 147
Platypus, 25
Polymerization initiator, 163
Pope, F.M., 59, 158, 159
Porcupine, 61, 73, 83
   See also Quill
Possum, 115, 134
Post-transcriptional gene expression, 121
Potential, 34, 40, 53, 54, 59, 61, 91, 92, 106, 113, 130, 131, 160–162, 165, 180, 186–188, 207, 210, 214
Praying mantis, 82
Primary cilia, 92
Primary follicle, *see* Follicle
Primate, 51
   See also Ape
Protein crosslinking, 206, 207, 209, 210, 214
   See also Protein-protein crosslinks
Protein-protein crosslinks, 174, 209, 211, 214
Proteoglycans, 115
Proteomic analysis, 211
Proto-macrofibril, 120, 123, 156, 158

Q
Quill, 54, 58, 61, 67, 73, 83
   See also Echidna; Porcupine

R
Rabbit, 10, 54, 75, 114, 115, 117, 142, 146, 156
Radicals, 208, 210–212
Radioactive cysteine, 117
Radio-label, 113, 114, 119
Random coil, 81, 191
Rat, 25, 104, 105, 112, 115, 119, 126, 129–131, 133–135, 137, 146, 147, 158
Reactive oxygen species, 95, 131
Redox, 210, 212–214
   See also Redox scoring
Redox scoring, 212–214
Reducing, 62, 67, 72, 127, 141, 175–177, 179, 191
Reptile, 3, 34, 39, 40, 58, 82
   See also Scale
Ribosome, 120, 122, 138, 145, 158, 190
Rod domain, 38, 60, 61, 63, 64, 174, 175, 177–179, 191
Rodent, 4, 15, 39, 40, 52, 53, 92, 101, 114, 133–135
Rogers, G.E., 72, 75, 84, 124, 156, 158, 160
Rosenbaum's model, 198

S
SAX, *see* Small angle x-ray diffraction pattern (SAX)
Scale, 3–5, 9, 34, 39, 40, 48, 51, 59, 67, 90, 91, 94, 100, 114, 123, 125, 135, 138, 141, 142, 144, 156
Scalp hair, 5, 10, 16, 42, 98, 103, 104, 106, 111, 112, 114, 120, 123, 125, 126, 130, 133, 135–138, 140, 143, 146, 148, 156, 157, 159, 208
Scanning transmission electron microscopy (STEM), 58, 65, 67
S-carboxymethylation, 22
Schizophyllan, 162
Sebaceous gland, 17, 42, 92, 98, 101, 103, 105, 148
Secondary follicle, *see* Follicle
Sheep, 4, 10, 24, 25, 28, 29, 40, 42, 43, 48, 54, 55, 59, 61, 74, 77–79, 81, 92, 98–101, 103–106
   Corredale, 117
   Lincoln, 54
   lustre mutant, 10, 54
   merino, 4, 5, 7–9, 11, 28, 54, 58, 75, 85, 103, 117, 147, 148, 176, 207
   Romney, 127
   See also Wool
Silk, 29, 81
Singlet oxygen, 208, 212
Skin, 5, 16, 17, 35, 38, 40, 53, 55, 90, 91, 94, 97–101, 111, 114–116, 144, 148, 163, 189, 207–209, 211, 214, 215
Sloth, 48, 53
Small angle x-ray diffraction pattern (SAX), 93, 122
   See also X-ray diffraction (XRD)
Snake-venom, 79, 80
Stem cells, 15–17, 91, 92, 94, 101, 105, 106, 111–114, 116, 117, 131, 132, 146
Stratum corneum, 94, 98, 113, 129, 137, 148, 207
   See also Epidermis; Skin
Stratum granulosum, 126
Stress-strain curve, 199
Stutter, 60, 179, 180
Sudoriferous, 94
Superoxide, 208

T
TAC, *see* Transient amplifying cell (TAC)
Tactoid, 121, 159–161, 163–167, 191, 192
Tail domain, 37–39, 60, 61, 63, 66, 67, 175, 179
Telogen, 16, 17, 101, 113
TEM, *see* Transmission electron microscopy (TEM)
Tensile strength, 58, 82, 208
Tetramer, 64, 124, 159, 162–164, 174, 189, 191
Tetrapod, 37, 40
Thermodynamic, 161, 163, 185–201
Thermoregulation, 98, 106
Tight junction, 126, 131, 137, 143, 146, 174
Tobacco mosaic virus, 161
Tobolsky, A.V., 164, 166
Tongue
   filiform papillae, 40, 42

Transcriptional factor binding sites, 38
Transcriptome, 55
Transglutaminase, 7, 12, 17, 125, 135, 140, 143, 174
    See also Isopeptide bond
Transient amplifying cell (TAC), 94, 111, 113, 114, 127
Transmission electron microscopy (TEM), 4, 10, 58, 64, 66, 67, 90, 103, 105, 119, 120, 123, 127, 130, 133, 135, 140, 141, 146, 148, 157, 160
Trichocyte keratin, see Keratin
Trichohyalin, 11, 12, 16, 17, 39, 43, 90, 91, 93, 94, 122, 132–134, 136–139, 143, 145, 147
Trichothiodystrophy, 84
Tubulin, 39
    See also Microtubule
Tyrosine, 9, 22, 23, 25–27, 29, 30, 38, 48, 49, 55, 59, 72–75, 81–85, 121, 139, 175, 208–211, 213
    See also Keratin associated protein (KAP)

U
ULF, see Unit-length-filament (ULF)
Unit-length-filament (ULF), 64, 159, 162–167, 191, 192

V
Vacuole, 12, 104, 120, 132, 133, 145, 147, 158
Van der Waals, 199
Vellus hair, 23, 24, 42, 106, 121
Vesicle, 93, 130–134, 140, 147
Vibrissae, 42, 104–106, 111, 112, 117, 119, 145, 148

Vimentin, 35–38, 59, 61, 158, 162, 165
Vomeronasal receptors water absorption, 51, 53
Vrentas-Vrentas model, 198

W
Whisker, see Vibrissae
Wool, v, 4–11, 22–24, 26–30, 48, 54, 55, 58–60, 66, 72–75, 77, 78, 81, 85, 90, 92, 94, 95, 99, 102, 103, 105, 106, 111, 112, 114, 116–133, 135–143, 145–148, 156–159, 165, 166, 173–176, 178, 180, 181, 192, 196, 197, 207–209, 211, 212, 215
    See also Sheep
Wooly hair condition, 145

X
Xanthan, 162
Xenopus, 40
X-ray crystallography, 81
X-ray diffraction (XRD), 61, 64–66, 91, 93, 122, 190
X-ray microbeam, 123

Y
Young's modulus, 199

Z
Zigzag, 42, 114, 146

Printed by Printforce, the Netherlands